Rhea County Court House, Dayton, Tennessee

THE WORLD'S MOST FAMOUS COURT TRIAL

TENNESSEE EVOLUTION CASE

THE LAWBOOK EXCHANGE, LTD.
Clark, New Jersey

ISBN-13: 9781886363311 (cloth)
ISBN-13: 9781616190569 (paperback)

Lawbook Exchange edition 2010

The quality of this reprint is equivalent to the quality of the original work.

THE LAWBOOK EXCHANGE, LTD.
33 Terminal Avenue
Clark, New Jersey 07066-1321

Please see our website for a selection of our other publications and fine facsimile reprints of classic works of legal history:
www.lawbookexchange.com

Library of Congress Cataloging-in-Publication Data

Scopes, John Thomas.
The world's most famous court trial : Tennessee evolution case : a complete stenographic report of the famous court test of the Tennessee Anti-Evolution Act, at Dayton, July 10 to 21, 1925, Including speeches and arguments of attorneys.
p. cm.
Originally published: Cincinnati: National Book Co., c1925.
ISBN 1-886363-31-5 (alk. paper)
1. Scopes. John Thomas—Trials, litigation, etc. 2. Evolution--Study and teaching—Law and legislation—Tennessee. 3. Bible and evolution. I. Tennessee. County Court (Rhea County) II. Title.

KF224.S3B7 1997
345.73'0288—dc21 97-38485

Printed in the United States of America on acid-free paper

THE WORLD'S MOST FAMOUS COURT TRIAL

TENNESSEE EVOLUTION CASE

A COMPLETE STENOGRAPHIC REPORT OF THE FAMOUS COURT TEST OF THE TENNESSEE ANTI-EVOLUTION ACT, AT DAYTON, JULY 10 TO 21, 1925, INCLUDING SPEECHES AND ARGUMENTS OF ATTORNEYS.

NATIONAL BOOK COMPANY
CINCINNATI

NATIONAL BOOK COMPANY
Cincinnati, Ohio

PUBLISHER'S NOTE

In presenting the material of this book we are moved by the great interest shown throughout the world in this most notable case.

It is unlike any other case ever tried, and we believe has an interest that will hold long after the individuals involved shall have passed away.

While some of the headings and sub-heads are ours, we have made no attempt at editing. We simply present the case from the court record as it was made from day to day.

In publishing this book we are indebted to many interested friends and especially to the Chattanooga Times for the use of their transcript copy.

We trust this work may find a hearty welcome from those who desire to know just what occurred at Dayton.

NATIONAL BOOK COMPANY.

* * * "Dayton is the center and the seat of this trial largely by circumstance. We are told that more words have been sent across the ocean by cable to Europe and Australia about this trial than has ever been sent by cable in regard to anything else happening in the United States. That isn't because the trial is held in Dayton. It isn't because a school-teacher has been subjected to the danger of a fine from $100 to $500, but I think illustrates how people can be drawn into prominence by attaching themselves to a great cause. Causes stir the world, and this cause has stirred the world. It is because it goes deep. It is because it extends wide, and because it reaches into the future beyond the power of man to see. Here has been fought out a little case of little consequence as a case, but the world is interested because it raises an issue, and that issue will some day be settled right, whether it is settled on our side or the other side."—*W. J. Bryan, in his last court speech,* *see page 316.*

TABLE OF CONTENTS

TABLE OF CONTENTS

CHAPTER I.

FIRST DAY OF DAYTON EVOLUTION TRIAL—FRIDAY, JULY 10, 1925.

The Court—The court will come to order. The Rev. Cartwright will open court with prayer.

The Rev. Cartwright—Oh, God, our divine Father, we recognize Thee as the Supreme Ruler of the universe, in whose hands are the lives and destinies of all men, and of all the world. We approach unto Thy province this morning, we trust with that degree of reverence that is due unto Thy supreme majesty, and with the consciousness of the fact that every good and every perfect gift comes down from Thee, Father of Lights.

We praise Thy holy and blessed name, that Thou hast made it possible for us to approach Thee at all times and in all places, at the throne of Thy divine grace, with the assurance that we shall find grace and help in our time of need.

We are conscious, our Father, that Thou art the source of our wisdom, and of our power. We are incapable of thinking pure thoughts or performing righteous deeds, unaided by Thee and Thy divine spirit, with the consciousness of our weakness and our frailty, and our ignorance, we come to Thee this morning, our Divine Father, that we may seek from Thee that wisdom to so transact the business of this court in such a way and manner as that Thy name may be honored and glorified among men, and we, therefore, beseech Thee, our Father, that Thou will give to the court this morning a sufficient share of the divine spirit as will enable the court to so administer its affairs as that justice may come to all and that God's standard of purity and holiness may be upheld.

We beseech Thee, our Heavenly Father, that Thou wilt grant unto every individual that share of wisdom that will enable them to go out from this session of the court, with the consciousness of having under God and grace done the very best thing possible, and the wisest thing possible. And to this end we pray that the power and the presence of the Holy Spirit may be with the jury and with the accused and with all the attorneys interested in this case.

Oh, God, in the midst of all, help us to remember that Thou art on Thy throne and that Thou knowest the secrets of our hearts, and that Thou art acquainted with the motive back of every act and thought; and may we also be conscious of the fact, our Heavenly Father, that there is coming a day in which all of the nations of the earth shall stand before Thy judgment bar and render an accounting for the deeds done in the body, and grant, our Father, that we may have kept in mind the great truth that we are amenable to God, and that Thou wilt search us, and that Thou wilt reward us according to our deeds.

Hear us in our prayers, our Father, this morning, for the cause of truth and righteousness, throughout the length and breadth of the earth, and Oh, God, grant that from the President of the United States down to the most insignificant officer thereof, that the affairs of church and state may be so administered that God may beget unto Himself the greatest degree of honor and glory.

Hear us in these our prayers. God help us to be loyal to God, and loyal to truth, and in the end of life's tremendous trouble, may we so have lived and so have wrought in this world, that we may be admitted into the grace of Thy kingdom and honor, and there, amongst the resplendent glories of a living God, offer praise to Thy glory and grace for ever more. Amen.

Judge Calls Case.

State of Tennessee vs. John Thomas Scopes.

The Court—Seat everyone you can, Mr. Sheriff, and those that can't get seats, let them stand around the wall.

The Court—Mr. Attorney-General, come right up here, please. Let me have my docket, Mr. Clerk.

9:22 A.M.—Mr. Attorney-General I am calling the case of the State vs. John Thomas Scopes.

The Attorney-General—May I have the papers, Mr. Clerk?

The Court—Hand the papers to the Attorney-General.

The Attorney-General—If the court please, in this case we think it is proper that a new indictment be returned.

The Court—Do you want a grand jury empaneled?

The Attorney-General—Yes, sir, and a new indictment.

The Court—Yes, sir.

The Attorney-General—This indictment has been returned by agreement on both sides, but both sides are anxious that the record be kept straight and regular, that no technical objection may be made to it in the appellate courts.

The Court—Very well.

The Court—Gentlemen: The lawyers that are interested in this case will please have their places behind the tables. Have you any further requests to make Mr. Attorney-General?

The Attorney-General—If the court please, some of the gentlemen interested in this case on both sides, of course, are not entirely familiar with our procedure. I understand the defense wants a little time to consult on some matters, an hour or an hour and a half.

The Court—That shouldn't interfere with the making up of a jury.

The Attorney-General—Not at all. I simply wanted to ask the court as a courtesy to them that they might have a recess for that length of time.

Judge Neal—There are a number of counsel on both sides from out of the state and I would like to have these men introduced to the court.

The Court—Yes, I will be glad to have them.

Visiting Lawyers Introduced.

Judge Neal—Gen. Stewart, I suggest that now would be the time to introduce the outside counsel.

Gen. Stewart—Mr. William Jennings Bryan and his son, both of whom need no introduction, are the only outside lawyers with the state.

The Court—Who are here for the defense?

Judge Neal—Mr. Darrow, Arthur Hays, Mr. Malone and Mr. Thompson.

The Court—Gentlemen: I desire to assure you that we are glad to have you. The foriegn lawyers for both the state and the defendent. I shall accord you the same privileges that are accorded the local counsel and assure you again that we are delighted to have you with us.

The Court—Now let's proceed to draw the jury, gentlemen.

The following grand jury was empaneled: J. B. Leuty, A. F. Odom, T. A. Odom, H. R. Thomas, R. M. Green, Lee Parham, L. N. Rogers, E. C. Byron, Dr. W. T. Green, T. H. Evans, John Rose, foreman; S. P. Hood, T. E. Benson.

After being duly empaneled and sworn, the usual oath being administered, the court gave the following charge to the grand jury:

Judge's Charge to Grand Jury.

Gentlemen of the grand jury, on May 25, 1925, John T. Scopes was indicted in this county for violating what is generally known as the anti-evolution statute. There is some uncertainty as to whether or not this indictment is valid, and, in order to avoid a possibility of it being invalid, I have determined to convene this grand jury for the purpose of reinvestigating these charges. I now use substantially the same charge I gave the first grand jury.

The statute, which it is alleged the said Scopes violated, is Chapter 27 of the acts of 1925, which makes it unlawful to teach in the universities, normals and all other public schools of the state, which are supported in whole or in part by the public school funds of the state, any theory that denies the story of Divine creation of man as taught in the Bible and teach instead thereof that man descended from a lower order of animals.

This act became the law in Tennessee on March 21, 1925.

This act in part reads as follows:

Section 1. Be it enacted by the general assembly of the state of Tennessee, that it shall be unlawful for any teacher in any of the universities, normals and all other public schools of the state, which are supported in whole or in part by the public school funds of the state, to teach any theory that denies the story of the Divine creation of man as taught in the Bible, and to teach instead that man has descended from a lower order of animals.

Since the act involved in this investigation provides that it shall be unlawful to teach any theory that denies the divine creation of man as taught in the Bible, it is proper that I call your attention to the account of man's creation as taught in the Bible, it is proper that I call your attention to the first chapter of Genesis, reading as follows:

Reads First Chapter of Genesis.

"In the beginning, God created the heaven and earth.

Second—"And the earth was without form, and void; and darkness was upon the face of the deep. And the spirit of God moved upon the face of the waters.

Third—"And God said, let there be light: and there was light.

Fourth—"And God saw the light, that it was good: And God divided the light from the darkness.

Fifth—"And God called the light day, and the darkness he called night. And the evening and the morning were the first day.

Sixth—"And God said let there be a firmament in the midst of the waters, and let it divide the waters from the waters.

Seventh—"And God made the firmament, and divided the watters which were under the firmament from the waters which were above the firmament; And it was so.

Eighth—"And God called the firmament heaven. And the evening and the morning were the second day.

Ninth—"And God said, Let the waters under the heavens be gathered together unto one place, and let the dry land appear, and it was so.

Ten—"And God called the dry land earth; and the gathering together of the waters called He seas: And God saw that it was good.

Eleventh—"And God said, let the earth bring forth grass, the herb yielding seed, and the fruit trees yielding fruit after his kind, whose seed is in itself, upon the earth: And it was so.

Twelfth—"And the earth brought forth grass, and herb yielding seed after his kind, and the tree yielding fruit, whose seed was in itself, after his kind; and God saw that it was good.

Thirteenth—"And the evening and the morning were the third day.

Fourteenth—"And God said let there be lights in the firmament of the heavens to divide the day from the night; and let them be for signs, and for seasons, and for days, and years.

Fifteenth—"And let them be for lights in the firmament of the heavens to give light upon the earth; and it was so.

Sixtenth—"And God made two great lights: The greater light to rule the day and the lesser light to rule the night: He made the stars also.

Seventeenth—"And God set them in the firmament of the heaven to give light upon the earth.

Eighteenth—"And to rule over the day and over the night and to divide the light from the darkness: and God saw that it was good.

Nineteenth—"And the evening,

and the morning were the fourth day.

Twentieth—"And God said, let the waters bring forth abundantly the moving creature that hath light, and fowls that may fly above the earth in the open firmament of heaven.

Twenty-first—"And God created great whales, and every living creature that moveth, which the waters brought forth abundantly, after their kind, and every winged fowl after his kind; and God saw that it was good.

Twenty-second—"And God blessed them, saying, Be fruitful, and multiply, and fill the waters in the seas, and let fowl multiply in the earth.

Twenty-third—"And the evening and the morning were the fifth day.

Twenty-fourth—"And God said, let the earth bring forth the living creature after his kind, cattle and creeping thing, and beasts of the earth after his kind: And it was so.

Twenty-fifth—"And God made the beasts of the earth after his kind, and cattle after their kind, and everything that creepeth upon the earth after his kind: and God saw that it was good.

Twenty-sixth—"And God said, Let us make man in our image, after our likeness: And let them have dominion over the fish of the sea and over the fowl of the air, and over the cattle, and over all the earth, and over every creeping thing that creepeth upon the earth.

Twenty-seventh—"So God created man in His own image, in the image of God, created He him; male and female created He them.

Twenty-eighth—"And God blessed them, and God said unto them, be fruitful and multiply, and replenish the earth, and subdue it; and have dominion over the fish of the sea and over the fowl of the air and over every living thing that moveth upon the earth.

Twenty-ninth—"And God said, Behold, I have given you every herb bearing seed, which is upon the face of all the earth and every tree, in which is the fruit of a tree yielding seed; to you it shall be for meat.

Thirtieth—"And to every beast of the earth, and to every fowl of the air, and to every thing that creepeth upon the earth, wherein there is life, I have given every green herb for meat; and it was so.

Thirty-first—"And God saw everything that he had made, and behold, it was very good. And the evening and the morning were the sixth day."

Therefore, the vital question now involved for your consideration is, has the statute been violated by the said John T. Scopes or any other person by teaching a theory that denies the story of the Divine Creation of man as taught in the Bible, and in Rhea County since the passage of this act and prior to this investigation.

If you find the statute has been thus violated, you should indict the guilty person or persons, as the case may be.

You will bear in mind that in this investigation you are not interested to inquire into the policy or wisdom of this legislation.

Both our state and federal governments are divided into three distinct and separate departments or branches and each has its functions and responsibilities independent of the other and there should be no interference, infringement or encroachment by the one upon the rights, duties, responsibilities and functions of the other.

The policy and wisdom of any particular legislation address itself to the legislative branch of government, provided the proposed legislation is within constitutional limitations.

Our constituton imposes upon the judicial branch the interpretation of statutes and upon the executive branch the enforcement of the law.

The statute involved in this investigation provided that a violation constitutes only a misdemeanor, but there are degrees involved in misdemeanors (not by expressed provision of statute, but in reality), as well as in felonies, and in the very nature of things I regard a violation of this statute as a high misdemeanor, and in making this declaration I make no reference to the policy or constitutionality of the statute, but to the evil example of the teacher disregarding constituted authority in the very presence of the undeveloped mind whose thought and morals he directs and guides.

To teach successfully we must teach both by precept and example.

The school room is not only a place to develop thought, but also a place to develop discipline, power of restraint, and character.

If a teacher openly and flagrantly violates the laws of the land in the exercise of his profession (regardless of the policy of the law) his example cannot be wholesome to the undeveloped mind, and would tend to create and breed a spirit of disregard for good order and the want of respect for the necessary discipline and restraint in our body politic.

Now, gentlemen of the jury, it is your duty to investigate this alleged offense without prejudice or bias and with open minds, and if you find that there has been a violation of the statute you should promptly return a bill, otherwise you should return "no bill."

You may proceed with your investigation.

Whereupon, the grand jury retired and court adjourned to 11 o'clock.

New Indictment Returned.

The Court—Call the grand jury, Mr. Clerk. Whereupon the clerk called the list of the grand jurors.

The Court—Have you a report to make, Mr. Foreman?

Mr. Rose—Yes, sir.

The Court—All right. Is this the final report, Mr. Foreman?

Mr. Rose—Yes, sir.

The Court—I thank you, Mr. Foreman. You gentlemen may be seated.

Gen. Stewart—Now, if your honor please, in No. 5231, I want to quash that indictment.

The Court—You want to move to quash the indictment? In No. 5231?

Gen. Stewart—Yes, sir.

The Court—Let the indictment be quashed. Draw the order, Mr. Attorney-General.

Gen. Stewart—Yes, sir, I will do that, judge.

The Court—Will you please change the number here? Mr. Clerk file this indictment and number it, please. Mr. Clerk number the indictment please and put in on my docket. And put a number on it.

Gentlemen and Mr. Attorney-General, I am calling now for trial Case No. 5232, the State of Tennessee vs. John Thomas Scopes.

Darrow Brings Up Question of Scientists' Testimony.

Mr. Darrow—Your honor, before that I want to have a little talk with the counsel on the other side and the court on the questions of witnesses here, before we do anything else. It is rather informal. Now we have arranged for a considerable number of scientists who will—who are all busy men and we do not want to take them away from their work any longer than we need to, so I thought we ought to get an idea of just how soon we would need them after we start.

The Court—Let me make an inquiry, colonel. You gentlemen are, perhaps, more familiar—you are more familiar with the lines of defense than I. How long do you think it might take to make up the jury? I will inquire from the attorney-general.

Gen. Stewart—Just a minute.

The Court—I just want to—

Gen. Stewart—Of course we cannot anticipate what we might have to contend with. I don't know whether any of these men might not qualify. If we do not have any trouble in the qualifications of the jurors so far as the state is concerned, it will only take a short time. By that I mean to say that it won't consume a day.

The Court—You say a day?

Gen. Stewart—So far as we are concerned it will take, perhaps, not a half a day to select the jury.

Mr. Darrow—Your honor, this case had a great deal of publicity, as the court knows, and in any case of this sort—I am not speaking of the locality, but in any locality, with all the publicity it has had, it is very hard to get impartial juries that the law prescribes, and we may get it quickly, but we feel, so far as the defense is concerned, we ought to have pretty full—a reasonable liberty of examination, to see that we do get as impartial a jury as it possible. As people generally have some general opinions on such subjects and I apprehend it might take some little time to get a jury.

The Court—Colonel, is there any reason why we should not proceed with making up the jury? When the jury is made the state, of course, introduces their proof first, then couldn't you notify your witnesses to be here after the jury is made?

Mr. Darrow—I think so. I assume that your honor and counsel on the other side will be fairly lenient with us at times, if we need it.

The Court—Sure, we will extend you any reasonable courtesy we can.

Mr. Darrow—We are going to try to co-operate with the court and do it expeditiously. Now I am not—I don't suppose the court has considered the question of competency of evidence. My associates and myself have fairly definite ideas as to it, but I don't know how counsel on the other side feel about it. I think that scientists are competent evidence—or competent witnesses here, to explain what evolution is, and that they are competent on both sides.

The Court—Colonel, when the jury is made I will expect you gentlemen—the lawyers for both sides—to outline your theories in an opening statement and in that way the court can have some ideas as to what the issues are going to be, and, of course, after the issues are made up and the evidence is offered, then the court will promptly rule as to the competency of any evidence that is offered.

Mr. Darrow—Of course, your honor, all I am doing at this time is because our witnesses are generally from a long distance. They get no pay for their time and are busy men, and I don't want to impose on them any more than I need to and, perhaps, if there is to be any question of competency of evidence that could be disposed of some time before we get them here.

The Court—Yes, we could. I take it you might raise the question by a motion, perhaps.

Mr. Darrow—Yes, we could raise it by agreement. I don't think that there is any disposition on the part of either of us to not be perfectly frank with each other about these matters.

The Court—Why not get an announcement from the state as to whether or not they are ready for trial and then I will call on you for an announcement and if you think you will be ready by tomorrow, some time soon, we could proceed in making the jury and when the jury is made, then, of course, if the defense asks for a little delay, I will give it.

Mr. Darrow—Well, just a minute now.

The Court—Yes, sir.

Whereupon the attorneys conferred informally in the presence of the court.

The Court—Have you any announcement for the state, Mr. Attorney-General?

Stewart Outlines State's Attitude.

Gen. Stewart—Yes, your honor. We have just been holding a conversation here for a few minutes, as has been evident. If the court please, in this case, as Mr. Darrow stated, the defense is going to insist on introducing scientists and Bible students to give their ideas of certain views of this law and that, I am frank to state, will be resisted by the state as vigorously as we know how to resist it. We have had a conference or two about that matter, and we think that it isn't competent as evidence; that is, it isn't competent to bring into this case scientists who testify as to what the theory of evolution is or interpret the Bible or anything of that sort. On the other hand, these gentlemen are just as earnest in their insistence that they are entitled to it. Now in order that we may not disqualify a number of jurors in the discussion of this matter, and further in order and for the purpose of accommodating these gentlemen, and accommodating perhaps the witnesses whom they anticipate bringing here, and whom they do not want to bring here if the court should hold that matter not competent, we have agreed to take that matter up out of order, but we prefer to select a jury before that is done. Now we are willing to take that up most any time, and we have agreed, if it meets with the approval of the court, that we adjourn until in the morning. These gentlemen are tired and they haven't gotten acclimated yet, and we are willing to give them a half a day.

The Court—May I inquire how many regular veniremen have you, Mr. Sheriff?

Mr. McKenzie (Gordon)—Twenty-nine, your honor.

The Court—Twenty-nine, excluding the grand jury?

Mr. McKenzie—No, sir.

The Court—Including the grand jury?

Mr. McKenzie—Yes, sir.

The Court—The grand jury would not be competent.

Mr. McKenzie—Sixteen.

The Court—Mr. Attorney-General, how many jurors would you anticipate we might need to make the panel? Mr. Attorney-General, I don't like to lose this afternoon. A great many people are here and I am willing to adjourn until 1:30, and I can have the sheriff to have us 100 men here at that time.

Gen. Stewart—Judge, these gentlemen, of course—I want to show all the courtesy I can to these visiting lawyers—these gentlemen have come in here on trains from a long distance last night, and they are tired and not feeling very well.

The Court—Well, it wouldn't require any great amount of energy to select a jury, would it?

Mr. Malone—Your honor, I think I am the only one who wanted it to go over until Monday, and since nobody else wants it, I believe we ought to continue and go on right now.

Mr. Darrow—I think we ought to have the afternoon on it.

The Court—Well, colonel, we will only hold about two hours, and then I will give you a good rest. Of course, I have a great regard for the lawyers, but I have some regard for others.

Mr. Darrow—Yes, I know; certainly you do have, but that doesn't seem that is hardly an unreasonable request, let it go over until morning.

The Court—I would prefer to proceed with getting the jury. I wouldn't expect you to enter on the trial this afternoon. What do you say, gentlemen, to ordering 100 extra jurors? What do you think for the state?

Mr. Darrow—May we get the court to tell us just what the law is as to—you say you have sixteen here. Suppose those are exhausted?

The Court—Well, you would be entitled, Col. Darrow, to exhaust those first, if you saw proper, and then the court would order an extra panel from bystanders, under the law. Usually by agreement of counsel in cases of this sort, we anticipate that we may perhaps need 100 men, and by agreement of counsel we would send the sheriff out and have 100 extra jurors summoned and brought in, and if we didn't get the panel out of that—get the jury out of that panel—we would send him out to get another panel.

Mr. Darrow—That is, drawn from a regular box?

The Court—No, sir, we have no regular jury box in this county—it is drawn if you want it—in a felony case it is drawn out of the box, yes, sir, if you require it, but in a misdemeanor it isn't.

Mr. Darrow—That is if you agree?

The Court—Yes, sir, if you rather, the names will go through a hat. It would be in the discretion of the court and it will be perfectly agreeable with me. I will give you any information, colonel, I can—anything you want to ask me.

Mr. Darrow—Thank you.

The Court—I do not mean that I know it all, but I will tell you anything I know.

Gen. B. G. McKenzie—If the court please, about the only thing I know is that Col. Darrow and I are the only two suspender men in the court room.

The Court—What do you say to ordering the extra jurors, summoning them here, and let the names go into a box and be drawn out until the jury is made?

Mr. Darrow—I suppose that would be the regular way, wouldn't it?

The Court—That would be regular, by agreement. Under the law strictly, you would be entitled to a panel—each side is entitled to three challenges in this case—peremptory challenges — that would be six, and twelve men would be eighteen on the regular panel. The regular panel under the law would be eighteen men, and usually, to save time, we put in more names and proceed until we get the jury.

Mr. Darrow—We have got local counsel that doesn't seem to be present. I think I ought to consult him about it.

The Court—Suppose I order 100 men to be here at 1:30 and we can take them out a panel at a time—eighteen at a time, and adjourn until 1:30 and we will proceed at that hour?

Mr. Darrow—Hadn't you better make that 2?

The Court—Court will adjourn until 1:30.

Thereupon court adjourned until 1:30 p. m.

AFTERNOON SESSION.

Court—Mr. Attorney-General, are you ready to proceed with the selection of this jury?

Gen. Stewart—Yes, sir.

Court—Are you ready, gentlemen?

Mr. Darrow—Yes, sir.

The Court—Do you want the names drawn from the box?

Mr. Darrow—Yes, sir.

The Court—Call them as they are drawn from the box.

Mr. Darrow—Just the panel?

The Court—Examine them, Mr. Darrow, when they are drawn from the hat.

Mr. Darrow—That's all right.

The Court—All right. Let the sheriff draw them?

Mr. Darrow—Yes, sir.

The Court—You may proceed, Mr. Sheriff.

Jury Is Selected.

Sheriff Harris—W. F. Roberson, number twelve.

Court—Come around, Mr. Roberson.

(The venireman was sworn by the court.)

Court—Are you a householder or freeholder in Rhea county?

Juror—Yes, sir.

Court—You are a householder?

Juror—Yes, sir.

Court—Have you formed or expressed an opinion as to the guilt or innocence of this defendant, John T. Scopes?

Juror—Well, to some extent, judge.

Court—What do you base that opinion on, Mr. Roberson?

Juror—Rumor.

Court—From some witness? Information from some witness, some of them who profess to know?

Juror—No, sir.

Court—General rumor?

Juror—Yes, sir.

Court—And do you think you can wholly disregard your opinion and go into the jury box, and try the case on the law and the evidence, and render a fair and impartial verdict?

Juror—Yes, sir.

Court—I think he is a competent juror, gentlemen. I will pass him to the state first.

Gen. Stewart—The state will take him.

Mr. Darrow—We have the right to examine him?

Court—Ask him anything you desire.

This talesman, W. F. Roberson, was examined as follows by Mr. Darrow, for the defense:

Q—What is your business, what do you do?

A—A farmer.

Q—Do you own a farm of your own?

A—I am a renter.

Q—What is your age?

A—I am 30.

Q—Have you ever given any special attention to this case?

A—Well, no more than just reading the newspapers.

Q—Are you satisfied that you could try it with perfect fairness to both sides?

A—Yes, sir, I think I could.

Mr. Darrow—All right, we will take him.

Court—All right. Have a seat, Mr. Roberson. Call the next man.

Sheriff Harris—Number 34, J. W. Dagley.

Court—Mr. Roberson, I intended to ask you, are you related by blood or marriage to Mr. Walter White, the prosecutor, or to John T. Scopes, the defendant?

Mr. Roberson—No, sir.

Court—Have you any interest at all in the case?

Mr. Roberson—No, sir.

The court swears Talesman J. W. Dagley.

Court—Mr. Dagley, are you a householder or freeholder of Rhea county?

Juror—Yes, sir.

Court—Are you related by blood or marriage, to Walter White, the prosecutor, or to John T. Scopes, the defendant?

Juror—No, sir.

Court—Have you formed or expressed an opinion as to the guilt or innocence of the defendant, John T. Scopes, on the charge of violating the antievolution statute?

Juror—Well, I can't hardly say that I have.

Court—Have you any fixed opinion, Mr. Dagley, any definite opinion as to his guilt or innocence?

Juror—No, sir.

Court—Have you heard rumors about the case?

Juror—Rumor.

Court—Have you talked to any person who is a witness in the case, or who professed to know the facts?

Juror—No, sir.

Court—You would wholly disregard any impression you have regarding to the matter, Mr. Dagley, and go into the jury box—the case wholly upon the law and the evidence and render an impartial verdict to both sides?

Juror—Yes, sir.

Court—He is a competent juror.

Gen. Stewart—We pass him to the defendant.

Examination by Mr. Darrow, for the defense:

Q—You are a farmer?

A—Yes, sir.

Q—Near here?

A—What is that?

Q—Do you live near here, near the town?

A—Twelve or fourteen miles.

Q—Have you lived in Tennessee most of your life?

A—I was born and raised here.

Q—In this community?

A—Yes, sir.

Q—Have you ever known anything about evolution, or read about it?

A—I have not.

Q—You don't know anything about it at this time?

A—No, sir.

Q—Are you a church member?

A—Yes, sir.

Q—Of what church?

A—Methodist.

Q—You have been for a good many years?

A—Yes, sir, a number of years.

Q—Have you ever heard it discussed in church?

A—No, sir.

Q—Did you ever hear your minister express himself on it?

A—No, sir.

Q—Did you ever hear your neighbors say what they thought about this case?

A—Well, no, sir, I don't know that I have.

Q—Did you ever hear anybody? I am not asking you now what you heard, but did you ever hear anybody say what they thought about it?

A—Well, not directly, I don't believe.

Q—Well, you have an opinion now, at this time, I believe you said?

A—No, sir.

Q—You have an opinion as to how this case should be decided at this time? I believe you said you did have? Did you?

A—I don't believe I did.

Q—Well, I might have misunderstood you. But you haven't, now, any opinion, one way or another?

A—Well, really, I haven't, no, sir.

Q—You haven't?

A—I haven't, no, sir.

Q—You don't know Mr. Scopes?

A—Do not. Outside of I have seen him here about town.

Q—You have never expressed an opinion as to what you thought ought to be done in this case, now?

A—I have not.

Q—You have a family, I suppose?

A—I have a family.

Q—Are your children going to school?

A—They go to school during school time.

Q—And at this time you haven't any idea about evolution at all?

A—No, sir, I have not.

Q—You don't know what it is, do you?

A—Well, I—I think I know what it is.

Q—Well, have you any prejudice against it?

A—Well, I don't believe I am competent to say. I understand it well enough; to say I have any prejudice either way—

Q—Well, you know your own mind, and we are entitled to a fair trial, by men who can be perfectly fair. You could tell whether you could be or not, couldn't you?

A—I think I would be fair, yes, sir.

Q—And you would give this man a fair trial, would you?

A—Yes, sir, I would.

Mr. Darrow—Have a seat.

Jim Riley, sworn by the court and examined on his voire dire:

Questions by the court:

Q—Mr. Riley, are you a householder or a freeholder in Rhea county?

A—Yes, sir.

Q—Are you related by blood or marriage to Walter White, the prosecutor, or to John T. Scopes, the defendant, in this case?

A—No, sir.

Q—Have you formed or expressed an opinion as to the guilt or innocence of the defendant?

A—No.

Q—And you have no definite opinion about it?

A—No, not anything about the facts at all; no, sir—not only just what I heard.

Q—Just rumor talk?

A—Yes, sir.

Q—You did not talk to any witness that undertook to tell you what the facts were?

A—No, sir.

Q—And you can go into the jury box and try the case wholly on the law and the evidence, disregarding any impression or opinion that you might have and render a fair and impartial verdict to both sides?

A—Yes, sir.

Court—Competent juror.

Mr. McKenzie—Pass him to you, colonel.

Questions by Mr. Darrow:

Q—Mr. Riley, you are a farmer?

A—Yes, sir.

Q—How far from Dayton do you live?

A—Just at the lower edge of town.

Q—You have lived in Dayton—you have lived here in this county for many years?

A—Yes, sir.

Q—Do you know Mr. Scopes?

A—I just know him—I just saw him once—just one time.

Q—Are you a member of any church?

A—Yes, sir.

Q—What one?

A—Baptist.

Q—You have been a member of the Baptist church for a long while?

A—Yes, sir.

Q—Do you know anything about evolution?

A—No, not particularly.

Q—Heard about it?

A—Yes, I have heard about it.

Q—Know what it is?

A—I don't know much about it.

Q—Have you any opinion about it—prejudice? Have you any prejudice against the idea of evolution? You understand my question—what I mean by prejudice, don't you? If you don't I will make it easier.

A—No, I have no prejudice.

Q—And you have heard that Scopes here has been indicted for teaching evolution?

A—Yes, sir, I have heard that.

Q—And you have no prejudice against it?

A—I don't know the man—wouldn't know him if I was to meet him out on the road at all—just saw him one time.

Q—I mean have you any prejudice on account of his having taught evolution, if he did teach it?

A—Well, I couldn't tell you about it because I don't know what he taught.

Q—Have you any feeling that it is a wrong teaching at ths time?

A—Well, I haven't studied very much about it.

Q—Ever talk to anybody about it?

A—None to amount to anything; no, sir.

Q—Ever heard anybody preach any sermons on it?

A—No, sir.

Q—Ever hear Mr. Bryan speak about it?

A—No, sir.

Q—Ever read anything he said about it?

A—No, sir; I can't read.

Q—Well, you are fortunate. You can be a perfectly fair juror, can you?

A—Yes, sir.

Q—And you will be if taken as a juror?

A—Yes, sir.

Court—What do you say for the state?

Mr. McKenzie—Sit down, Mr. Riley.

Court—Have a seat, Mr. Riley.

No. 20, J. P. Massingill, duly sworn by the court and examined on his voire dire, testified as follows:

Q—Mr. Massingill, you are a householder or freeholder in Rhea county?

A—Householder.

Q—Mr. Massingill, are you related by blood or marriage to Walter White, the prosecutor, or John T. Scopes, the defendant?

A—Not at all that I know of.

Q—Have you formed or expressed an opinion as to the guilt or innocence of the defendant in this case?

A—From rumors and newspapers—of course, I read. I don't know anything about the evidence.

Q—You haven't talked with any person who professed to know the facts?

A—No, sir.

Q—Have you read any detailed account of the charge, Mr. Massingill, undertaking to give the details of the charge and what the evidence was?

A—Yes, sir, I have read a sketch of it.

Q—Did you read what the evidence was, given before the magistrate's court, or wherever it has been tried, or not?

A—I don't recall.

Q—Now, Mr. Massingill, could you go into the jury box and wholly disregard any impression or opinion you have?

A—Yes, sir.

Q—And try the case wholly on the law and the evidence, rendering a fair and impartial verdict to both sides?

A—I think so; yes, sir.

Court—He seems to be competent, gentlemen.

Mr. McKenzie—Pass him to you, Colonel.

Mr. Darrow—Your honor, if we exercise peremptory challenges, must we do it as we go along?

Court—Yes, sir, you have three.

Mr. Darrow—That is a different practice to what I am familiar with. In Illinois you can do it at any time.

Court—No, do it as we go along.

Mr. Darrow—If that is true, of course, you never know which one to challenge.

Court—Yes, I can see the difference, but the practice is different here.

Mr. Darrow—May I consider just a moment on the last one?

Court—Yes, sir, you may.

Mr. Darrow—May I ask Mr. Riley one question, your honor?

Court—Yes, sir.

Mr. Darrow's question to Mr. Riley:

You said you couldn't read. Is that due to your eyes?

Mr. Riley (A) No, I am uneducated.

Q—That is because of your eyes?

A—I say I am uneducated.

Q—Have your eyes bothered you?

A—No, I am uneducated.

Q—You have worked always on a farm?

A—Not all the time. I have worked in the mines a good deal of the time.

Q—Whereabouts?

A—Right up here at Nelson's.

Q—How long did you work in the mines?

A—Some four or five years.

Q—When did you leave the mines?

A—Well, it has been twenty years ago.

Gen. Stewart—I presume, of course, that the defense know, since they ask about the peremptory challenges, that they have three.

Mr. Darrow—Yes, sir, I had already found that out.

Court—What do you say to Mr. Massingill?—for the state?

Mr. McKenzie—I pass him to you, Colonel.

Questions by Mr. Darrow:

Q—What is your business?

A—I am a minister.

Q—Whereabouts?

A—How is that?

Q—Where?

A—I live in Rhea county.

Q—What part of it?

A—I live in the second district of Rhea county, twenty miles north of this place.

Q—Where do you preach?

A—I preach over the county in the rural sections.

Q—You mean you haven't any regular church?

A—I have. I am pastoring four churches—have four appointments.

Q—Ever preach on evolution?

A—I don't think so, definitely; that is, on evolution alone.

Q—Now, you wouldn't want to sit on this jury unless you were fair, would you?

A—Certainly, I would want to be fair; yes, sir.

Q—Did you ever preach on evolution?

A—Yes. I haven't as a subject; just taken that up; in connection with other subjects. I have referred to it in discussing it.

Q—Against it or for it?

A—I am strictly for the Bible.

Q—I am talking about evolution, I am not talking about the Bible. Did you preach for or against evolution?

A—Is that a fair question, judge?

Court—Yes, answer the question.

A—Well, I preached against it, of course! (Applause).

Q—Why, "of course?"

Court—Let's have order.

Mr. Darrow—Your honor, I am going to ask to have anybody excluded that applauds.

Court—Yes, if you repeat that, ladies and gentlemen, you will be excluded. We cannot have applause. If you have any feeling in this case you must not express it in the courthouse, so don't repeat the applause. If you do, I will have to exclude you.

Q—You have a very firm conviction—a very strong opinion against evolution, haven't you?

A—Well, some points in evolution.

Q—Are you trying to get on this jury?

A—No, sir.

Q—Have you formed a strong conviction against evolution?

A—Well, I have.

Q—You think you would be a fair juror in this case?

A—Well, I can take the law and the evidence in the case, I think, and try a man right.

Q—I asked if you think you thought you could be a fair juror?

A—Yes, sir.

Q—Have you heard about Mr. Scopes?

A—Yes, sir; yes.

Q—You have heard that he is an evolutionist, haven't you?

A—Yes, sir, I have heard that.

Q—And in your opinion he has been teaching contrary to the Bible?

General Stewart—If your honor please, I except to that. The question involved here will be whether or not—not, I apprehend if Mr. Scopes taught anything that is contrary to the Bible—that isn't the question. He has asked him whether or not he has prejudged the guilt of the defendant.

Court—He has a right to know that.

Gen. Stewart—The man has already stated to him that he had no opinion in the case.

Mr. Darrow—Do you think he would be a fair juror in the case?

Gen. Stewart—Yes, I do, if he says so.

Mr. Darrow—I don't.

Court—I think the lawyers have the right to get all the information they can on the subject, and I will treat both sides alike.

Mr. Darrow—What was that question? (question read).

Court—You may answer that.

A—Yes, sir.

Q—You have that opinion now?

A—I have no opinion to convince me otherwise.

Court—Questions by the court:

Q—Have you, in your mind now, Mr. Massingill, a fixed opinion that he has taught a theory contrary to the theory of the Bible as to the creation of man?

A—Yes, sir.

Q—Would that have any weight with you or any bearing with you in the trial of this case if you were selected as a juror?

A—I think I am fair and honest enough to lay aside things and give a man justice.

Q—Could you disregard any opinion you have and go in the jury box and render a fair verdict to both sides regardless of any impression you now have?

A—The opinion I have is from the public press and what I heard. Of course, I could surrender that.

Q—You don't know whether it was true or not? What I want to be sure of is this, if you were selected on the jury, could you go in the box and wholly disregard any impression or opinion you have and try the case wholly on the law and the evidence, rendering a fair verdict to both sides?

A—You mean in regard to this particular case?

Q—In regard to the charges here?

A—Sure, I would do that, too.

Court—You may proceed, gentlemen. He seems to be competent.

Mr. Darrow—You now have an opinion that evolution is contrary to the Bible and that my client has been teaching evolution; as you stand there now, that is your opinion?

A—From the information I have in regard to his teaching.

Q—That is your opinion now, isn't it, as you stand there now?

A—Sure it is.

Q—You could change it if you heard evidence enough to change it on?

A—Yes, sir.

Q—Otherwise you couldn't?

A—I have no right to; I don't think.

Mr. Darrow—I challenge for cause.

Court—Well, I want every juror to start in with an open mind. I will excuse you, Mr. Massingill.

J. H. Harrison (29), called and sworn, upon examination testified:

Examination by court:

Q—Are you a householder and freeholder in this county?

A—Yes, sir. I claim my age, too old, I don't want to sit on the jury.

Q—How old are you?

A—Sixty-six.

Q—Claiming exemption on account of your age?

A—Yes, sir.

The court—You may be excused.

W. D. Taylor (14), sworn and examined on the voir dire, testified:

Questions by the court:

Q—Mr. Taylor, are you a householder and freeholder of this county?

A—Yes, sir; householder.

Q—Are you related by blood or marriage to Walter White, the prosecutor, or John T. Scopes, the defendant?

A—No, sir.

Q—Have you formed or expressed an opinion as to the guilt or innocence of this defendant?

A—Well, I have to a certain extent.

Q—Have you talked to any witness, Mr. Taylor?

A—No, sir; I have not talked to any witness.

Q—Have you talked to any person who professed to know the facts?

A—Yes, I was present during part of the preliminary, I heard a part of the lawyer's talk. I never heard any of the evidence.

Q—You heard some of the argument in the preliminary?

A—Yes, sir.

Q—Mr. Taylor, here is the important thing I am going to ask you, could you wholly disregard any impression that you might have as to his guilt or innocence, and go into the jury box and try this case wholly on the law and the evidence?

A—Yes, I think I can.

Q—And any impression that you have now, would it have any influence on your verdict, do you think?

A—No, sir; I think I could.

Q—You could wholly disregard that?

A—Yes, sir.

Q—Eliminate it from your mind?

A—Yes, sir.

Q—And try the case as if you never heard it before?

A—Yes, sir.

Examination by Gordon McKenzie, Esq.:

Q—Now, Mr. Taylor, you could go into the jury box, and before you went into the jury box you could disregard any opinion you might have and give the defendant a fair trial?

A—You mean before I went into the jury box?

Q—Yes, before you went into the jury box you could disregard any opinion you might have and give the defendant a fair trial?

A—Yes, sir.

Q—In other words, it would not take any evidence to remove the opinion that you have, would it, Mr. Taylor?

A—Well, I have never heard any, that is, I never heard any witness; all I have heard is what I have read.

Q—And it would not take any evidence as I understand you to remove any opinion you might have?

A—Well, that is not strong enough to require that, I don't think.

Q—And you could give the defendant a fair and impartial trial?

A—Yes, sir.

Mr. McKenzie—I pass him to you, Colonel.

Re-examination by the court:

Q—What you mean, is, what you have read, you do not know whether it is true or not?

A—Well, I don't know much about it, just what I have read about it.

Q—Are you sure if you were chosen on the jury, when you sit down in the jury box, you could go in there with an open mind without any leaning or bias or prejudice against either side; could you do that?

A—Yes, sir.

The court—All right.

Examination by Mr. Darrow:

Q—Mr. Taylor, what is your business?

A—Farmer.

Q—How far from Dayton do you live?

A—Ten miles, east.

Q—You have been a farmer here for a good many years?

A—All my life, yes.

Q—Born in Tennessee?

A—Yes, sir.

Q—A member of any church organization?

A—What say?

Q—Are you a member of any church organization?

A—Yes, sir.

Q—What one?

A—Methodist Episcopal, South.

Q—Methodist Episcopal, South, that is what we call the Southern Methodist?

A—Yes, sir.

Q—You were present at the preliminary hearing of Mr. Scopes?

A—I was in town that day, heard Col. Neal, a part of his talk and a part of the other side.

Q—You just heard the lawyers talk?

A—Yes, sir.

Q—You would not pay much attention to that anyway?

A—I could not hear them, I was not close enough. I would pay attention to what I could hear, but I could not hear it, though.

Q—You have heard about evolution, I suppose?

A—Oh, yes, I have heard of evolution all my life.

Q—Have you read any of it?

A—No, sir; I never read on evolution at all.

Q—Did you ever hear anybody talk against it or for it?

A—Well, I have heard lots of talk against it, and some talk for it, whether either one knew what they were talking about, I don't know. They might have been like me, did not know.

Q—You have not any opinion as you stand there now, as to whether it is a true doctrine or a false one?

A—No, I do not think I have. I could hear the evidence in the case and then decide.

Mr. Stewart—I submit that is not a proper interrogation, whether evolution is true or not. The correct test is whether or not he has an opinion that the defendant is guilty or not guilty.

Mr. Darrow—I was going to follow with that.

The court—Go ahead.

Mr. Darrow—I did not get up to that.

Mr. Stewart—My objection is, I do not think the other is proper.

The court—Yes, go ahead.

Q—(Mr. Darrow) You have not any opinion now as to whether Mr. Scopes ought to be convicted or discharged?

A—I do not know what he taught. I do not know anything about it, only what I have read.

Q—You have no prejudice against evolution?

A—No, sir; as far as I know evolution, I have not.

Q—You have no prejudice against Mr. Scopes, one way or the other?

A—No, sir.

Q—Mr. Taylor, you would not sit on a jury without you thought you could be perfectly fair?

A—I try to be fair wherever I am at.

Q—And you think your mind is in such shape that you could be perfectly fair?

A—Yes, sir.

Q—You went to the public schools here I suppose?

A—Not in this county.

Q—In Tennessee, I suppose?

A—Yes, sir.

Q—How far did you go?

A—When I went to school they did not grade like we do now, I went through high school.

Q—They taught science in the schools?

A—No, sir.

Q—Do you take many newspapers?

A—I take one, that is, one outside of the county, I take the county paper.

Q—Do you take a religious paper?

A—Yes, sir; I take a religious paper.

Q—How?

A—Yes, sir, one church paper.

Q—Have you read anything about evolution in it?

A—Yes, something about it. I have not read anything about evolution in a church paper. I see the headlines, but have not read it.

Q—Did you ever hear anybody speak on it?

A—Yes, sir; I have heard them speak, preach on it, that is, ministers of the gospel preach on it.

Q—What is that?

A—I have heard ministers of the gospel preach on it one time.

Q—Your own church?

A—Yes, sir; in my own church.

Q—Well, did you form any opinion on that account?

A—I don't know that he told me any more than I knew about it.

Q—Now, you say you are sure you will be fair of mind, I will not ask any more questions, are you sure?

A—Yes, sir.

Mr. Darrow—All right.

The court—What do you say for the state?

Gen. Stewart—Sit down.

Tom Jackson (23), being duly sworn and examined on the voir dire, testified:

Examination by the court:

Q—Are you a householder and freeholder in Rhea county?

A—Yes, sir.

Q—Are you related by blood or marriage to Walter White, the prosecutor, or John T. Scopes, the defendant?

A—No, sir.

Q—Have you formed or expressed an opinion as to the guilt or innocence of this defendant?

A—Well, I do not know. I expect I have.

Q—Have you talked to any witness that professes to—

A—No, sir.

Q—Any person that professes to give the facts, Mr. Jackson?

A—No, sir.

Q—Do you think you have got an opinion in your mind as to whether or not he is guilty or innocent, a definite opinion?

A—No, sir, I don't know about that.

Q—You have no fixed opinion either way?

A—No, sir.

Q—You have not heard any of the proof?

A—No, sir.

Q—Just heard rumors?

A—Just rumors and newspaper reading, yes.

Q—Newspaper accounts?

A—Yes, sir.

Q—Did you read any account of the evidence that may have been given in the preliminary?

A—I don't think so.

Q—Just read the newspaper comments?

A—Yes, sir; just comments.

Q—You do not know whether they knew what the facts were or not?

A—No, sir.

Q—Don't know?

A—No, sir.

Q—Now, Mr. Jackson, could you go in the jury box, and wholly divest yourself of any impression you have and go in the jury box and try the case wholly upon the law and the evidence, being fair to both sides?

A—Yes, sir.

Q—You think you could?

A—Yes, sir; I think I could.

Q—Now, if you were chosen, Mr. Jackson, on the jury, could you just make up your mind before you hear any proof taken, whatever you read, you do not know whether it was true or not, would you go in the jury box with an open mind without leaning either way?

A—Yes, sir.

Q—You think you could, before hearing any proof?

A—Yes, sir.

The court—All right.

Mr. McKenzie—You live up near Spring City?

A—Yes, sir.

Mr. McKenzie—Pass him to you, Colonel.

Examination by Mr. Darrow:

Q—You are a farmer?

A—Yes, sir.

Q—You have lived most of your life in Tennessee?

A—Well, sir, I have been all over the United States, I suppose, almost.

Q—Where else have you lived?

A—Oh, I have not lived, I have traveled around, just from one state to another.

Q—What is your kind of work?

A—Farmer.

Q—When you traveled around, was that just to see the country, or working?

A—No, sir. I was in the United States army.

Q—And when did you get out of the army?

A—1912; December, 1912.

Q—When did you go in?

A—Oct. 9, or October, 1909.

Q—1909?

A—1909.

Q—What is your age now?

A—Forty-three.

Q—You went to school here in Tennessee?

A—Yes, sir.
Q—Now, whereabouts do you live now?
A—Spring City.
Q—How far is that from here?
A—It is about sixteen or eighteen miles.
Q—And what do you do now for a living?
A—Farming.
Q—A farm of your own?
A—Yes, sir.
Q—Do you belong to the church?
A—Yes, sir.
Q—What one?
A—Methodist, Southern Methodist.
Q—You have belonged to that church most of your life, I suppose?
A—No, sir; six or seven years.
Q—Join it here?
A—I belong at Washington.
Q—Now, have you ever heard anything about evolution?
A—I have read about it, yes, sir.
Q—Do you remember what you read it in?
A—Newspapers.
Gen. Stewart—Talk louder, please.
Q—(Mr. Darrow) If you will face the court, I will hear and my friend will hear you over there. All of us will hear. You have read in the newspapers, anywhere else?
A—I do not know. I may have read in magazines, something like that, just through reading.
Q—You do read magazines, do you?
A—Yes, sir.
Q—Do you take any magazines?
A—No, sir; not now.
Q—Have you ever heard anybody talk about it, make speeches or sermons?
A—No, sir; not public speeches.
Q—You have heard this case talked about?
A—Yes, sir.
Q—By people around town or people in your town?
A—Yes, sir.
Q—You have heard them say what they thought about it?
A—Well, some I have, yes, sir.
Q—Well, now I am not asking what you thought about it, but have you ever expressed your opinion as we call it?
A—Well, I don't know, I might have done so.
Q—How?
A—I might have done so. I don't know.
Q—Well, if you had you don't remember, is that right? If you have, you do not remember. If you know —
A—I don't.
Q—What?
A—I don't know, I expect I have. It is the general talk all over the community, all over the county, all over the country since this came up.
Q—You need not tell me now what you said, but do you remember now what you said?
A—No, sir.
Q—Well, have you any opinion now as to what should be done in this case? You need not tell what it is, but have you an opinion?
A—Do you mean as to the guilt or innocence?
Q—Yes.
A—No, sir.
Q—Have you any opinion as to what ought to be done if you are a juror, I mean, at this time?
A—No, sir.
Q—Have you any opinion on evolution at all?
A—I have, yes.
Q—How long have you had it?
A—Well, I have been almost since —
Gen. Stewart—We except to that, if the court please?
The court—I am not sure whether that would be competent or not, Colonel. I think the question is whether or not he has an opinion as to the guilt or innocence of this defendant.
Mr. Darrow—That is my—
The court—Of course, for your own information, I might allow you to ask about that, that you might determine whether or not you would wish to peremptorily challenge him.
Mr. Darrow—Yes. Have you got a strong opinion one way of the other on evolution?
A—Yes, I have my opinion on evolution, yes.
Q—Do you know where you got it?

A—I got it from the Bible.

Q—Have you any opinion from what you have heard, whether Mr. Scopes taught evolution?

A—No, sir; I do not know anything about that.

Q—You have a prejudice against evolution, have not you?

A—Well, to some extent, I suppose I have.

Q—And against teaching it?

A—Yes, I am against teaching evolution—evolution of man, not evolution of the mind.

Q—Of the man?

A—Of the man, yes.

Q—But not of the mind?

A—No, sir.

Q—Do you think you would be a fair juror in this case, where Mr. Scopes is charged with teaching evolution?

A—I don't know, I would do the best I could.

Q—I think I know what kind of a mind you have got. I think you want to be perfectly fair. Mr. Scopes is here charged with teaching evolution. You have told us about your opinion on evolution. Now, you can tell better than the lawyers on either side, and better than I, or better than the judge, as to whether you think you would be a perfectly fair man to try Mr. Scopes.

The court—Do you think you could be fair?

A—Yes, sir.

The court—All right.

Mr. Darrow:

Q—If you were unlucky enough to be a defendant, would you think you would get a perfectly fair trial from one who feels as you do?

A—Yes, sir.

Q—Have you any feeling or prejudice against a man because he believes in evolution?

A—No, sir.

Q—Or because he disagrees with you in religious matters?

A—No, sir; that is his own affair.

Q—What?

A—That is his own affair; no, sir.

Mr. Darrow—May we have a minute for consultation?

Q—If you were on a jury, would you care what anybody else thought about it, so long as you did what you thought was right?

A—Yes, sir. I would just do whatever I thought was right and would be what I would do, if I thought I was right, I would still be right, I would stay right.

Mr. Darrow—We will excuse him.

The court—Excused by the defendant.

R. L. Gentry, being examined on his voir dire, testified as follows:

Examination by the court:

Q—Mr. Gentry, raise your right hand please. Do you solemnly swear you will make true answers to all such questions as may be asked you in the present inquiry?

A—Yes, sir.

Q—Are you a householder and freeholder of Rhea County?

A—Yes, sir.

Q—Are you related to Walter White, the prosecutor in this case?

A—No, sir.

Q—Have you formed or expressed an opinion as to the guilt or innocence of the defendant in this case?

A—No, sir.

Q—You don't know anything of the facts, Mr. Gentry?

A—Not only what I have heard.

Q—Do you know whether that is true or not?

A—Only what I have seen in the papers.

Q—Sometimes you don't know whether everything you read in the papers is true or not?

A—No, sir.

Q—You have no fixed or definite opinion, present ideas as to his guilt or innocence?

A—No, sir.

Q—Could you go in the jury box and try the case according to the law and the evidence?

A—Yes, sir.

The court—Competent juror.

Gen. Stewart—Ask him about his relationship to Mr. Scopes.

The court—I did that.

Gen. Stewart—You asked him about the prosecutor, but not about Scopes, the defendant.

The court—Are you related to the prosecutor or to J. T. Scopes, the defendant?
A—No, sir.
Mr. McKenzie—You can ask Prof. Gentry, Col. Darrow. We pass him to you.
Examination by Mr. Darrow:
Q—Where do you live?
A—I live about two miles from here.
Q—Farm?
A—Yes, farmer and teacher.
Q—How is that?
A—Farmer and teacher in the public school.
Q—Been teaching in the public schools here in Tennessee?
A—Yes, sir, in Rhea county.
Q—How long?
A—About twenty years. I came here in 1901, over twenty years.
Q—Do you own your own farm?
A—Yes, sir.
Q—How large a farm?
A—Right about 106 acres.
Q—And for how many years have you been teaching and farming?
A—Several years.
Q—Are you still teaching?
A—Yes, sir.
Q—Winters?
A—I teach in the fall. Start in August and teach the winter months.
Q—You have a family, I take it?
A—Yes, sir.
Q—Belong to some church?
A—Belong to the Baptist.
Q—Been a Baptist for a good many years?
A—Well, about twenty-five years.
Q—Never belonged to any other church?
A—No, sir.
Q—Common school you teach in?
A—Yes, sir.
Q—Do you teach in the high school?
A—No, sir.
Q—Do you read the papers?
A—Yes, sir.
Q—And magazines?
A—Yes, sir.
Q—Do you go to church in town or out where you live?
A—Go out where I live and come to town some.
Q—You visit around a little?
A—Yes, sir.
Q—Well now, you have, of course, read and thought some about the theory of evolution, have you?
A—Yes, sir; I read the books and taught evolution.
Mr. Darrow—Read that answer.
Mrs. McCloskey (court reporter):
A—Yes, sir, I read the books and taught evolution.
Mr. Darrow:
Q—How long have you been reading?
A—I don't know; a long time.
Q—I don't ask you to be exact.
A—I started when I was a school boy and read those books. I have read them off and on all my life.
Q—Still read them?
A—Yes, sir, I have and read them once in a while when I want to refer to something. I have read them all my life.
Q—Well, I am asking your opinion, but you have an opinion, haven't you?
A—Of course a person would have an opinion about such as that if he had thought and read about it.
Q—You could not give Mr. Scopes a fair trial?
A—Yes, sir, I think so.
Q—You know pretty well?
A—Well, I know I could.
Q—Probably I oughn't to ask you this question, but if you were sitting here as a juror, of course, you know how important that is, don't you?
A—Yes, sir.
Q—You know how important this matter is, and that it has caused a great deal of discussion?
A—Yes, sir.
Q—And you know there are people who feel pretty strong on both sides?
A—Yes, sir, I know that.
Q—And you are a school teacher?
A—Yes, sir.
Q—Do you think it would embarrass you in your position as school-teacher, embarrass you any as a juror?
A—Well, not as I know of any.
Court—Would it have any weight with you in the jury box, Mr. Gentry?
A—No, sir.

Mr. Darrow:

Q—Whatever it meant to you you would decide it the way you think it ought to be?

A—According to the evidence, yes, sir.

Q—Did you ever hear anybody speak against evolution?

A—Yes, sir.

Q—Here?

A—Yes, sir.

Q—Did you ever hear Mr. Brown speak against it?

A—No, sir.

Q—Some of the ministers here?

A—I heard this fellow Martin. I have heard him several times speak against it. I have heard several preachers speak against it.

Q—Did you ever hear anybody speak for it?

A—Yes, sir.

Q—I am not asking you now to express any opinion. I am trying to find out whether you have one. Have you read the Bible lately?

A—Yes, sir, I have read the Bible a great deal.

Q—Have you ever yourself formed any opinion from the theory of evolution that it is in conflict with the Bible? I am not asking you to tell what that opinion is, but have you formed any?

Gen. Stewart—We except to that.

Court—I think he is asking that to see whether or not he should use a peremptory challenge. That is my ruling.

Mr. Darrow—Your Honor ruled my way and I am satisfied. I don't believe there can be any offense in this case unless a jury should find from the evidence that the teaching of the origin of man, as taught by Mr. Scopes, was in conflict with the Bible theory.

Court—I anticipate that question will arise and I have to pass on it later on, perhaps.

Mr. Darrow—I didn't want the court to get set on it. I am satisfied.

Court—I am not going to get set on anything but a chair right now.

Mr. Darrow—You may have to get set on something else later.

Court—Yes.

Mr. Darrow—I want to register with your honor that I consider I have the right to challenge for cause, although I have no such idea now.

Court—Yes, sir.

Mr. Darrow:

Q—Have you read the various books and magazine discussions that bear upon the question of whether there is a conflict between—

A—No, I haven't read very much of that.

Q—Have you read any of them?

A—I have read a little magazine here, a little while ago, called The Conflict, that is the name of it, and another one called the Present Fruit, that says that the Bible and evolution are contrary. They can't go together.

Q—You have been reading evolution for thirty or thirty-five years, have you, more or less?

A—Yes, sir, something like thirty years.

Q—And have you been reading the Bible that long?

A—Yes, sir.

Q—You have settled in your own mind without reading the little magazine as to whether you think they are in conflict?

A—Yes, yes, sir.

Q—That is something you read lately, that Conflict?

A—Yes, sir, a few days ago.

Q—You didn't study evolution under Martin?

A—Under who?

Q—Under Martin?

A—No, sir.

Q—I will just ask you this question. Do you think you would be a perfectly fair juror?

A—Could I have—

Q—Yes?

A—Yes, sir.

Q—That is all.

The Court—What do you say for the state?

Mr. McKenzie—Have a seat, professor.

The Court—Have a seat, professor. Call the next man.

J. C. Dunlap, being examined on his voir dire, testified as follows:

Examination by the court:

Q—Do you solemnly swear you

will true answers make to all questions asked you touching your qualifications as a juror?

A—Yes, sir.

Q—Are you a householder or freeholder of Rhea County?

A—Yes, sir.

Q—Are you related by blood or marriage to the prosecutor, Walter White, or Mr. Scopes, the defendant?

A—No, sir.

Q—Have you formed or expressed an opinion as to the guilt or innocence of Scopes, the defendant?

A—To some extent, yes, sir.

Q—Have you heard any of the proof?

A—No, sir.

The Court—Have you talked with any person who professed to know the facts?

A—I had a conversation with Mr. Scopes way back there. My knowledge of it, just as to his action. I don't know what he taught.

Q—Did he undertake to tell you what the facts were in this case, Mr. Dunlap?

A—No, sir.

Q—That was back before he was arrested or charged with this offense?

A—That was, I believe it was just after he was indicted.

Q—Mr. Dunlap, have you any definite or fixed opinion in your mind as to the guilt or innocence of the defendant?

A—No, sir.

Q—Have not?

A—No, sir.

Q—Did you know whether any of the remarks you heard were true or false, did you have any definite information as to that?

A—No, sir.

Q—Could you go in the jury box, Mr. Dunlap, and wholly disregard any impression or opinion you have and try the case wholly upon the law and evidence and return a fair verdict to both sides?

A—Yes, sir.

Q—You say you could?

Q—Could you go in the jury box before hearing any proof with an open mind, without any leaning or bias to either side?

A—Did you understand when I said I heard it that I know what he had taught?

The Court—Yes.

A—What he taught I don't know.

Q—Independent of that could you go in the jury box without any bias or leaning, and take your seat in the jury box without hearing any proof and be absolutely fair to both sides?

A—Yes, sir.

The Court—Competent juror.

Examination by Gordon McKenzie:

Q—Mr. Dunlap, have you been on the regular panel in the last two years?

A—No, sir.

Q—Let me see if I understand you. You make the statement that you knew that Mr. Scopes was teaching this in high school, is that right?

A—I knew that he taught this textbook, yes, sir.

Q—And you have read the textbook?

A—No, sir.

Q—Have you read a portion of the textbook it is claimed Prof. Scopes did teach?

A—No, sir.

Q—You never read that?

A—No, sir.

Q—As I understand you, you have got an opinion like the rest of them?

A—Just as I said. I know he taught, and that is far as my knowing goes.

Q—Of course your opinion is made up on what you have heard that he had taught?

A—No, I don't know what he has taught. All I know—

The Court—You just know he is a school-teacher.

Mr. McKenzie:

Q—What is it then you do know in conflict that he has taught?

A—I don't know a thing he has taught. All I know is what the newspapers claimed he taught, evolution.

Q—Now, then, from that did you form an opinion?

A—No, sir.

Q—Didn't form any at all?

A—No, sir.

Q—As to the guilt or innocence of Prof. Scopes?

A—No, sir; that is what I mean.

Q—You talked to Prof. Scopes in regard to the case?

A—At that time, yes, sir.

Q—And heard him say as to what he taught?

A—No, sir.

Q—What did he talk to you in regard to what he taught?

A—He just made the statement, someone asked him—I was in the conversation and I don't remember what the conversation was, someone asked him if he taught evolution and he made the remark that he taught what was in the textbook. That was way back there.

Q—Still you didn't form an opinion?

A—I say I am sure he taught what was in that book. I am confident of that.

Q—You still didn't form any opinion about it, after he said he had taught that?

A—Not as to his guilt or innocence as to this indictment, no, sir.

Mr. McKenzie—I will pass him to you, Colonel. Just a second, please.

Q—Mr. Dunlap, do you know where F. E. Robinson Company place is, where it started?

A—Yes, sir.

Q—I will ask you if you haven't participated in a number of arguments down there time and again in regard to this case?

A—Oh, yes.

Q—And after you have taken the affirmative or negative of this question, then you want the court to understand you have no opinion even though you have argued it down there at the drug store?

A—Mr. McKenzie, the point I am trying to make is if he is guilty of teaching evolution as the law defines it, someone smarter than I am will have to tell you. I know he taught school in the state and we have argued it and everybody else.

Q—As I understand you, Mr. Dunlap, if the statute should say that this man should not teach anything contrary to the Divine Creation of man as taught in the Bible and this textbook that you have spoken of teaches something contrary to that, then you would still have to have additional evidence before you could arrive at it, as to whether or not he was innocent or guilty?

Mr. Darrow—That question is argumentative.

Mr. McKenzie—I am trying to ascertain what his opinion is in regard to the matter.

The Court—I rather think your question was a little involved, Mr. McKenzie. Ask him again, I didn't get that.

Mr. McKenzie:

Q—I say if this law should state that no theory shall be taught that conflicts with the story of the creation of man as taught in the Bible and then this textbook teaches that man evolved from a one-cell animal, then you think it would take still additional proof?

A—No, sir, if that testimony would come up in the trial my mind would be made up.

Mr. McKenzie—I submit, if the court please, he would not be competent.

The Court—I understand the attorney-general insists that—

Gen. Stewart—I understood this gentleman here to say he had a conversation with Mr. Scopes, in which Scopes told him that he taught evolution and if that is true I think that would disqualify the man as a juror.

The Court—Is that what you said, Mr. Dunlap?

A—I really don't remember what was said, whether he said that or not. I am under the impression he said he taught what was in the textbook.

Q—Is that all he said, you think?

A—That is all I remember.

Gen. Stewart—I understood him to say he told he had taught evolution.

Mr. Darrow—May I object to the question as to whether he told him he taught evolution? They have insisted that under the question here, the question as to whether a man came from some lower form, that that is in conflict with the Bible and it has got to be proven by them to make the case here.

Gen. Stewart—That is not in order here. I don't think that should be argued before these prospective jurors.

The Court—I think you are correct about that.

Q—Mr. Dunlap, have you any fixed opinion in your mind at all as to the guilt or innocence of this defendant? You know what he is charged with?

A—I have not as far as the indictment is concerned.

Q—In any way have you any fixed opinion that he is guilty or not guilty? I don't want to know what your opinion is, without any modifitions.

A—I couldn't say he was guilty without some more evidence.

Q—Have you any fixed opinion now in your mind either way is what I want to know?

A—No, sir.

Q—None at all?

A—No, sir.

Q—You think anything you have heard said about it would have any influence on your verdict at all?

A—Yes, what I have heard would have a little influence on my verdict. The point I am trying to make is this, I can't be fair about it.

Q—That is what I want you to say?

A—I believe his statement that he taught evolution. As it was set down in that textbook. Now if that is a violation and it breaks the law, I have got a fixed opinion.

Q—Do you think now what you have heard or what you have read or what you know might have any weight with you in the jury box?

A—Yes, it is bound to.

The Court—I will excuse you, Mr. Dunlap.

W. A. Ault, on the examination on his voir dire, testified as follows:

Examination by the court:

Q—Do you solemnly swear that you will true answers make to all questions asked you touching your qualifications as a juror?

A—Yes, sir.

Q—Are you related by blood or marriage to Walter White, the prosecutor or J. T. Scopes, the defendant?

A—No, sir.

Q—Are you a freeholder or householder of Rhea county?

A—Yes, sir.

Q—Have you formed or expressed an opinion as to the guilt or innocence of the defendant?

A—No, sir.

The Court—Let's have order in court. I am going to have to exclude someone if you don't keep quiet, and if we put you out you will have to stay out.

Q—Mr. Ault, have you heard something of the case?

A—Yes, sir.

Q—Could you go in the jury box and wholly disregard whatever you have heard and try the case and be fair to both sides?

A—Yes, sir.

Mr. McKenzie—Pass him to you, Colonel.

Examination by Mr. Darrow:

Q—Mr. Ault, are you a married man?

A—Yes, sir.

Q—You are a merchant here?

A—Yes, sir.

Q—Have been for a good many years?

A—Yes, sir.

Q—Born in Tennessee?

A—Yes, sir.

Q—Member of any church?

A—Yes, sir.

Q—What one?

A—Baptist.

Q—Been a member a long time?

A—Yes, sir.

Q—Did you ever hear about evolution?

A—Yes, sir, I have read about it.

Q—You have a definite opinion about it?

A—To a certain extent. I believe evolution is progress or whatever you want to call it.

Q—Is that all the belief you have about evolution?

A—Yes, sir.

Gen. Stewart—We except to all that. I don't think it is proper to go into what he believes about it.

The Court—Not except for his information.

Mr. Darrow—Have you any positive opinion as to whether man is the development of a lower section?

A—Yes, sir.

Q—Very decided on that, aren't you?

A—Yes, sir.

Q—You have heard a good many people talk about it?

A—Yes, sir.

Q—Heard Mr. Bryan talk about it?

A—Yes, sir; not on just that subject. I have heard him talking on evolution at the banquet; that is the only time I heard him.

Q—Did you hear anybody else talk about it?

A—Yes, sir. I have heard it on both sides. I think I heard you talk some about it.

Q—Didn't hear me talking about evolution, did you?

A—A little bit.

Q—You didn't believe it if you did, did you?

A—I don't fall out with a man on what he thinks.

Q—You have a very definite and fixed opinion about that question, haven't you?

A—As to a man coming from a lower order?

Q—Yes.

A—Yes, sir.

Q—And you have an opinion as to whether Mr. Scopes taught it, haven't you?

A—No, sir, I haven't. I didn't know about what he had taught. I didn't know he was teaching.

Q—You have heard about it?

A—Heard about him teaching, but didn't know what he taught.

Q—You have heard what he taught?

A—Yes, sir.

Q—You have heard it stated frequently, haven't you?

A—Yes, sir.

Q—Have you formed any opinion about Mr. Scopes' guilt or innocence in this case now?

A—I have not, Colonel.

Q—No opinion of any sort?

A—Not as to guilt or innocence. I have an opinion as to evolution.

Q—Would that opinion prejudice you any in this case?

A—Not a bit.

Mr. Darrow—We will take him.

Mr. McKenzie—Have a seat.

Mr. Darrow—Wait a minute, your honor.

The Court—I thought you said you would take him.

Mr. Darrow—No, not for some time, your honor.

Q—Do you think you would be a fair juror to Mr. Scopes?

A—Yes, sir.

Mr. Darrow—We will excuse him.

No. 2, Will Weir.

Examination by the court:

Q—Mr. Weir, are you a householder or freeholder of Rhea county?

A—Yes, sir.

Q—Are you related to the prosecutor, Walter White, or the defendant, John T. Scopes?

A—No, sir.

Q—Have you formed or expressed an opinion as to the guilt or innocence of this defendant?

A—Yes, sir.

Q—From what, Mr. Weir?

A—I am a teacher myself, and have been teaching that book. I have read it very carefully since this case came up, have studied it very closely so as to understand it if it was necessary for me to teach it.

Q—You have a very definite opinion as to his guilt or innocence?

A—Yes, sir, I have.

Q—Your opinion would have some weight with you in the jury box?

A—I am afraid it would, sir.

Court—You may be excused.

No. 6, J. R. Thompson.

Being duly sworn, was examined as follows by the court:

Q—Capt. Thompson, are you a householder or freeholder in this county?

A—Both.

Q—Are you related to Walter White, the prosecutor, or John T. Scopes, the defendant, by blood or marriage?

A—No, sir.

Q—Have you formed or expressed an opinion as to the guilt or innocence of this defendant?

A—I don't think I have, Judge.

Q—Well, have you any definite or

fixed opinion, captain; a fixed opinion about—?

A—As to whether he is guilty or innocent?

Q—As to whether he is guilty or innocent?

A—No, sir.

Q—Captain, could you go into the jury box and try this case free from passion or prejudice, divesting yourself of any opinion you may have, and try this case according to the law and the evidence?

A—I think I could, sir.

Court—He is a competent juror.

J. G. McKenzie—We pass Capt. Thompson to you.

Examination by Mr. Darrow:

Q—You have lived here a good many years?

A—Yes, sir. I have lived in this county all my life. I haven't been here all my life; I was born in this county and raised here.

Q—You are a United States marshal?

A—Yes, sir, I was, for six—or five years during Wilson's administration.

Mr. Darrow—That doesn't prejudice you with me.

Q—Where were you stationed?

A—Knoxville.

Q—All your life you have lived at Dayton?

A—I didn't live at Dayton; I lived about the center of the county, General.

Q—You aren't a farmer, are you?

A—I own a farm; I am no farmer.

Q—That is different. Are you in any other business?

A—No, sir, not at this time.

Q—Do you know Mr. Scopes?

A—I do not. I hardly know the man by sight. I have seen him; I have seen him on the streets since I have been here, but as to knowing him, I don't.

Q—I presume that you belong to the church?

A—I do. I am not a good member, not as good as I ought to be.

Q—Of what church?

A—Methodist.

Q—Do you work at it very hard?

A—Well, no, sir; not as hard as I ought to.

Q—Is that church here at Dayton?

A—No, sir; it is up in the country.

Q—You go sometimes?

A—I beg your pardon?

Q—You go to church sometimes?

A—Yes, I do.

Q—Your wife probably goes more than you do.

A—More than I do.

Q—Well, now, do you read much?

A—I am not an extensive reader, outside of magazines and newspapers; I am not a book reader.

Q—You are not a book reader?

A—No, sir.

Q—Do you take a number of magazines?

A—No, sir, I can't say that I do; I read a great many magazines, but am not a subscriber.

Q—Have you ever heard evolution argued?

A—Yes, I have read that a good deal, and also in the papers.

Q—Now, Mr. Scopes is charged with violating the law. Have you ever given much, if any, attention to the question of evolution?

A—I never have.

Q—That is one of the things you have not studied?

A—No, sir.

Q—You haven't any opinion about it at the present time?

A—Well, I couldn't say that I have no opinion. I have never—it is a question I have made no study of.

Q—So your opinion would not be worth much?

A—No, I don't think it would be, General.

Q—Most of us have opinions on everything?

A—That is true.

Q—And a good many things of which we don't know a thing about?

A—And ought to know more than I do.

Q—Well, you know yourself; you do not know enough about it to form an opinion at this time, a decided opinion; that is an important question you haven't studied, and on it your opinion would not be worth much?

A—I don't think my opinion on evolution would be worth very much to the court or to any court or on the outside generally because it is a question I have never studied, General; I have never made a study of it.

Q—You are perfectly competent to try this case here with fairness, aren't you?

A—Why, I would be on that subject, yes.

Q—You wouldn't want to work in such a case?

A—I had not rather work at all.

Q—You would listen to the evidence and—

A—I certainly would try to listen to it as much as I could. I don't know how much attention I would pay to the lawyers.

Q—Well, perhaps you are right about that, but you would try to inform yourself?

A—I would, yes, sir.

Q—Sincerely and honestly?

A—Yes, sir.

Q—Anyhow, if you were a juror you would want to do right and get it right?

A—I would do that; yes, sir; I wouldn't want to do anything else.

Q—You think you are in a frame of mind to do what is right in this case, don't you?

A—I think so, yes, sir.

Q—And do your duty, would you?

A—I would try.

Q—Anyhow, if you thought you would not be perfectly fair you would tell us, wouldn't you?

A—I would.

Q—You haven't anything against Mr. Scopes?

A—And nobody else.

Q—So you wouldn't hold out on him?

A—No, sir.

Q—You haven't heard anybody talk about evolution?

A—General, it has been talked about, especially in this section, since this case came up. I have heard it talked about pro and con, especially since this case came up in this county.

Q—There has been something about it since this case came up?

A—Seems so, yes, sir.

Q—Do you know whether you have heard anybody talk about it who knew anything about it, that you know of?

A—I don't think I have heard anybody talk about it except just generally. I haven't mixed up with the farmers, and the reason I don't know any more about it than I do is perhaps they didn't know much more about it than I did.

Q—That is probably right. Now, let me ask you a little more.

A—Yes, sir; glad to have it.

Q—You are a church member. Are you much of a Bible student?

A—No, sir.

Q—You don't pretend to be very much posted on the Bible, do you?

A—I do not.

Q—And if it was necessary for you to have kept posted, you would not have permitted it to prejudice you one way or another?

A—I have no prejudice whatever.

Mr. Darrow—I can see no reason why I should not take you for a juror. Of course, they would rather not have you on the other side; we are not prejudiced.

A—That is to be left up to you.

Q—Well, you think you can decide it without prejudice?

A—I wouldn't be willing to go into the jury box unless I could.

Q—But you are willing to go in?

A—I prefer not to go.

Q—We understand that. But you think you could be perfectly fair as a juror?

A—Yes, sir.

Q—Unless you would, you would tell me so readily and openly?

A—I do think so, yes, sir.

Mr. Darrow—All right; have a seat.

Venireman No. 1, W. B. Smith, was duly sworn and examined as follows by the court:

Q—Are you a householder or freeholder in this county?

A—Yes, sir.

Q—Are you related to the prosecutor, Walter White, or the defendant, J. T. Scopes, by blood or marriage?

A—No, sir.

Q—Have you formed or expressed an opinion as to the guilt or innocence of the defendant?

A—A kind of one, yes, sir.

Q—The main thing I want to know is whether you have a definite, fixed opinion as to his guilt or innocence?

Have you done so?

A—I don't think so.

Q—You have not?

A—No, sir.

Q—Have you heard of the case?

A—Yes, sir.

Q—Heard some rumor about it?

A—Yes, sir.

Q—You haven't heard any of the proof?

A—No, sir.

Q—Has any one undertaken to detail to you what the facts were?

A—I don't think so; no, sir.

Q—Now, Mr. Smith, if you were a juryman, in the jury box, could you go into the box and try the case wholly on the law and the evidence, and render a fair verdict to both sides?

A—I think so.

Court—Competent juror.

Mr. J. G. McKenzie—Ask 'Squire Smith, Col. Darrow.

Talesman—I have been on the regular panel in less than two years.

Mr. Darrow—You have been on the regular panel in less than two years?

Talesman—Yes, sir.

Mr. Darrow—I suppose this entitles him to be excused?

Court—No, sir; it entitles you to challenge him for cause.

Mr. Darrow—Hadn't I better find out whether I like him, first, Judge?

Examination by Mr. Darrow, for defense:

Q—Mr. Smith, do you know anything about evolution?

A—I do not; no, sir.

Q—You would like to find out, would you?

A—I ain't wanting nothing about it.

Q—Are you a member of the church?

A—Yes, sir.

Mr. Darrow—Speak a little louder.

A—Yes, sir.

Q—Of what church?

A—Baptist.

Court Officer—You will have to quit talking over there.

Court—I indorse what you say, captain.

Q—Are you a farmer?

A—Yes, sir.

Q—You really work at it?

A—What did you say?

Q—Do you run your own farm?

A—Yes, sir, but I haven't this year. I have rented it; got it rented out.

Q—You live near here, do you?

A—Nearly six miles.

Q—I don't blame you for being weary. Are you a regular church attendant?

A—About once a month.

Q—That is regular, isn't it? Did you ever hear any preacher talking about evolution?

A—Yes, sir.

Q—Did you find out anything?

A—I don't think I did; I don't know; I didn't know what they were talking about.

Q—You have never heard anything about it, and don't pretend to know anything about it? Is that right?

A—That is right, yes, sir.

Q—You haven't any opinion one way or another?

A—No, sir.

Q—And you don't know Mr. Scopes here, do you?

A—No, sir.

Q—You haven't anything against him, then?

A—No, sir.

Q—You haven't any opinion in the case?

A—No, sir.

Q—Have you been much of a Bible reader?

A—I have read the Bible some.

Q—You have never read it clear through?

A—I guess not.

Q—You have never given a great deal of attention to it, have you? To reading it, I mean.

A—I have read the Bible right smart.

Q—How is that?

A—I have read the Bible right smart, yes, sir.

Q—Did you ever have any opinion, or try to have any opinion on whether the Bible was against evolution or not?

A—I never gave it any thought that way.

Q—You never gave it any thought that way?

A—No, sir.

Q—You have thought about both of them, to be sure, haven't you?

Is there any reason why you could not be a fair juryman?

A—Which?

Q—There is not any reason why you could not be a fair juryman, is there?

A—No, sir, I believe not.

Q—Did you ever hear many people talk about this case?

A—Yes, I have heard a heap of talk about it in the past three or four months.

Q—Have you paid much attention to it?

A—Yes, I guess I have.

Q—You have heard talk on both sides, haven't you?

A—Yes, sir.

Q—You haven't been able to make up your mind yet, have you?

A—I have not, no, sir; I haven't decided myself.

Q—What?

A—I have not decided myself, except I have heard people talking about it.

Q—You don't know whether anybody who talked about it knew anything about it, do you?

A—No, sir.

Q—Nobody but the lawyers know anything about it? You just let it go and paid no attention to it? You haven't any prejudice? How is that? Do you know what your neighbors think about it?

A—I do not, no, sir.

Q—Do you care?

A—No, sir.

Q—It doesn't make any difference to you what anybody says?

A—No, sir.

Q—You would do what you thought was right, would you?

A—Yes, sir.

Mr. Darrow—I think you would, too. You are a juror.

Court—What do you say, gentlemen?

J. G. McKenzie—Have a seat.

Mr. Darrow—I wasn't through.

Court—I beg your pardon; I thought you said, "You are a juror."

J. G. McKenzie—I thought so, too, or I would have waited.

Mr. Darrow—I agree, but the state had not challanged, as I understand it. Your practice is different.

Venireman J. T. Leuty was duly sworn and replied as follows to questions asked by the court:

Q—Are you a householder or freeholder in this county?

A—Yes, sir.

Q—Are you related to Walter White, the prosecutor, or to J. T. Scopes, the defendant?

A—No, sir.

Q—Have you formed or expressed an opinion as to the guilt or innocence of this defendant?

A—No, sir.

Q—Have you no opinion about his guilt or innocence?

A—Rumor.

Q—Just due to rumor?

A—Yes, sir.

Q—You have had no definite information?

A—No, sir.

Q—If chosen on the jury, could you go into the box without prejudice or bias either way, and try the case on the law and the evidence?

A—Yes, sir.

Court—He is a competent juror.

J. G. McKenzie—Col. Darrow, did you accept 'Squire Smith?

Mr. Darrow—Oh, yes.

Examination by J. G. McKenzie:

Q—Mr. Leuty, you say you have been hearing about this case?

A—No, sir, just talk.

Q—And the first fact discussed was in regard to the arrest of Mr. Scopes?

A—Well, I think that—

Q—When he was arrested?

A—Yes, sir.

Q—And of course everybody formed an opinion, and naturally would? That's right?

Mr. Darrow—Did he answer that?

A—No, I didn't exactly form an opinion or anything about it.

Q—Did you form any opinion at all? You didn't form any opinion at all?

A—No, sir; I didn't hear any evidence in this case, and didn't form any opinion at all.

Q—You didn't form any opinion from what you heard other people say?

A—No, sir.

Q—And haven't an opinion now?

A—No, sir.

Q—You have not been on the regular jury panel in the last two years?
A—No, sir.
Q—You have no suit against you up here?
A—No, sir.
Mr. J. G. McKenzie—We pass him to you.
Examination by Mr. Darrow:
J. G. McKenzie—I want to ask one question: What church do you belong to?
A—None.
Mr. McKenzie—What orders?
A—K. P.
(Examination resumed by Mr. Darrow):
Q—Have you ever been a member of a church?
A—No, sir.
Q—How long have you lived here?
A—All my life.
Q—What is your business?
A—Well, I am a kind of a farmer now.
Q—Have you ever been in any other business?
A—Yes, sir; I have been clerking in a store.
Q—Here in Dayton?
A—Sir?
Q—Here in Dayton?
A—No, sir; I live in Rhea Springs.
Q—That is in this county?
A—Yes, sir.
Q—You have never studied evolution?
A—No, sir.
Q—Are you much of a reader?
A—I read some. I used to read a great deal.
Q—Books?
A—Yes, and magazines and newspapers. Used to read books.
Q—You used to read books. And you went to school here, I suppose, rather than where you live now?
A—I went to the public schools in Rhea county.
Q—Did you ever hear anybody talk about evolution?
A—Oh, well, I have heard it talked about when they got this question up.
Q—I didn't get your answer.
A—I say I have heard them talking about evolution since this question has been up.
Q—They never talked about it before down here, did they?
A—Well, they might in a general way, but people never paid much attention to it.
Q—Well, you have not heard it talked about; nobody else has talked about it, and all the information you have has been since this case came up?
A—Yes, sir.
Q—You have not any prejudice against the doctrine or idea of evolution?
A—No, sir.
Q—Have you ever heard of Mr. Scopes?
A—I have heard of him, yes, sir.
Q—You don't know him?
A—No, sir.
Q—You are not prejudiced against him?
A—No, sir.
Q—You have never made any comment on this case?
A—No, sir.
Q—You will be perfectly fair in dealing with it?
Q—I just want to ask you this—you are a farmer, now?
A—Yes, sir.
Q—You have a family, I suppose?
A—Yes, sir.
Q—You don't know what your neighbors think about this case?
A—I suppose some of them have thought about it.
Q—You wouldn't care what they thought if you were on this jury?
A—No, it wouldn't make any difference to me if I was on this jury.
Q—If you were on the jury it would not make any difference to you what your neighbors thought?
A—No, sir.
Mr. J. G. McKenzie—Challenge by the state.
The Court—Mr. Leuty, we will excuse you.
Mr. Darrow—Have they got a right to do that?
The Court—Colonel, perhaps you don't understand our practice.
The Court—They examine a juror. They pass him to you, and you can examine him and say that you pass him back; then they have the right to challenge him. They

have a right to pass him back and then you take him or reject him. That is our practice.

Mr. Darrow—I thought they were trying to put something over on us.

The Court—No, if they tried to I would not let them.

Mr. Darrow—Don't let them.

Venireman No. 6, Jess Goodrich, being duly sworn, was examined as follows by the court:

Q—Mr. Goodrich are you a householder or freeholder of Rhea county?

A—Yes, sir.

Q—Are you related to Mr. Walter White, the prosecutor, or Mr. J. T. Scopes, the defendant, in this case?

A—Yes, sir.

Q—Have you formed or expressed an opinion as to the guilt or innocence of this defendant?

A—No, sir.

Q—You have not?

A—No, sir.

Q—Have you heard of this case, Mr. Goodrich?

A—Yes, sir, I have heard rumors of it.

Q—You didn't hear any evidence?

A—No, sir.

Q—You have no bias or leaning or prejudice either way, you say?

A—No, sir.

Q—You think you would be absolutely free from prejudice?

A—Yes, sir.

Q—And could try the case on the law and the evidence?

A—Yes, sir.

J. G. McKenzie—Pass him to you, Colonel.

Mr. Darrow—You mean we may examine him?

The Court—You can ask him any question you wish and pass him back to him.

Examination by Mr. Darrow:

Q—What is your name?

A—Goodrich.

Q—What?

A—Goodrich.

Q—How do you spell it?

A—G-O-O-D-R-I-C-H.

Q—What is your business?

A—Shipping clerk.

Q—What is it?

A—Shipping clerk in a wholesale house.

Q—Here in Dayton?

A—Yes, sir.

Q—Have you been in that business long?

A—A year and a half.

Q—What did you do before that?

A—Sold goods for ten years or so.

Q—Here in this town?

A—In this county.

Q—That was in a store, working as a clerk?

A—No, not inside.

Q—Where was that?

A—Eleven miles from here, north.

Q—Did you ever farm here?

A—Farm? Oh! I have done a little side farming.

Q—Are you a member of the church?

A—Yes, sir.

Q—Of what church?

A—The Christian church.

Q—That is what some of us up north call Disciples?

A—Well, yes.

Q—Or Campbellites?

A—Yes.

Q—Is there a church of that sort here?

A—Yes, sir.

Q—I haven't been around long enough to know about it. Has your minister, so far as you know, taken any hand in this evolution case?

Mr. J. G. McKenzie—Colonel, would you mind letting us in on your conversation over there, for we can't hear a word of it.

Mr. Darrow—I asked him whether or not his minister had taken any hand in this evolution case.

Q—Have you been a member of the Christian church for a good while?

A—About eight years.

Q—That was the first church you joined?

A—Yes, sir.

Q—You are a regular attendant, are you?

A—Yes, sir.

Q—Did you ever hear anything about evolution?

A—A good deal.

Q—In what way have you heard about it?

A—Oh, just hearing them talking about it here, and reading the papers.

Q—How long since you have been hearing about it?

A—Just since this came up.

Q—You never heard of it before?

A—Yes.

Q—But you paid no attention to it before?

A—No, sir.

Q—And didn't pay much attention to it since?

A—No, sir.

Q—Just listened to what they said, without attempting to form any opinion on it? Is that right?

A—Yes, sir.

Q—Have you heard talk on both sides?

A—I heard much talk about it.

Q—Anybody say anything to you about it?

A—I have not heard it, except just what is rumored around; I haven't heard it discussed much.

Q—You really have not taken any interest in it?

A—No, sir.

Q—Well, you had no interest, one way or the other?

A—No, sir.

Q—Or any prejudice in the matter?

A—None whatever.

Q—No prejudice against the defendant?

A—No, sir.

Q—Do you know Mr. Scopes?

A—When I see him, yes, sir.

Q—You have never known him in any other way?

A—No, sir.

Q—You have nothing against him?

A—No, sir.

Q—You know of no reason why you could not be perfectly fair as a juror here?

A—No, sir.

Q—If you have had any prejudice about it before, you would throw it aside?

A—Yes, sir.

Q—You will be perfectly fair about it?

A—Yes, sir.

Q—Are you a married man?

A—Yes, sir.

Q—Have you any children?

A—No, sir.

Q—You would not care what anybody else thinks about this case, would you?

A—No, sir.

Q—You would do what was right in it?

A—Yes, sir.

Mr. Darrow—We will accept him.

J. H. Bowman (28) being duly sworn and examined on the voir dire, testified as follows:

Question by the court:

The Court—I want to announce here that the court's hours will be from 9 in the morning to 11:30 and from 1:30 in the afternoon to 4:30.

Mr. Bowman, are you a householder or freeholder of Rhea county?

A—Yes, sir.

Q—Are you related to Walter White, the prosecutor, or the defendant John T. Scopes?

A—No, sir.

Q—Have you formed or expressed an opinion as to the guilt or innocence of this defendant?

A—I have not.

Q—You have not?

A—I have not. No, sir.

Q—You have not heard any proof?

A—No, sir.

Q—You have not talked to any person who professed to know the facts?

A—I believe not.

Q—And you could go into the jury box and try the case wholly on the law and the evidence free from any prejudice passion or bias either way?

A—Yes, sir.

The Court—He is a competent juror.

Mr. J. G. McKenzie—Pass him to you.

Mr. Darrow—If I am right about this we will just pass him back.

The Court—If you do not care to examine him. You do not care to examine him?

Mr. Darrow—I will a little.

Examination by Mr. Darrow:

Q—Where do you live, Mr. Bowman?
A—I live in Graysville.
Q—What is your business?
A—Well, I am here as a farmer. I have been working at cabinet work until this summer.
Q—In the cattle business?
A—No, sir; cabinet.
Q—A cabinet maker?
A—Yes, sir.
Q—What has been your business, generally?
A—Farming.
Q—Where did you learn the trade of cabinet maker?
A—At Dayton.
Q—You have lived at Dayton the most of your life?
A—What?
Q—You have lived at Dayton the most of your life?
A—I have lived at Graysville the most of my life.
Q—How far is that from here?
A—Five miles from Dayton.
Q—Have you a farm of your own?
A—Well, not entirely my own.
Q—Well, I don't mean to be inquisitive about it, but you do not live on a rented farm?
A—No, sir; I live in my own home.
Q—Are you a member of the church?
A—Yes, sir.
Q—Of what church?
A—The Methodist church.
Q—Have you been a Methodist for quite a while?
A—I became connected through my father and mother, and I have been a good while.
Q—You were born, then, in the church?
A—Yes, sir.
Q—How often do you go; pretty regular?
A—No, sir.
Q—Is that Southern Methodist?
A—No, sir, M. E.
Q—Did you ever know anything about evolution?
A—Nothing especially.
Q—You never paid any attention to it especially?
A—No, sir.
Mr. Darrow—You will have to talk a little louder so we can hear you.
Q—You never paid any attention to it especially?
A—Well, no, I never did read on it much; I just paid a little attention to it since this trial came up.
Q—You never heard it spoken of at all before the trial came up?
A—Well, I have heard it spoken of; and I have read books about it.
Q—Have you ever heard any addresses on it?
A—Well, I have one, probably?
Q—When was that?
A—About two weeks ago.
Q—Who was that ?
A—I am not sure, but I think W. J. Abernathy.
Q—Well, that did not make any difference in your opinion?
A—No, sir.
Q—You really have no special opinion on the truth or falsity of the theory of evolution; is that right? At this time?
A—Well, I don't know whether I could say I don't have any opinion or not.
Q—Well, let me put it a little different. I guess, or I reckon we all have opinions, whether they are good for anything or not, but we have got to have opinions. Have you got any opinion that would influence you any if you were trying to find out the truth here, or would you lay it aside and try for the truth?
A—I am always open to truth.
Q—Well, you have not made an investigation, and you are not sure what you know about it, so far, have you?
A—Well, I really don't know.
Q—Have you been quite a Bible reader?
A—Well, I read the Bible a good deal.
Q—Well, you do read it; have you read it all through?
A—No, sir.
Q—Have you tried to discover whether there is anything conflicting between the Bible and the doctrine of evolution?
A—No, sir.
Q—You have not any information on that?

A—No, sir.
Q—And no opinion?
A—No, sir.
Q—Have you had it fixed strong enough to affect your judgment in a case?
A—No, sir.
Q—That is right, is it?
A—Yes, sir.
Q—You have no fixed opinion on that question, anyhow, have you?
A—I don't really understand your question.
Q—I asked you whether you had any opinion on the subject, on the conflict between the Bible and evolution?
A—Why, no, I am sure about that.
Q—And your mind is open to what you may hear? Is that right?
A—Yes, sir.
Q—You can be fair, if you are put on the jury, can't you?
A—I think so.
Q—You are not prejudiced against a man because he is an evolutionist?
A—No, I don't hold prejudice against any one.
Q—You will be fully fair as a juror?
A—Yes, sir.
Q—To the best of your knowledge?
A—Yes, sir.
Mr. Darrow—We will take him.
Mr. J. G. McKenzie—Sit down, Mr. Bowman.
Mr. Bill Day being duly sworn by the court and examined on his voir dire, testified as follows:
Questions by the court:
Q—Mr. Day, you are a householder or freeholder of Rhea county?
A—Yes, sir.
Q—Are you related by blood or marriage to Walter White, the prosecutor, or to John T. Scopes, the defendant?
A—No, sir.
Q—Have you heard of the case?
A—Yes, I have heard of it.
Q—You haven't heard any of the proof?
A—No, sir; I haven't read it.
Q—You have no bias, or leaning, or prejudice, either way?
A—No, sir.
Q—Haven't read any of the proof?
A—No, sir.
Q—You would go in the jury box and try the case wholly upon the law and the evidence, being fair to both sides?
A—Yes, sir.
Court—Competent juror.
Mr. McKenzie—You may ask Mr. Day, Colonel.
Court—Colonel, he is unloading on you again.
Mr. Darrow—What is your business—farmer?
A—I have been a farmer.
Q—Really work at it?
A—No, sir, I don't.
Q—Rent your farm?
A—Yes, sir.
Q—You live in the city here?
A—No, sir, not here.
Q—Where do you live?
A—Spring City.
Q—That is a bigger place?
A—No, it is a little place.
Q—You have heard a lot about this case?
A—Yes, I have heard a little about it.
Q—You think you can be a fair juror here?
A—I think so.
Q—You wouldn't say so if you couldn't, would you—you wouldn't say that you could?
A—No, siree, I wouldn't.
Q—That is what I mean. If you thought you couldn't you would say you wouldn't.
A—I would do what I think was right.
Q—Do you belong to the church, do you?
A—Yes, sir.
Q—Which one?
A—Baptist.
Q—Baptist?
A—Yes, sir.
Q—Been a Baptist always?
A—Yes, sir.
Q—Where is your church?
A—Yellow Creek.
Q—Well, I have never been down there. Is that near here?
A—It is about eighteen miles of here—fifteen or eighteen.
Q—Has a minister ever talked to you about evolution down there?

A—I haven't heard him. The fact of the thing, I haven't been down there—I have been sick.

Q—Been going anywhere else?

A—Yes, sir.

Q—Whereabouts?

A—Spring City.

Q—Baptist?

A—Yes, and others.

Q—How is that?

A—Yes, and other churches there.

Q—Did you ever hear any of the preachers talk evolution?

A—No, I don't know as ever I have in the pulpit.

Q—Have you out of the pulpit?

A—How is that?

Q—Have you outside of the pulpit?

A—Yes, I have heard people talk. I don't know just who I have heard —very often hear somebody say something about it.

Q—It wouldn't necessarily make you for or against a thing because a preacher said so, would it?

A—No, sir.

Q—You reserve your right to decide for yourself, don't you?

A—Yes, sir.

Q—Your own conscience and your own judgment?

A—Yes, sir.

Q—Have you ever been interested in evolution?

A—No, I have never paid very much attention to it to tell you the truth.

Q—Ever read a book on it?

A—No, I have never read no book on it.

Q—Never read a lecture on it?

A—I noticed a little in the newspapers.

Q—Never went to hear anybody speak on it?

A—No, I haven't. I haven't been interested in it that much.

Q—You have no prejudice against it?

A—No, sir.

Q—You are willing to try to find out what the truth is, if you get interested?

A—Yes, sir.

Q—Been quite a Bible reader all your life?

A—I haven't read it like I ought to.

Q—How is that?

A—I haven't read it like I should have.

Q—Well, you never have studied the Bible to see whether there is anything against evolution in it or not?

A—No, I haven't studied it as I should.

Q—Well, I don't know about that. That is, have you—you have not studied it enough to find out whether it is against evolution or not?

A—No, sir.

Q—You haven't any opinion on that?

A—No, sir.

Q—There really isn't any reason why you would not be perfectly fair to our client?

A—Perfectly fair.

Q—You probably heard people talk about this case—neighbors and friends?

A—Yes, sir.

Q—They probably didn't know any more about it than you do—they might say more about it without knowing?

A—I haven't been interested in it.

Q—If you were a juror in this case you wouldn't care what anybody thought about it—you would do what you thought?

A—I would do what I think; yes, sir.

Mr. Darrow—We will take the juror.

Mr. McKenzie—Have a seat, Mr. Day.

H. A. Davis was called and did not respond.

F. S. Collins was called and did not respond.

R. L. West, being duly sworn by the court, and examined on his voir dire, testified as follows:

Questions by the court:

Q—You are a householder or a freholder in Rhea county?

A—Yes, sir.

Q—Are you related by blood or marriage to Walter White, the prosecutor, or John T. Scopes, the defendant?

A—No, sir.

Q—Have you formed or expressed an opinion as to the guilt

or innocence of the defendant?

A—No, sir.

Q—Have you any leaning or prejudice either way?

A—Well, I don't know that I have.

Q—Well, could you go into the jury box and be perfectly fair to both sides and try the case wholly on the law and the evidence?

A—Yes, sir.

Court—Competent juror.

Questons by Mr. McKenzie:

Q—What church are you a member of?

A—Baptist.

Mr. McKenzie—Pass him to you, Colonel.

Questions by Mr. Darrow:

Q—You are a Baptist.

A—Yes, sir.

Q—How long have you been a Baptist?

A—About eighteen years.

Q—Were you a member of the church before?

A—No, sir.

Q—Do you go pretty regularly?

A—Well, I haven't—I haven't been for a little while. I used to go pretty regularly.

Q—Have an automobile?

A—No, sir.

Q—Well, you are a farmer?

A—Yes, sir.

Q—Always been a farmer?

A—Most of the time; yes, sir.

Q—Ever work at anything else?

A—Yes, sir.

Q—What?

A—Carpenter trade.

Q—Where abouts?

A—I worked in Ohio.

Q—How long did you work there?

A—I worked about three years.

Q—Belong to the union?

A—Yes, sir.

Q—Then you came to Dayton, Tenn.?

A—Hear this—I don't live in Dayton.

Q—Well, you think you could be a fair juror here?

A—Yes, sir—I don't know.

Q—Heard much about the case?

A—Well, I have heard a right smart about it; yes, a little all around.

Q—Heard much about evolution.

A—Not until this came up—not very much.

Q—Never knew about it until this?

A—Well, nothing to amount to anything.

Q—Have you read any books on it?

A—No, sir.

Q—Ever listened to any speeches?

A—No, sir.

Q—Ever hear your minister make any speeches?

A—Well, no, I don't think so.

Q—On evolution?

A—No, sir.

Q—If he did, you would probably make up your mind for yourself, wouldn't you?

A—Yes, sir.

Q—You don't pretend to know much about it?

A—No, sir.

Q—And you haven't any opinion that you call an opinion, have you?

A—Oh, I don't know whether I could say I don't have any opinion about it or not—not at all.

Q—Well, you don't think it is worth much, do you?

A—The opinion?

Q—On evolution?

A—Well, I don't know about that.

Q—Where did you get it?

A—Well, the opinion I have—you mean whether it is true or untrue?

Q—Yes.

A—Why, nothing more than only just rumors of what I have heard talk and the newspapers.

Q—Do you think that gave you an opinion of whether evolution is true or not?

A—No, I couldn't say that it did.

Q—You do know that it s quite an interesting question, don't you?

A—Well, I don't know about that. I don't know just what it is.

Q—Well, do you think it is a question upon which you could form an intelligent opinion without some study?

A—Yes, sir; I think so.

Q—You think you could form it without any study.

A—Well, I don't know about not studying it. I thnk I would have to study something about it or know something about it, or hear something about it.

Q—Do you think you have heard or studied enough about it to know or have an opinion of any values on it?

A—No, I haven't.

Q—That is, if evolution was of any importance in the case, you would want to learn the truth of the case the best you could about it, wouldn't you?

A—Why, sure.

Q—Your mind would be open for anything you could receive?

A—Yes.

Q—And just as much open to one side or the other?

A—Yes.

Q—You haven't any desire to be wrong or not to learn just what the truth is?

A—To be right.

Q—You want to know what the truth is? There are lots of things that may be important that we don't any of us study enough to form an intelligent opinion, I suppose, and this is one of the things that you have regarded in that way?

A—Yes, sir.

Q—Have you been quite a Bible reader?

A—Well, I have read the Bible some, not so awful much—I have read it some.

Q—Have you read it to find out whether there is any conflict between the Bible and evolution?

A—No, I never reached that point; I don't know as I have. I don't know whether I am just following up what you mean.

Q—All right, I will make it plainer. I know you will tell me, if you have—you haven't any opinion that you yourself think of any value on evolution?

A—No, sir.

Q—I don't mean casual opinion like a man may form on a thing without any study, but I mean something that is substantial and amounts to something—you haven't any such opinion?

A—No, sir.

Q—You haven't any opinion as to whether evolution as you understood it would be contrary to the Bible, if you have studied the Bible carefully or not, have you?

A—Well, I have a slight opinion on that line, but not—I don't know—I couldn't say whether—as I already have told you, I don't understand what is meant by evolution, or really what—I don't know just exactly what your idea is.

Q—Have you paid any special attention to what the Bible says about how man came?

A—Yes, sir, I have.

Q—Is that from your reading or what you heard?

A—From my reading.

Q—Well, what is your judgment as to whether you would be a fair and impartial juror in this case and can decide it without any opinions or bias?

A—Yes, sir.

Q—You think you can?

A—Yes, sir.

Q—Did you ever read much about the Bible and how it came into being, outside of the Bible itself—did you ever read much about that?

A—Let's see that question.

Q—Did you ever read much about the Bible outside the Bible itself?

A—Well, yes, I have read something.

Q—Ever made any study of it outside the reading of the book—outside of the Bible?

A—What do you mean?

Q—Read what people have written about it—it's history and all that?

A—Yes, I have read some—books, you mean, on the Bible?

Q—Yes.

A—Yes, I have read some few; not much though—very little.

Q—And do you think you have no fixed opinion as to whether evolution is contrary to the Bible?

A—Well, I don't know whether I could say that or not.

Q—You mean you don't know whether it is contrary or not?

A—Yes, sir, that is what I mean.

Q—What I am getting at—have you any opinion on the subject or are you ready for argument—open for it?

A—I am open for it.

Q—You are a man who wants to find out the truth?

A—I certainly do.

Q—And want to do right and would not be influenced by any consideration outside of getting at what the truth is?

A—No, sir.

Q—If you are a juror here and the question of evolution is put up to you, would you try to find out whether evolution is true amongst other things, won't you?

A—Yes, sir.

Q—At this time you have no prejudice against Mr. Scopes?

A—No, sir.

Q—No desire to convict him?

Gen. Stewart—What was that question, Colonel?

Court—Whether he had any desire to convict him. He said he had no prejudice against him and no desire to convict him.

Mr. Darrow—Desire, I said.

Mr. Stewart—We want to except to that question unless he makes it explicit.

Court—The witness says he has none.

Venireman West—If I understand the question, I am not disposed to convict him unless he is guilty—the truth is what I am for.

Mr. Darrow—The question was perfectly competent.

Court—I am allowing you to go right on.

Q—You understand that he is presumed to be innocent don't you, in this case?

A—Yes, sir.

Q—And every presumption goes to his innocence—every doubt goes to his innocence—every reasonable doubt you entertain, and if you hear no evidence you would acquit him of course?

A—Yes, sir.

Q—He is presumed to be innocent like everybody else in court and you have no desire to find him guilty?

A—No, sir, if he is not guilty.

Q—You have no desire to find him guilty?

Gen. Stewart—I except to his arguing with the prospective juror. I don't think he has a right to do that and deliver a lecture to him on what he should do or not do.

Q—What do you say—whether you are accepted or not—you know, before you can convict him—convict anybody—you must find that he is guilty beyond a reasonable doubt?

A—Yes, sir.

Q—And you would not convict anybody unless the evidence convinced you that he was guilty beyond a reasonable doubt?

A—No, sir, according to the law and the evidence.

Mr. Darrow—We will take him.

W. P. Ferguson, examined on the voir dire, being duly sworn and examined, testified:

Questions by the court:

Q—Are you a householder and freeholder in this county?

A—Freeholder.

Q—Are you related by blood or marriage to Walter White, the prosecutor, or John T. Scopes, the defendant?

A—No, sir.

Q—Have you formed or expressed an opinion as to the guilt or innocence of the defendant in this case?

A—I cannot say that I have.

Q—You say you cannot say that you have?

A—No, sir.

Q—Have you heard any of the evidence, Mr. Ferguson?

A—I do not remember that I have heard the evidence, just what I saw in the papers.

Q—Just what you have read?

A—Yes.

Q—You do not know whether it was true or not?

A—No, sir.

Q—Now, have you definitely made up your mind in any way, at any time, Mr. Ferguson, as to whether he is guilty or not guilty?

A—Well, sir, I could not say.

Q—You mean you cannot say that you have?

A—No, sir.

Q—Have you any fixed opinion, now? Any definite opinion now that he is guilty or not guilty?

A—No, sir.

Q—Could you go in the jury box, Mr. Ferguson, and try the case wholly on the law and the evidence, disregarding anything you have heard or know about it?

A—I think I could, yes, sir.

Q—You can go in there and free from any leaning or bias before you hear any proof, without any leaning in either way?

A—I think so.

The Court—I think you could. He seems to be competent.

Mr. McKenzie—Pass him to you, Colonel.

Examination by Col. Darrow:

Q—Where do you live?

A—I live in this county.

Q—Where?

A—I live out in the Third district in this county.

Q—How far from Dayton?

A—Two and a half miles.

Q—A farmer?

A—Yes, sir.

Q—Own a farm and run it yourself?

A—Yes, sir.

Q—Has that been your business for a long time?

A—All my life.

Q—Are you a member of the church?

A—Yes, sir.

Q—Baptist?

A—Yes, sir.

Q—And you have been a Baptist for a long time?

A—Yes, sir.

Q—A pretty regular church attendant?

A—Constant.

Q—Sir?

A—Yes, sir.

Q—Go every Sunday?

A—Not every Sunday.

The Court—Talk louder, please.

A—Not every Sunday.

Q—Do you read much?

A—Yes, sir, a right smart.

Q—What do you read, books?

A—Well, mostly, no I read the Bible some. I read newspapers.

The Court—Louder. They complain they cannot hear you.

A—My voice there seems to be something the matter. Yes, my voice, I read the Bible some, newspapers mostly. I don't read much books.

Q—Magazines?

A—Well, some.

Q—Did you ever read about evolution?

A—Well, nothing only just what I have seen in the papers.

Q—What papers do you take?

A—Well, I am taking the Tri-Weekly Constitution, an Atlanta paper, now, I have taken the Chattanooga News, I am not taking it now.

Q—Any church papers?

A—Not now, I have taken the Baptist Reflector, but not now.

Q—Ever hear anybody talk on evolution?

A—Yes, sir; some.

Q—Who have you heard?

A—Well, it has been general talk, since this case came up.

Q—Ever hear anybody lecture on it?

A—No, sir.

Q—Hear any preachers talk about it?

A—Yes, sir; some.

Q—In church?

A—Yes, sir.

Q—Your preacher?

A—Yes, sir. Yes.

Q—Do you think he is an authority on evolution; do you?

A—Well, I don't dispute it at all, anything he said, no.

Q—He talked against it, didn't he?

A—Well, I think so, yes, sir.

Q—Well, you know he is against it, don't you, are you against it too?

A—Well, if evolution is what I have heard, I would have to say I am.

Q—What you have heard from the preacher?

A—Well, preachers and others, just talking.

Q—You are against evolution as you understand evolution?

A—Yes, sir.

Q—And you think it is against the Bible?

A—I think so.

Q—You have that opinion now?

A—That is the way I understand it.

Q—It would take evidence to change it?

A—Yes, sir.

Q—A good deal of it, would not it?

A—Well, I suppose it would.

Mr. Darrow—Challenge him for cause.

Mr. Stewart—Challenge for cause.

The Court—Let me ask him? Yes.

Q—Would that opinion of evolution have any weight wth you in determining whether or not this defendant is guilty or innocent?

A—No, sir.

Q—It would not?

A—No sir.

The Court—I do not want to prejudge a question that I may have to pass on tomorrow. I believe I will excuse Mr. Ferguson.

Mr. Stewart—We except to that, if the court please.

The Court—I want to hear you on that, before I pass on it, it looks like it is up to them.

Mr Stewart—We have some authorities we will be glad to submit to the court. Of course, if your honor is to excuse a man, we will not do that.

The Court—That is the very question you are to argue before me tomorrow. I take it, at great length.

Mr. Stewart—No, sir, a different matter. Of course, if a man is subject to a challenge by the defendant because he believes the Bible conflicts with the theory of evolution as he understands it, if that gives them a ground to challenge for cause, then, for the converse reason the state would have ground to challenge for cause and the result would be everybody on earth who could be brought here, would be challenged.

The Court—Let me see, the statute says it shall be unlawful to teach any theory that conflicts with the story of the Divine creation of man, as taught by the Bible.

Mr. Stewart—Yes, the result is, the defense will challenge every man who does not believe in evolution, if the court's ruling is correct. That would given the state the right to challenge every man who does believe in the theory of evolution; we would have the same right to challenge that the defense would. The result is, everybody who was capable of having an opinion at all, would be subject to challenge by one side or the other.

Mr. Darrow—I think, your honor, that statement is hardly correct. If you can find one that believes in it we will promptly challenge him.

J. S. Wright, being duly sworn and examined on the voir dire testified:

Examination by the court:

Q—Are you a householder or freeholder of this county?

A—Yes, sir.

Q—Are you related to Walter White, the prosecutor, or John T. Scopes, the defendant?

A—No, sir.

Q—Have you formed or expressed an opinion as to the guilt or innocence of this defendant?

A—No, sir.

Q—You have no opinion as to the guilt or innocence, at all, Mr. Wrght?

A—No, sir.

Q—You have heard of the case?

A—Yes, sir.

Q—You have not heard any of the proof?

A—No, sir.

Q—Have no bias or leaning either way?

A—No, sir.

Q—Have not studied evolution at all?

A—No, sir.

Q—So you have no opinion in any way about it?

A—No, sir.

The Court—Competent.

Mr. McKenzie—What church do you belong to, Mr. Wright?

A—Belong to the Baptist.

Mr. McKenzie—Pass him to you, Colonel.

Examination by Col. Darrow:

Q—You are a farmer?

A—Yes, sir.

Q—Live near here?

A—No, sir; live at Spring City.

Q—How far away is Spring City, five miles about?

A—Yes, sir; about sixteen.

Q—Have a farm of your own?

A—Yes, sir.

Q—You lived in this county most of your life, I presume, or all of it?

A—Most all my life.
Q—How long have you been a Baptist?
A—Sir?
Q—How long have you been a Baptist?
A—About eight years.
Q—You are a regular attendant?
A—Yes, sir.
Q—On church, I mean?
A—I attend church.
Q—Have you ever belonged to any other before that?
A—No, sir.
Q—You told the court you did not know much about evolution; that is correct, I suppose, never paid much attention to it?
A—No, sir.
Q—You are a Baptist, did you ever hear your preacher say anything about evolution?
A—Well, I do not believe I have heard the preacher where I belong at say anything about it.
Q—If you did hear your preacher speak about it, you would still think you had a right to use your own judgment on the question, regardless of your preacher?
A—Sure, I would.
Q—You would make up your mind for yourself, would you?
A—What is that?
Q—You would still make up your own mind for yourself.
A—Yes, sir; sure.
Q—You have not any opinion one way or the other as to whether evolution is correct doctrine or not, or a correct theory?
A—I do not know that I have ever read any or studied any.
Q—And you would not form an opinion on a subject without some study.
A—No, sir.
Q—Did you ever hear anybody talk about it?
A—Yes, sir; I have heard it talked about.
Q—Lately?
A—Yes, sir.
Q—Since this case came up?
A—Yes, sir.
Q—But, you have not given enough attention to it yet to form an opinion?
A—Well, no, I have not.
Q—Have you heard both sides talk about it?
A—Well, yes, I have heard it talked from both sides.
Q—Have you been much of a bible reader?
A—Yes, I read the Bible.
Q—Have you not formed any opinion as to whether evolution is in conflict with the Bible or not?
A—I never did read anything about it.
Q—And if evolution should cut any figure in this case you would try to find out, amongst other questions, whether it was true, whether it should be taught, and make up your mind on those points yourself, so far as your opinion goes?
A—Yes, sir.
Q—You know perfectly well whether you can give this defendant a perfectly fair trial, don't you?
A—I think I could.
Q—Are you sure of it? You would not sit in the jury box if you did not think you could?
A—I do not think I would.
Q—You will give him a fair trial?
A—Yes, I think I could.
Q—You won't care what anybody else thinks if you are a juror, you will do what you think is right?
A—Yes, sir.
The Court—Take him, Colonel?

Darrow Asks First Juror if He is a Church Member.

Col. Darrow—Judge, there is one, the first juror, I did not ask the question as to whether he belongs to a church. I will not challenge him, but I would like to ask him.

The Court—Do you take this man?

Col. Darrow—Yes, sir.

W. F. Roberson, the first juror, recalled, testified:

Col. Darrow—Are you a member of the church?

A—No, sir.

Col. Darrow—That is all.

Court—That makes the jury. Now, did you want to read your indictment, Mr. Attorney-General, and have the jury sworn tonight?

Gen. Stewart—Your honor, it has been discussed here that perhaps an adjournment to Monday would be asked by the defense, and not seriously objected to by the state.

Court—Let me ask a question. Is there some preliminary matter to be threshed out before the court?

Gen. Stewart—Yes, there will in all probability be. That is the matter with reference to the competency of evidence that will be introduced by the bringing here of these scientists, that was mentioned this morning.

Court—Might not it be better for me to hear you in the morning so they will have the advantage over Sunday to arrange for witnesses or not?

Gen. Stewart—I take it that they can get them here on short call. That is wholly immaterial to me.

Mr. Neal—May it please your honor that is a matter that is very serious to the defense. The defense would like tomorrow, we would like extremely the opportunity of having a consultation and conference.

The Court—Is the state insisting on court tomorrow?

Gen. Stewart—Your honor, we realize these gentlemen have not had an opportunity to confer very much together. While I would prefer to go right on with the trial and get the matter disposed of, we think we understand their position, and want to be courteous to them, and do not seriously object to it, that is the opinion of all of us.

The Court—In other words, you agree to it?

Gen. Stewart—Yes, your honor, if you see fit.

The Court—What do you say, Colonel?

Col. Darrow—We in the time have not had a chance to be together and take up the different subjects, as we are from the different cities.

The Court—Would you rather present the legal questions to the court Monday morning instead of Saturday?

Col. Darrow—We would like to do that.

Gen. Stewart—Tomorrow is Saturday, the last day of the week.

Mr. Hicks—I believe if we could get together and discuss the matters, we would be able to eliminate matters in that way, and could save time.

The Court—I believe you could save time.

Mr. Hicks—I think so.

Gen. Stewart—Your Honor, I prefer that they not be sworn and the indictment read until Monday morning. If the jury is sworn, if any of the jury got sick, or something happened, I think it might be better to leave it open. It makes no particular difference, except it affects the record.

The Court—Whether sworn or unsworn, gentlemen of the jury, you have been selected here as the jurors to try the issues in this case and it would be highly improper for any of you gentlemen to talk to any person about the facts of this case, or allow any person to talk in your presence about the facts of the case, and if any person or persons want to begin a discussion in your presence about these issues, it is your duty to say to them you are on the jury. You have to pass on the issues and it is not proper for them to discuss the issues in your presence. If they persist in such conduct report to me and I will deal with them as the law directs.

You gentlemen will not be kept together in a case of this character. I prefer that you gentlemen not attend any meeting or any debate or any service where these issues would probably be discussed by any person. What I want you to do is to keep your minds open and free from anything that might prejudice your opinion or your minds in any manner, so you can take up and try these issues absolutely without prejudice or bias, and try the case solely upon the law and the evidence. So, I give you that instruction.

Darrow Insists on Jury Being Sworn

Col. Darrow—I want to insist on their being sworn, tonight! this jury has been accepted by both sides, they are under no obligation as jurors until they are sworn. I think it is their duty to be sworn and accepted.

Gen. Stewart—Of course, when they are sworn, they are only sworn to truly try the issues of the case. That is all the oath covers.

Mr. Darrow—But, it means jeopardy.

Gen. Stewart—I see, but they are only sworn to well and truly try the issues, that covers more than that; suppose one juror becomes sick in the morning, have to select a man to take his place Monday.

Mr. Darrow—I will let you sweat about it.

Mr. Stewart—We would have to enter a mistrial, is all there is to it.

Mr. Neal—What we want is to put the man in jeopardy.

The Court—Let me state to you gentlemen, lawyers who are not familiar with our practice. Before the jury is sworn, the issues are made up, and in making up the issues, I would expect you both to outline your theories in an opening statement. That would take some thirty minutes, now, to read the indictment and have the opening statements from both sides. That would make a little late adjournment.

Mr. Darrow—May I save the record? I presume you are entirely right, but may I make my request to have them sworn?

The Court—Yes, let the record show that.

Mr. Darrow—You understand I do not know.

The Court—Yes, I am giving you the benefit of all the information I can.

Mr. Darrow—Yes, sir.

The Court—Gentlemen, you will heed my instructions, I am sure. Now, by agreement of counsel for both sides, I rather think we will have to let the court go over until Monday morning. They think they can save time. All the other jurors that have been here will be discharged, except the twelve gentlemen. I can appreciate these issues are profound and the lawyers are in need and entitled to have opportunity to make such investigation as they see proper.

Court thereupon adjourned until 9 o'clock Monday morning, July 13, 1925.

Judge John T Raulston

John T. Scopes
Defendant

Clarence Darrow
of Defense Counsel

CHAPTER II.

SECOND DAY'S PROCEEDINGS—MONDAY, JULY 13, 1925.

Court opened with prayer by the Rev. Moffett. Oh, God, our Father, Thou Who are the creator of the heaven and the earth and the sea and all that is in them. Thou Who are the preserver and controller of all things, Thou who wilt bring out all things to Thy glory in the end, we thank Thee this morning that Thou doest not only fill the heavens, but Thou doest also fill the earth. We pray Thy blessings upon this Court this morning. We pray that Thy blessings might guide the presiding judge, that he may give wise decisions in his conduct of this case. We pray that Thou would bless the jury, each member of it, as they shall hear and receive testimony, that they may be able to receive it and make a decision according to the law and the evidence in the case. We pray Thee, our Father, that Thou would bless the lawyers on each side of this case, that each one of them singly and individually shall have nothing before their minds, but each one shall do his duty that justice may be done. We pray Thee that Thou wouldst bless the principles in this case, that Thou wouldst bless those in the court and those on the outside to the ends of the earth. Bless these newspaper men as they take reports and interpret the facts throughout the world. Our Father, we pray Thee that Thy blessings may so overshadow and that Thy spirit may so direct and that Thy spirit may so guide and that the highest ideals of justice and righteousness and truth may prevail in this court in its decision for the good of men and for Thy glory, we ask in the name of our Lord and Saviour, Jesus Christ. Amen.

Jurors Called.

The Court—Open court, Mr. Sheriff.

Mr. Sheriff—We will have to have order in the courtroom. Call the jury Mr. Clerk. Answer to your name, gentlemen. (List of jurors was thereupon called.)

The following corrections were made by the jurors as to their several initials.

W. G. Day, R. F. West, J. W. Riley.

The Court—The jury is all present. Are you ready to proceed Mr. Attorney-General?

Gen. Stewart—Is the defendent present?

Mr. Neal—Yes, sir.

Mr. Neal—Before the jury is sworn we want to call attention to our motion to dismiss and quash the indictment which has been filed.

The Court—I think, Dr. Neal, that the indictent should be read first and then when I call on you to plead you may present your motion. Are you ready to read the indictment General?

Gen. Stewart—Your honor, we want to interrogate one of the jurors.

The Court—Very well, which one?

Gen. Stewart—Mr. Gentry, Prof. Gentry.

The Court—You want the rest of the jury to retire?

Gen. Stewart—Yes, sir.

The Court—Mr. Sheriff, take the jury please sir, for the present, except Mr. Gentry. Let's have order in the courtroom. Where is my policeman that had the gavel here the other day?

Spectator—Right over there.

Judge Demands Order.

The Court—Come over here, Mr. Rice, I wish you would keep order please and if they don't do what you say I will put them out. Gentlemen, we cannot proceed in the courtroom, as many people as there are without absolute order, so if any person persists in being disorderly in the courtroom they will be removed from the courtroom by the officers. I give you warning and I hope you will take this warning and heed it and that no person has to be removed from the courtroom. You want to ask Mr. Gentry some questions?

Mr. Darrow—Just a minute. We want to object. The juror has been passed and accepted and we want to object to any further interrogation.

The Court—The juror has not been sworn and I think either side has a right to interrogate any juror they see proper.

Mr. Darrow—We want to save our exception.

Gen. Stewart—This interrogation, of course, is no reflection on Prof. Gentry.

The Court—You might state why you make this inquiry.

Gen. Stewart—The reason we make it—we make this inquiry to definitely determine as to Mr. Gentry's expression of opinion. It has come to our ears that he had perhaps expressed an opinion and I just wanted to interrogate him about that.

The Court—An opinion as to the guilt or innocence of the defendant?

Gen. Stewart—Yes, sir.

Mr. Darrow—We want to save our objection anyway.

The Court—Yes, sir, that will be overruled.

Stewart Questions Juror.

Gen. Stewart—Have you made any expression of opinion as to the guilt or innocence of Scopes?

Mr. Gentry—I don't know anything about it only what I have read in the papers, not a thing.

Gen. Stewart—Did you make the statement at any time that Mr. Scopes ought not to be convicted?

Mr. Gentry—No, I don't know that I did. I don't remember a thing about it.

Gen. Stewart—You have nothing in mind now?

Mr. Gentry—No, sir, not a thing in the world.

Gen. Stewart—There is no reason why you would not be willing and could not hear the evidence in this case and return your verdict on the evidence alone?

Mr. Gentry—Not a thing.

Gen. Stewart—That is all we care to ask. We just wanted to verify the report we heard.

The Court—Do you want to interrogate the juror, colonel?

Mr. Darrow—No, sir.

Gen. Stewart—There is no reflection on him at all.

The Court—Mr. Gentry, you have an absolutely open mind, no prejudice or leaning or bias either way?

Mr. Gentry—I haven't any.

The Court—None at all.

Mr. Gentry—No, sir.

The Court—And can try the case wholly upon the law and the evidence?

Mr. Gentry—Yes, sir.

The Court—Let the jury be brought back please.

The Court—Let the jury come in. I don't like for the jury to come in under the ropes, Mr. Sheriff, but come over the ropes.

Mr. Attorney-General, are you ready to proceed?

Gen. Stewart—Yes, your honor.

The Court—Very well, sir. Prepare the indictment.

Gen. Stewart—Mr. Clerk, give me the indictment, please, sir. One of the jurors is not in.

A Voice—He will be in in just a minute.

The Court—One of the jurors is not in.

A Voice—He will be in in just a minute.

The Court—You may read the indictment, gentlemen.

Gen. Stewart—State of Tennessee, County of—

The Court—Wait a minute. Is the other juror in?

Gen. Stewart—No, sir.

The Court—Who is the other juror?

A Voice—Riley. He will be in in just a minute.

Mr. Darrow—The jury has not been sworn either?

The Court—No. We make out the issues before we swear the jury.

Mr. Darrow—Will your honor explain the procedure of this court? I am not familiar with it.

The Court—We make up the issues and then swear the jury to try the issues as joined, as joined.

Mr. Darrow—You don't mean by a statement on both sides?

The Court—No, sir. I mean by the reading of the indictment, and your plea.

Mr. Darrow—I understood it was a little different the other day.

The Court—Yes.

Gen. Stewart—Your honor, the defense has notified us of the filing of a motion to quash. Before reading the indictment we want to say that we want that properly disposed of.

The Court—Wouldn't that come when I call upon them to plead, Mr. Stewart, or not? I can proceed either way.

Gen. Stewart—The practice has been to dispose of that even before the jury is sworn.

The Court—I mean to dispose of that before the jury is sworn.

Gen. Stewart—Our practice has been to dispose of that even before the jury was empaneled.

The Court—Suppose you read the indictment first?

Indictment Read.

Gen. Stewart (Reading)—
State of Tennessee,
County of Rhea.

Circuit Court.
July Special Term, 1925.

The grand jurors for the state aforesaid, being duly summoned, elected, empaneled, sworn, and charged to inquire for the body of the county aforesaid, upon their oaths present:

That John Thomas Scopes, heretofore on the 24th day of April, 1925, in the county aforesaid, then and there, unlawfully did wilfully teach in the public schools of Rhea county, Tennessee, which said public schools are supported in part and in whole by the public school fund of the state, a certain theory and theories that deny the story of the divine creation of man as taught in the Bible, and did teach instead thereof that man has descended from a lower order of animals, he, the said John Thomas Scopes, being at the time, or prior thereto, a teacher in the public schools of Rhea county, Tennessee, aforesaid, against the peace and dignity of the state.

A. T. STEWART,
Attorney-General.

The Court—What is your plea, gentlemen?

Mr. Neal—May it please your honor. We make a motion to quash the indictment, and we would like simply to present the motion, possibly read it, and then with a very brief explanation, if any, ask your honor to reserve judgment on that until later in the trial.

Gen. Stewart—That would not be the practice at all. We would insist on the disposition of the motion before we proceed at all.

The Court—Under the practice, if they insist upon it, I would have to pass upon your motion before I go further.

Mr. Neal—We want to get it in the record, with the reading and a brief statement.

The Court—I will hear your motion.

Mr. Neal—Where is your motion? Have you it, general, over there?

Gen. Stewart (Handing document to counsel)—

Defendant Moves to Quash.

The defendant moves the court to quash the indictment in this case for the following reasons:

First—(a) Because the act which is the basis of the indictment, and which the defendant is charged with violating is unconstitutional and void in that it violates Sec. 17, Article II of the constitution of Tennessee.

Sec. 17. Origin and frame of bills. Bills may originate in either house, but may be amended, altered or rejected by the other. No bill shall become a law which embraces more than one subject, that subject to be expressed in the title. All acts which repeal, revive or amend former laws shall recite in their caption, or otherwise, the title or substance of the law repealed, revived or amended.

(b) In that it violates Sec. 12, Article XI of the constitution of Tennessee:

Sec. 12. Education to be cherished; common school fund, poll tax, whites and negroes, colleges, etc., rights of—knowledge, learning and virtue being essential to the preservation of republican institutions, and the diffusion of the opportunities and advantages of education throughout the different portions of the state, being highly conducive to the promotion of this end, it shall be the duty of the general assembly in all future periods of the government to cherish literature and science. And the funds called the common school fund and all the lands and proceeds thereof, dividends, stocks and other property of every description whatever, heretofore by law appropriated by the general assembly of this state for the use of common schools, and all such as shall hereafter be appropriated shall remain a perpetual fund, the principal of which shall never be diminished by legislative appropriations; and the interest thereof shall be inviolably appropriated to the support and encouragement of common schools throughout the state, and for the equal benefit of all the people thereof; and no law shall be made authorizing said fund or any part thereof to be diverted to any other use than the support and encouragement of common schools. The state taxes derived hereafter from polls shall be appropriated to educational purposes, in such manner as the general assembly shall from time to time direct by law. No school established or aided under this section shall allow white and negro children to be received as scholars together in the same school. The above provisions shall not prevent the legislature from carrying into effect any laws that have been passed in favor of the colleges, universities or academies, or from authorizing heirs or distributees to receive and enjoy escheated property under such laws as may be passed from time to time.

(c) In that it violates Sec. 18, Article II of the constitution of the state of Tennessee:

Sec. 18. Of the passage of bills. Every bill shall be read once on three different days, and be passed each time in the house where it originated, before transmission to the other. No bill shall become a law until it shall have been read and passed, on three different days in each house, and shall have received, on its final passage, in each house, the assent of a majority of all the members to which that house shall be entitled under this constitution; and shall have been signed by the respective speakers in open session, the fact of such signing to be noted on the journal; and shall have received the approval of the governor, or shall have been otherwise passed under the provisions of this constitution.

(d) In that it violates Sec. 3, Article I of the constitution of Tennessee:

Sec. 3. Right of Worship Free—That all men have a natural and indefeasible right to worship Almighty God according to the dictates of his own conscience; that no man can of right, be compelled to attend, erect or support any place of worship, or to maintain any minister against his consent; that no human authority can, in any case whatever, control or interfere with the rights of conscience; and that no preference shall ever be given, by law, to any religious establishment or mode of worship.

(e) In that it violates Section 19, Article I of the constitution of Tennessee:

Sec. 19. Printing presses free; freedom of speech, etc., secured. That the printing presses shall be free to every person to examine the proceedings of the legislature, or of any branch or officer of the government, and no law shall ever be made to restrain the right thereof.

The free communication of thoughts and opinions is one of the invaluable rights of man, and every citizen may freely speak, write and print on any subject, being responsible for the abuse of that liberty. But in the prosecutions for the publications of papers investigating the official conduct of officers, or men in public capacity, and the truth thereof, may be given in evidence; and in all indictments for libel the jury shall have the right to determine the law and the facts under the direction of the court, as in other crimnal cases.

(f) In that it violates Section 8, Article I of the constitution of Tennessee:

Sec. 8. No man can be disturbed but by law. That no man shall be taken or imprisoned, or disseized of his freehold, liberties, or privileges, or outlawed, or exiled, or in any manner destroyed or deprived of his life, liberty or property but by the judgment of his peers or the law of the land.

(g) In that the act and the indictment and the proceedings herein are violative of Section 9, Article I of the constitution of Tennessee:

Sec. 9. Rights of the accused in criminal prosecutions. That in all criminal prosecutions, the accused hath the right to be heard by himself and his counsel; to demand the nature and cause of the accusation against him, and have a copy thereof, to meet the witnesses face to face, to have compulsory process for obtaining witnesses in his favor, and in prosecutions by indictment or presentment, a speedy public rial, by an impartial jury of the county in which the crime shall have been committed, and shall not be compelled to give evidence against himself.

(h) In that the act, prosecution and proceedings herein violate Section 14, Article I of the constitution of Tennessee:

Sec. 14. Crimes punished by presentment, etc. That no person shall be put to answer any crimnal charge but by presentment, indictment or impeachment.

(i) In that the act violates Section 8, Article II of the constitution of Tennessee:

Sec. 8. General laws only to be passed; corporations only to be provided for by general laws. The legislature shall have no power to suspend any general law for the benefit of any particular individual, nor to pass any law for the benefit of individuals, inconsistent with the general laws of the land; nor to pass any law granting to any individual or individuals rights, privileges, immunities or exemptions other than such as may be, by the same law, extended to any member of the community who may be able to bring himself within the provisions of such law. No corporation shall be created, or its powers increased or diminished by special laws; but the general assembly shall provide general laws, for the organization of all corporations hereafter created, which laws may, at any time, be altered or repealed; and no such alteration or repeal shall interfere with or divest rights which have become vested.

(j) In that the act violates Section 2, Article II of the constitution of Tennessee:

Sec. 2. No person to exercise powers of more than one department. No person or persons belonging to one of these departments shall exercise any of the powers properly belonging to either of the others, except in the cases herein directed or permitted.

Second—(a) That the indictment is so vague as not to inform the defendant of the nature and cause of the accusation against him.

(b) That the statute upon which the indictment is based is void for indefiniteness and lack of certainty.

Third—(a) In that the act and the indictment violate Section 1 of the Fourteenth amendment of the constitution of the United States:

Mentions Fourteenth Amendment of U. S. Constitution.

Sec. 1. Art. XIV. All persons born or naturalized in the United States, and subject to the jurisdiction thereof, are citizens of the United States and of the state wherein they reside. No state shall make or enforce any law which shall abridge the privileges or immunities of citizens of the United States. Nor shall any state deprive any person of life, liberty or property, without due process of law, nor deny to any person within the jurisdiction the equal protection of the laws.

Mr. Neal—Now, may it please your honor, we would prefer to have you reserve judgment, if the state will permit and the argument in connection with this question until the whole case, the evidence will be of enlightening character both to your honor and the jury and our intention, unless the state insists, was simply to read the indicment and then allow it to remain—

Mr. Darrow—Read the motion.

Mr. Neal—I mean read the motion and allow your honor to pass upon it later we think the whole evidence in the whole case will be enlightening, and I say particularly perhaps to your honor, and your honor will be in much better position to decide these issues after our whole case rather than hearing an argument this morning, no matter how elaborate.

The Court—What course do you want to pursue, Mr. Attorney-General?

Gen. Stewart—We want the matter disposed of at this time, yes, sir.

Mr. Neal—As I understand, we would have the right to make an explanatory statement and then the Attorney-General make his argument. and we to make the final argument?

Gen. Stewart—Yes, that is right.

The Court—Yes, you would have the right to open and close, take the affirmative of the argument and state your position.

Mr. Neal—The only thing we want to understand is we have the right to close the argument.

Mr. McKenzie—To open and close.

Mr. Neal—May it please your honor, I am simply going to run through and explain our attitude or view. One of my associate counsel will make the final argument.

The Court—Yes.

Mr. Neal—May it please your honor, it is useless for us to stress right at the beginning hour, that your honor has the power, not only power but the duty, to pass on the constitutional matters. A great deal of misunderstanding exists in regard to that matter. A great many people, I think a great many lawyers, seem to unconsciously have the understanding that the appelate courts have that power alone, to pass on the unconstitutionality of statutes; but I am sure your honor is not deceived in the matter. As was said in the great case of Meador vs. Madison, it is the very essence of judicial functions to determine what the law is, and to determine what the law is, necessarily requires the determination of it constitutionality. I am sure it is not necessary for us to pause to explain to your honor, that it is not only your power but your sworn duty to support the constitution of the United States and of the state of Tennessee.

The Court—It is not necessary to argue that point.

Mr. Neal—So, while I do not expect to read all the motion, it is a very brief explanation of our idea, appealing to that particular section of the constitution, naturally and logically the first objection we make to this statute, is to call attention to that well known provision of our constitution, at least well known to Tennessee lawyers, in regard to the caption and the substance of the bill.

I do not think I exaggerate, may it please your honor, when I say probably four-fifths of the law which the Tennessee supreme court has ultimately held unconstitutional, the constitutionality has been based upon this particular provision. They have praised it highly. They have not looked upon it as purely a technicality, but looked upon it as a matter which is very important, to hold the legislature's hands, within the provisions of this particular law. Now, we will not take your honor's time in explaining why this provision, and why our courts have praised it so highly, but it is there, and I am sure your honor is familiar with it.

Now, coming to the application of this particular law. We will just mention our contention in this respect and not elaborate. The act commences, "An act inhibiting the teaching of the evolution theory in our universities, normals and all other schools of Tennessee which are supported in whole or in part by public school funds." There is the caption speaking of evolution. When we get to the body of the act: "Be it enacted by the general assembly of the state of Tennessee that it will be unlawful for any teacher in any university, normal or other school in Tennessee supported in whole or in part by public school funds, to teach any theory" any theory—not the theory, not the one contemplated by the legislative mind in the caption, but when we get to the body of the act, which must be responsive in every way to the caption, there is adversity of the act which the act is attempting to make a misdemeanor.

Passing from the first objection, the second objection, in that it violates Section 13, Article II of the constitution of Tennessee. I will not read the part which is rather lengthy, but only the particular provision we have in mind, when we say this particular provision conflicts with the statute:

"Knowledge, learning and virtue, being essential to the preservation of republican institutions, and the diffusion of the opportunities and advantages of education throughout the different portions of the state, being highly conductive to the promotion of this end, it shall be the duty of the general assembly in all future periods of this government, to cherish literature and science."

That is, in this very part of the constiuion is carried with its grant of power, the mandatory duty to cherish science.

Now, may it please your honor, we will have evidence, and now we think simply by appealing to your judicial knowledge, we can show that not only can the legislature not cherish science, but in no possible way can science be taught or science be studied without bringing in the doctrine of evolution, which this particular act attempts to make a crime. Whether it is true or not true, all the important matters of science are expressed in the evolution nomenclature. It would be impossible, if Tennessee wanted to, to strip from modern expressions of science, or announcements of science the evolutionary theory, and therefore, we think this act attempts to cut out of the very provision of the constitution upon which our common school system is based the very purpose for which this power was given.

Now, that will be elaborated a little later.

In that it violates Section 18, Article III of the constitution of Tennessee, in regard to the passage of bills. We will not stress that. We thought possibly some defect might be found, but some other speaker will explain in regard to that, that is with regard to the regularity of the procedure of the legislature at the time this particular bill was passed.

Now, may it please your honor, we come to the most sacred provision of the constitution of Tennessee, and with your honor's permission, I would like to read that.

The Court—Yes.

Mr. Neal—(Reading) "That all men have a natural and indefeasible right to worship Almighty God according to the dictates of their own conscience; that no man can of right be compelled to attend, erect or support any place of worship, or to maintain any minister against his consent; that no human authority can, in any case whatever, control or interfere with the rights of conscience; and that no preference shall ever be given by law, to any religious establishment or mode of worship."

Now, may it please your honor, we do not for one moment in this case question the right of the state of Tennessee, through proper legislative enactment or through administrative authority, to supervise and control its schools. We think, of course, the curriculum in that school must be fixed by some authority, that authority may be a local authority, municipal authority, it may be a country authority or may be a state authority. It may, as I say, fix that through administrative councils, tribunals and committees, or it may be by legislative enactment. But, may it please your honor, we insist, that in exercising this power, it is limited by the express provisions of the constitution, itself.

And, therefore, we contend, and in my humble judgment this is the most important contention of the defence, that in exercising this power, it cannot exercise it so as to violate this great provision of the constitution in regard to religious liberty, in regard to the prevention of any establishment of any particular religion or of any particular church. Our contention, to be very brief, is that in this act there is made mandatory the teaching of a particular doctrine that comes from a particular religious book, and to that extent, it places the public schools of our state in such a situation, in regard to particular church establishments, that they contravene the provisions of our constitution. Now, may it please your honor, that will be elaborated on later by some of my associates.

In that it violates Section 19, Article I of the constitution of Tennessee in regard to printing presses and expression of opinion. I will not read that.

Stewart Asks Retirement of Jury.

Gen. Stuart—It has occurred to me, perhaps, that if we are going to elaborate this argument don't you think you would, perhaps, ask the jury to retire?

Mr. Darrow—I object to the jury retiring.

Gen. Stewart—You don't object?

Mr. Darrow—We do object.

Gen. Stewart—It don't make any difference whether you do or not. It is a matter that addresses itself to the court. I ask your honor to let the jury to retire.

Mr. Neal—State why? The jury has got to be the judge of the law and the facts in this case, and this is up to this jury.

Gen. Stewart—You are not here under a plea of not guilty, and the case is not before the jury.

Mr. Neal—We are here with our motions before the jury, and we have got a right to state our motion, since the jury will be the judge of the law and the facts. We will have to go over it again anyway, and it is the same matter that we will present in the opening statement.

Gen. Stewart—There is no issue before the jury. There is nothing for the jury to consider. There is no issue before the jury.

Mr. Neal—Then what is the harm in having them here? It is the same jury that will try the case.

Gen. Stewart—That is the harm in having them here. I ask your honor to let the jury be discharged. I don't want to invade their province. I don't want anything said here that might handicap them in rendering a verdict on the evidence that will be presented to them. I think right now we are getting on dangerous territory, and I think we might invade some of the jury's rights in this case.

Mr. Neal—The jury is the judge of the law and the facts.

Gen. Stewart—Oh, that is all foolishness.

Mr. Neal—They ought to hear anything that the court has a right to listen to.

The Court—This matter that is being presented now, is purely a matter for the court to pass on. The jury has no jurisdiction to pass on this question. The jury in the final analysis are the judges of the law and the facts when the case is presented to them properly. And I think if you gentlemen are going to discuss matters that are vital to the issues in this case, before the court, it is in the discretion of the court to have the jury retire?

Mr. Thompson—Before you make a statement on that, may I make a suggestion? Of course this question of whether or not the jury retires is discretionary with the court.

The Court—Absolutely so.

Mr. Thompson—That makes first, then the inquiry in what way it can possibly prejudice the jury to hear a discussion of it if the attorney-general cannot state in what way the jury can be prejudiced, why should the court exercise its discretion by having the jury retire?

Gen. Stewart—I understand your honor had already decided the proposition?

Judge Retires Jury.

The Court—Mr. Officer, you may let the jury go. I know we are safe to let the jury be excluded. If they stay, there might be some discussion that might invade their province.

Mr. Darrow—Your honor, we will go right over it on the opening statement again, in a few minutes?

Mr. Neal—The same statement, in the same way, to the same jury.

The Court—It may become necessary for the court to make inquiries from you gentlemen, during the arguments from which the jury might infer that the court had certain opinions as to the facts and so the court will be more at ease with the jury not present.

Mr. Darrow—We will be less at ease.

The Court—Let the jury retire.

(Whereupon the jury retired from the courtroom.)

Mr. Neal—May it please your honor?

The Court—You may proceed, Judge Neal.

Mr. Neal—The last clause was the clause in regard to freedom of communication, thought and opinion. One of the fundamental rights of men. Every citizen may freely speak and write on any subject, being responsible for the abuse of that liberty. We think that particularly refers to libel.

The Court—To which?

Mr. Neal—The abuse of that liberty granted by the latter clause evidently applies to libel, and we think that then there is the freedom of expression of opinion regardless of the site, whether the site of it is in a schoolhouse, or store, or street, or building, or any place—the freedom of expression of a man's ideas and a man's thoughts, limited only by his responsibility under libel law.

In that it violates Section 8 of Article I of the constitution of the state of Tennessee—which is Section 8—which is the great section in our constitution which corresponds to the section of the great section in the fourteenth amendment—the first section of the fourteenth—that no man shall be taken or imprisoned or disseized of his freehold and liberties or privileges or outlawed or exiled or in any manner destroyed or deprived of his life, liberty or property, but by the judgment of his peers or the law of the land. We will refer to that later when we come to the final section which has to do with the federal constitution. By numerous decisions the law of the land, as your honor knows, is the same thing as due process of law.

The Court—Yes, sir.

Mr. Neal—In that the act and the indictment and the proceedings herein are violated Section 9, Article I of the constitution of Tennessee and that is that in all crimnal prosecutions the accused shall have the right to be heard by himself and his counsel. Now this is the vital part—to demand the nature and cause of the accusation against him and have a copy thereof, to meet the witnesses face to face, etc. Now our contention, may it please your honor, is that this crime which they have attempted to define—the crime in this act—the definition is so indefinite that it is absolutely impossible for the defense to know exactly the nature of its charge—of the charge. Now if there is one thing that is fundamental to crimnal law, it is that the crime must be defined with sufficient particularity, not only in the indictment, but in the statute, so that the court, the individual, everyone, may know whether this particular individual has violated that particular command of the state or not. Now we think that the act in many particulars, especially in attempting to make a crime of teaching of certain doctrines in the Bible, which we think you now can take due judicial knowledge of, and which we hope later to present evidence in regard to, a doctrine in the Bible is so indefinite that every man that reads the Bible will have a different interpretation as to exactly what that theory of creation is and how it is possible, may it please your honor, for the state of Tennessee to make a crime that which every individual—and the individuals are millions—would arrive at a different idea as to exactly what the offense is.

Next that the proceedings herein violates Section 16 of Article I of the constitution. We contend that this act is so indefinite that there cannot possibly be trained an indictment based upon the law, therefore this piece of paper which the distinguished attorney-general has filed here as an indictment does not come within the meaning of such, on account of its indefiniteness and the statute on which it is based, and therefore violates this particular provision of the constitution in that the act violates Section 8, Article 11 of the constitution. The legisature shall have no power to suspend any general law for the benefit of an individual, inconsistent with the general laws of the state, contemplating such laws, nor to pass any law carrying to an individual or individuals any grants, immunities or privileges other than such as may be by the same law extended to any member of the community.

Now, we contend, may it please the court, and this is one of the most serious and one of the many serious contentions of the defense that this particular law lacks uniformity, that it must be, if you can defend it at all, an exercise of the police power of the state, and the crimnal jurisdiction of the state which most writers classify under the head of police power. Here is a mandatory provision of our constitution, that these laws must be general and uniform.

Now this law tends to say that which is an offense if committed in the high schools would be no offense if committed up here on the streets and highways or in public halls of our state.

Suppose, may it please the court, the legislature of Tennessee should attempt to say that it is murder in one part of your town and not murder the other part of town. We do not think that would violate any more the spirit or provision of this law than does this act, in that the act violates Section III of the constitution of Tennessee, no person or persons belonging to one of these departments shall exercise any of the powers particularly belonging to the either of the other, except in the case herein directed, or permitted.

Now, may it please your honor, under that particular objection this statute is so indefinite, it fixes no definite time, as we noticed a moment ago, just one aspect of it, one particular aspect as we understand it, what parts of the act must be committed under this law. The act violating the story of the Divine creation set out in the Bible or the other that man is descended from lower animals. You have just as many interpretations of the particular offense there as individuals who read the Bible.

Now the act being so indefinite, if it is made definite and specific, it would force that upon the court. Your honor would have to assume legislative powers and attempt to make specific what the legislature left indefinite, and that is the reason.

Now, may it please your honor, that is the first section of our objection. The other two sections are very brief. The second section is that the indictment is vague as not to inform the defendant of the nature and cause of the accusation against him. We have been speaking about the law; we have said that the indictment is too vague, that these gentlemen have simply said Mr. Scopes taught evolution, simply followed the statute, or attempted to follow this, if it can be followed, which I doubt very seriously, but I think the learned attorney-general has made a very strenuous effort to follow the statute with all its indefiniteness, but we do not think that is sufficient; we think that the indictment should set out just exactly what our defendant was supposed to have taught. My associate will emphasize, that particular part of our motion. Secondly, that the statute upon which the indictment depends is void for indefiniteness and lack of certainty, which we have stressed all through this hurried statement of ours, which will also be stressed by my associate.

Now, if your honor pleases, we come to the third and last section of our motion to dismiss; that the act and the indictment violates Section 1 of the Fourteenth Amendment of the constitution of the United States. Now, will your honor bear with me and let me read that?

The Court—Yes, sir. Take your course.

Religion Not Proper Subject for Legislation.

Mr. Neal: (Reading) I want to say that our main contention after all, may it please your honor, is that this is not a proper thing for any legislature, the legislature of Tennessee or the legislature of the United States to attempt to make and assign a rule in regard to. In this law there is an attempt to pronounce a judgment and conclusion in the realm of science and in the realm of religion. We contend, may it please your honor, that was not the purpose for which legislatures were created; under our system they were created for very definite, limited purposes, to lay down rules of conduct, rules of conduct that the framers of our constitution made a very definite, very precise and a very narrow line within which these rules of conduct should be drawn. But the great domain of opinion, the great realm of religion, the framers of our constitution, not that they regarded it unimportant, but that they regarded it so important that no power, legislative or court, would attempt to lay down and assign a rule to bind conscience and the minds of the people.

Now, may it please your honor, we have been met constantly and this is my concluding word, we have been met constantly by the assertion if you don't like this law, have it repealed. The bitter tragedy and humor of such a remark to us, we know, of course, that we cannot have this law repealed; we grant you that the legislature spoke for the majority of the people of Tennessee, but we represent the minority, the minority that is protected by this great provision of our constitution, that that man that hollers out to us the assertion that we should have his law repealed is either ignorant or has only contempt for this great provision of the constitution that was made to protect one sole individual or a dozen or a thousand.

Mr. Hays—If your honor please.

Gen. Stewart—Your honor, we have the right to speak.

The Court—Gentlemen, who of you will argue? We want all of you if you want to be heard.

Mr. Neal—Just Mr. Hays and Mr. Darrow will follow.

The Court—Mr. Hays, I will hear you now.

Hays Argues.

Mr. Hays—There are only a few phases of the argument of Judge Neal to which I wish to address myself. I should like to direct the court's attention to the indefiniteness of the indictment as drawn. Mr. Scopes is charged in the caption of the act with one thing and in the body of the indictment it is put in another way. It is a good deal like charging a man with murder and trying him for another offense. I believe this act is indefinite in many respects. I will pay my respects to the phase of it which I consider most indefinite. A man could not tell whether he is commiting a crime. It is not clear what is meant by the word "teach." Suppose during my next half hour I expound the theory of the divine creation. Have I violated the law? I presume our teachers should be prepared to teach every theory on every subject. Not necessarily to teach a thing as a fact. There are many hypotheses about which the world is talking. And we desire to know the facts. I can conceive a law as bad that would provide that we could not repeat the story of divine creation as taught in the Bible. It should not be wrong to teach evolution, or certain phases of evolution, but not as a fact. That is quite a different proposition. Even with all the discussion about this law, which has been talked about all over the United States, if I were a teacher in the schools of Tennessee I would not be able to tell whether I, in explaining to my children the facts concerning the theory of evolution, and the facts concerned in teaching the theory of divine creation in the Bible, whether I would know when I was violating this law.

I direct your honor's attention to the fact that a law cannot stand unless it is definite enough for a man to know when he is commiting crime. And if we are to teach this or not to teach that. We must know whether or not the making of a particular statement is a crime. If it means that we cannot teach certain things, it should be definitely stated. If it means that you cannot explain a certain theory that should be stated plainly, or whether either or both of them can be expounded.

And the last point to which I wish to address myself, is to consider this act under the police powers of the state. The only limitation on the liberty of the individual is in the police power of the state. The preservation of public safety and public morals falls under this head. The determination of what is a proper exercise of the police power is under the jurisdiction and supervision of the court.

Now, as to whether a law is reasonable or unreasonable under the police power of the state, I have taken the liberty of drafting a law, which it seems to me would be constitutional if this law is constitutional. I have entitled this, "An act prohibiting the teaching of the heliocentric theory in all the universities, normals, and all other public schools of Tennessee which are supported in whole or in part by the public school funds of the state, and to provide penalties for the violation thereof.

Hays Drafts a Law With Death Penalty as a Comparison.

Sec. 1—Be it enacted by the general assembly of the state of Tennessee that it shall be unlawful for any teacher in any of the universities, normals and all other public schools in the state which are supported in whole or in part by the public school fund of the state to teach any theory that denies the story that the earth is the center of the universe, as taught in the Bible, and to teach instead, that the earth and planets move around the sun.

Sec. 2—Be it further enacted that any teacher found guilty of a violation of this act shall be guilty of a felony, and upon conviction shall be put to death.

Sec. 3—Be it further enacted that this act take effect from and after its passage, the public welfare requiring it.

Now, my contention is that an act of that sort is clearly unconstitutional in that it is a restriction upon the liberties of the individual, and the only reason Your Honor would draw a distinction between the proposed act and the one before us is that it is so well fixed scientifically that the earth and planets move around the sun. The Copernican theory is so well established that it is a matter of common knowledge. I might say that when the Copernican theory was first promulgated, he was under censure of the state. The book was published in Hamburg and Copernicus was banished from the state. And Georgiana later fell under the displeasure of the inquisition, and was put to death, and because of that theory Galileo, too, incurred the displeasure of the inquisition. The only distinction you can draw between this statute and the one we are discussing is that evolution is as much a scientific fact as the Copernican theory, but the Copernican theory has been fully accepted, as this must be accepted.

Law Under Police Power Must Be Reasonable.

My contention is that no law can be constitutional unless it is within the right of the state under the police power, and it would only be within the right of the state to pass it if it were reasonable, and it would only be reasonable if it tended in some way to promote public morals. And, Your Honor, and you, gentlemen of the jury, would have to know what evolution is in order to pass upon it. And I feel that it would be in the interest of justice for your honor to reserve a decision on this motion until after the case is in; then you can determine more definitely whether this comes within the police power of the state. If it is unreasonable, if it is not necessary, or does not conserve the public morals, it is not within the police power. To my mind, the chief point against the constitutionality of this law is that it extends the police powers of the state unreasonably and is a restriction upon the liberty of the individual.

The Court—Have you a brief, Mr. Hays?

Mr. Hays—I shall have it.

The Court—I should like to see it.

Mr. Hays—The reason I suggested that Your Honor reserve your decision on this, is that it is in the interest of justice that you do so until the case is in.

The Court—I cannot proceed until I have a plea of not guilty.

Mr. Hays—We are asking that you proceed, and ask that you reserve your decision until the case is developed. We are ready to proceed.

The Court—I will hear you, General.

Gen. Stewart—We will only have two arguments. Gen. McKenzie will make the first argument.

Gen. McKenzie—May it please Your Honor, I have been very much interested in the remarks of distinguished adversary counsel and by the remarks from the entire array in the case. Upon the first proposition, may it please your honor, that the indictment is not sufficient; it has been passed on by the supreme court of our state too often, and this indictment is in the language of the statute. Under the laws of the land, the constitution of Tennessee, no particular religion can be taught in the schools. We cannot teach any religion in the schools, therefore you cannot teach any evolution, or any doctrine that conflicts with the Bible. That sets them up exactly equal. No part of the constitution has been infringed by this act. Under the law we have the right to regulate these matters. Col. Neal in his argument has admitted this. Now, the distinguished gentleman, Mr. Hays, got up some indictment by which he was to hang somebody. That was not at all a similar case to this act; it has no connection with it; no such act as that has ever passed through the fertile brain of a Tennessean. I don't know what they do up in his country. It has been held by the supreme court that the Tennessee legislature has the right to arbitrate and to judge as to how they shall proceed in the operation of the schools. They have provided school funds and say that they shall not be diminished in any way, shape, form or fashion, and the Tennessee legislature is the proprietor of the schools and directs the handling of the school funds.

The Court—General, there was some insistence that the caption did not conform with the requirements of the law.

Gen. McKenzie—Your Honor, that is their caption.

The Court—That is their objection to it. What is their obligation?

Gen. McKenzie—I could not say as to that.

Mr. Neal—The caption sets forth a bill touching the theory of evolution and the body of the bill says any theory of evolution.

General McKenzie Charges Interference by Foreign Lawyers.

Gen. McKenzie—The object of the restriction is to give notice to the legislature that they should prevent surprise and fraud in the enactment of laws. However, they are to be construed liberally. In Railroad vs. Tennessee this is fully explained. Another thing, you do not construe these statutes according to their technical sense, unless it is a technical statute; you construe them in common ordinary language, and give them an interpretation like the common people of this state can understand. You do not need experts to explain a statute that explains itself. Under the law you cannot teach in the common schools the Bible. Why should it be improper to provide that you cannot teach this other theory? This indictment says that this is what he did; and that he was a school teacher, employed by a school supported wholly or in part by the public school funds of the state of Tennessee. Now, if the court please, in the construction of a statute, it has to be construed in common ordinary language. In the construction of a statute we don't have to send out and get some fellow to construe it for us.

Mr. Neal—Is the general discussing our motion, or the admissibility of evidence?

Gen. McKenzie—I am replying to the extensive speech of the gentleman over there on evolution, and, incidentally, to your argument. The rule of construction in these matters is in favor of the statute and every doubt must be solved so as to sustain it where that can be done and its constitutionality maintained. You do have to look to the interpretation of the titles as well as to the acts. The questions have all been settled in Tennessee, and favorable to our contention. If these gentlemen have any laws in the great metropolitan city of New York that conflict with it, or in the great white city of the northwest that will throw any light on it, we will be glad to hear about it. They have many great lawyers and courts up there.

Says Sixteen-Year-Old Boy Could Understand Law.

The United States supreme court has also sustained our contention in this matter. As to the scientific proposition, the words employed in the constitution or a statute are to be taken in their natural and popular sense, unless they are technical legal terms, in which event they are to be taken in their technical sense. But this is not such a statute. This is not a statute that requires outside assistance to define. The smallest boy in our Rhea county schools, 16 years of age, knows as much about it as they would after reading it once or twice.

Mr. Malone—We object to this argument. The motion before the court does not involve the discussion of the admissibility of evidence. We are discussing the constitutionality of this indictment on a motion to quash. And I would like to say here, though I do not mean to interrupt the gentleman, that I do not consider further allusion to geographical parts of the country as particularly necessary, such as reference to New Yorkers and to citizens of Illinois. We are here, rightfully, as American citizens.

The Court—Col. Malone, you do not know Gen. McKenzie as well as the court does. Everything he says is in a good humor.

Mr. Malone—I know there are lots of ways of saying—

The Court—I want you gentlemen from New York or any other foreign state, to always remember that you are our guests, and that we accord you the same privileges and rights and courtesies that we do any other lawyer.

Mr. Malone—Your Honor, we want to have it understood we deeply appreciate the hospitality of the court and the people of Tennessee, and the courtesies that are being extended to us at this time, but we want it understood that while we are in this courtroom we are here as lawyers, not as guests.

Gen. McKenzie—Your Honor, we have the very highest regard for these distinguished lawyers. I will admit that I have no respect for their opinions that have been advanced as to the law, and do not believe it to be the law—that I have the right to say in the legal form. But, so far as wanting to insult or hurt the feelings of either one of these various gentlemen, that is not my intention. I have been reading from our supreme court opinion. I do not know whether they have any respect for that or not.

Now, then, the distinguished gentleman remarked in regard to the police power of the state. Our supreme court said that this can be classified as either the exercise of power under the power of the legislature or under the police power, either one they want, against the state. And our supreme court said that the police power of the state and of the government has never been defined. The United States supreme court in 128 U. S. said the same thing. So, it don't seem to be so very restricted.

Police Power Never Defined.

In determining whether the statute enacted under the police power and discriminating between particular classes of persons, is reasonable, the courts have no power to pass upon the statute with a view to determining whether it will ultimately redound to he public good, or counteract to natural justice or equity, because these expressions are solely for the legislature. But the function of the courts is merely to decide whether it has any real tendency to carry into effect the purpose designed in the act, ultimately the protection of the public safety, the public health or the public morals. There can be no question, as we view it, as to the constitutionality of the act, or the validity of the indictment.

It serves notice on the defendant of what? That you were employed to teach in the public schools of Rhea county, that you taught a theory that is contrary to the record given by the Holy Writ as to the creation of man, and I insist it defines its own self. It does not need any construction. Instead, you taught that a man descended from a lower order of animals, just in the language of the statute. There can be no question on that ground.

Sue Hicks—I do not want to take up much time of Your Honor, because I think the most of their exceptions, I think that all of their exceptions are not valid, and I think the most of them are not worth considering, but I would like to say a word or two on one or two of the assignments made, that my colleagues have overlooked.

Now, further on the question of education and science, literature and science, I would like to say this —that the constitutional convention had in mind when they made that clause that the great public school fund should be preserved and not directed to any other purpose, and that is the main intention of the constitutional convention.

I will go on and read right here in part, I want to read from the case of Leeper vs. State, a particular excerpt from it, which has not been quoted, that Your Honor has not seen:

"We are of the opinion that the legislature under the constitutional provsion may as well establish a uniform system of schools and a uniform administration of them, as it may establish a uniform system of criminal laws and of courts to execute them."

Then, it goes on and says under the police powers that they have the right to do that, and then further it says: The court not only upholds the right of the legislature to pass this new police power, and also under the inherent right of the state to control its schools. They have two grounds on which to pass the act, if they think the teaching of evolution is harmful to the children of the state, to the future citizens of the state, upon the ground of police power, they may pass the act. They do not have to consider whether it is harmful, if, in their own judgment, they want to pass the act regulating the schools, because they are the supreme head of the schools, and they can regulate the schools as any other part of the regulations might be had. They can pass the law under the inherent powers vested in them, and that has nothing to do with the police powers.

Taking up another exception or two, the right of religious worship, "that all men have a natural and indefeasible right to worship Almighty God according to the dictates of their own conscience," that seems to me as perfectly ridiculous to say when a state employs a teacher, and he is employed under men appointed by the legislature by their acts, it is perfectly ridiculous to me to think that when they employ that teacher that he can go in and teach any kind of doctrine he wants to teach, and yet be violating that act of free speech, but they say they cannot do that, it would be violating it, if they did. Suppose a teacher wanted to teach architecture in a school when he has been employed to teach mathematics. Suppose he is employed to teach arithmetic to the class which the uniform textbook commission has adopted, and by the way, the uniform textbook commission, as Your Honor knows, has been established by the legislature. Suppose that instead of teaching arithmetic this teacher wants to teach architecture. Under their argument they say that they cannot control him and make him teach that arithmetic in that school. They go on and say that his religious worship is hindered thereby. The teaching in the schools has nothing whatever to do with religious worship, and as Mr. McKenzie brought out, he can preach as he wants to on the streets—his religious rights—but cannot preach them in school. I think that about covers all their exceptions that are worth while to mention.

The Court—Have you a copy of that brief for the state?

Mr. Hicks—Yes, sir, we can get it for you, Your Honor.

Court—Well, I will see it later. Any other counsel? Mr. Haggard? Gen. Stewart?

Gen. Stewart—Yes, sir.

Court—If you gentlemen would prefer the court will now adjourn for dinner in about twenty-five minutes.

Gen. Stewart—It is ten minutes after eleven according to my time.

Court—The court will adjourn at 11:30 and I wouldn't want to break into your argument.

Gen. Stewart—Well, I couldn't finish in twenty minutes. It will take thirty or forty minutes, I think. Of course, I want to read some authorities.

The Court—Well, I want to say to both sides, gentlemen, these issues are too profound for the court to guess at. I want briefs from both sides. If you have briefs I want you to file them with me. If you haven't any briefs, I will ask that you prepare them hurriedly.

Mr. Neal—May it please Your Honor, we had contemplated that possibility — especially Mr. Hays more than myself—we had contemplated that these proceedings would be more or less informal.

Mr. Hays—We will promise Your Honor to furnish the brief.

Mr. Neal—We contemplated the brief will come later. We contemplated your decision coming later, but if your decision is coming now we will very quickly have in your hands the brief.

Court—Any one else for the state besides Gen. Stewart? Anyone else to argue besides you?

Gen. Stewart—No, sir; that is all we will have. I want to make a few—

Court—Except you?

Gen. Stewart—I wanted to argue a little.

Court—I say, except you.

Gen. Stewart—That is all except I wanted to make an argument on the proposition.

Court—I said any other lawyer except you. The defense seems possibly to have misconstrued the procedure and I wouldn't want to break into your argument, so having these things in view I think the court will adjourn until 1 o'clock and then I want any authorities you have.

Mr. Hicks—I want all the witnesses that are in the courtroom to answer to their names and meet me right outside just as we go out—I want to see if you are here—in Judge McKenzie's office over there.

Read list of witnesses as follows:

Frazier Hutchison, James Benson, Howard Morgan, Richard Gill, Rose Cunningham, Mara Stout, Harry Shelton, Orville Gannoway, Charles Stokeley, Gregg Kyle, Elsie Farrar.

Court—Court will adjourn until 1 o'clock.

MONDAY AFTERNOON SESSION.

The Court—Call the court to order.

The Court—I will hear you, Gen. Stewart.

Gen. Stewart—Your Honor, may I—

The Court—Proceed without your coat.

Gen. Stewart—Yes. sir.

The Court—I wish you would this afternoon take up these different rounds as they are stated in the motion.

Gen. Stewart—Yes, sir, that is my purpose, Your Honor. Now if the court please, in this motion to quash as Your Honor has requested I will take up each—undertaking to state our position or theory on each assignment of each section of the constitution upon which they base this motion.

Stewart Answer Defense on Motion to Quash.

The first assignment is with reference to the origin and frame of the bill and cites Section 17, Article 11 of the constitution of Tennessee, which has been read, but the part underscored I take it is the part that is most material, Dr. Neal, so I will leave the other alone and address what remarks I shall make solely to that part that is indicated from the citation that they insist more seriously upon. This they underscore. "No bill shall become a law which embraces more than one subject, that subject to be expressed in the title." Now if Your Honor please, the constitution of the—as I understand their position, they say the caption doesn't correspond with the body of the act.

The Court—Yes, sir.

Gen. Stewart—The constitution of the state of Tennessee I have here, Your Honor. I have also most of these matters briefed, which brief I will present to Your Honor. I cannot read from the book. I have here the annotated constitution of Tennessee, Shannon's annotation, and under this, reading from the annotations under this section, among other things I want to call the court's attention to this. "A general title to an act is one which is full and comprehensive and covers all legislation germaine to the general subject stated. A title may cover more than the body, but it must not cover less. It need not index the details of the act, nor give a synopsis thereof." Citing Railroad Company vs. Burns, 11 Cates, and Green vs. State, 13 Cates. In this case if the court please—where is the copy of that act?

Mr. McKenzie—The law?

Gen. Stewart—Yes, sir.

Mr. Darrow—I will lend you my copy.

Gen. Stewart—We have one here, I thank you.

The copy of the acts says this: "An act prohibiting the teaching of the evolution theory in all the universities, normals and schools of this state which are supported in whole or in part by the public school funds of the state, and provides the penalties for violation thereof. Section 1, Be it enacted by, the general assembly of the state of Tennessee, that it shall be unlawful for any person in any of the universities, normals, and all other public schools of the state which are supported in whole or in part by the public school funds of the state, to teach any theory that denies the story of the divine creation of man as taught in the Bible and teach instead that man has descended from a lower order of animals." If anything, your honor, the caption to this act is broader than the title. The caption of the act states the legislature's conception of the evolution theory, that is, that it states in words —in so many words—that this act shall prohibit the teaching of the evolution theory and the body of the act —I mean to say states the legislature's conception of the theory of evolution—that is the particular part they undertake to prohibit teaching. Now if anything, your honor, the caption of this act is broader than the title—broader than the body. It covers the evolution theory. It may be said that there are many theories of evolution but it refers in the body of the act to one particular theory of evolution which the legislature certainly had in mind when they passed the law. It has been repeatedly held by our courts that it does not invalidate the act if the caption is broader, or shall be broader than the body of the act—that doesn't invalidate it at all. All that is necessary under our law, is that the caption of the act and the body of the act shall be germaine one to the other. The caption of the act shall simply state enough to put the legislature on notice when the caption is read as to what they are passing—what they, the legislature, are passing upon. This, if your honor please, undertakes to deal with only one thing, and that is to prohibit the teaching in the public schools of Tennessee the evolution theory, that is the particular evolution theory that man descended from a lower order of animal. I don't think, your honor, that that can be seriously considered. I have several cases here—a number of citations I can read to your honor, but I know, of course, that your honor has had a number—or some questions presented to you a number of times and are familiar with the general principles.

The Court—You are insisting that if the caption is broader than the body of the act that it doesn't invalidate the act?

Says Caption is Broader Than the Act.

Gen. Stewart—Our insistence is that the only objection that could be made is that the caption is broader than the act and it is well settled in Tennessee that that would not invalidate it.

Mr. Darrow—There is no question but the caption is broader than the act, but the act can be broader than the caption. I think that is something different.

Gen. Stewart—The caption cannot be broader than the act and then the act in turn broader than the caption. I don't understand that.

Mr. Darrow—We understand that the caption may be broader than the act.

The Court—Without affecting the validity of the act?

Mr. Darrow—Yes, but the act cannot be broader than the caption, or cannot include something that is not in the caption and two subjects cannot be included.

Gen. Stewart—No, that is true, there cannot be, and certainly if the caption of the act is broader than the body of the act, then the body of the act could not be broader than the caption. That could not be true both ways. If the caption is broader than the body, then there couldn't be two subjects within the body of the act, but there are not two subjects in the body of the act. I understand their insistence, your Honor, to be that in order to violate this act it must be necessary first to teach, by specific reference to the story of divine creation in the Bible, that that is untrue—that the story of divine creation is untrue, and to say at the same time that instead of that the story of man's creation by evolutionary process is true. I understand that to be their insistence and about all I would care to remark—to say in remarking to that, would be this, that we have a rule of construction in Tennessee which prohibits the court from placing an absurd construction on the act and that certainly would be an absurd construction. Now the next assignment if the court pleases is that

Section 12, Article 11 of the constitution of Tennessee, and they point to that part of the constitution which makes it the duty of the legislature to cherish literature and science. Now, your honor, there is a case of Green vs. State of Tennessee, which to my mind settles that proposition thoroughly. This brief was prepared in accordance with another motion that was filled and I will have to lose some time in looking through it, because the chronological order in this is different than from the other.

The Court—Have you the books here?

Gen. Stewart—The books? I have some of the cases here. Most of them are just quoted from, your Honor. This case Green vs. State of Tennessee, says this: It is cited a number of times in various reports and decisions and they quote from Judge White in a dissenting opinion, in dissenting on the particular point in question—dictum you might call it—that is what they do call it, but, nevertheless, it is an authority in which he states—

The Court—Judge White, of the Supreme Court of the United States?

Gen. Stewart—No, sir, of the Supreme Court of Tennessee.

Mr. Malone—General, can you give us the citation?

Gen. Stewart—I lost it in my brief case. Here it is. Here is the footnote of the annotation here in the volume of the constitution. (Read ing the constitutional provision making it the duty of the legislature to cherish literature and science.) That is merely a direction to the legislature, but, nevertheless, it indicates the popular feeling on this question. That was the comment Judge White made in his dissenting opinion in 5 Humpheys, 215.

"Cherish Literature and Science" Merely Directory.

To cherish literature and science. The constitution maks it the duty of the legislature to cherish literature and science, but this is our position in following that reasoning that is merely directory to the last legislature and constituting as to that, and is stated in the opinion of Judge White, that indicates the popular feeling of the people, that they realize the importance of education, they realize the importance of literature, they realize the importance of scientific investigation, and they say to the legislature through the constitution, that they should cherish literature and science.

Now, that, if your honor pleases, is merely directory to the legislature. Being so, the legislature has a right to exercise its discretion in placing its discretion on that when they speak to us through the statute.

And that, your Honor, disposes of the matter.

The Court—Was that case disposed of by Judge White rendering a dissenting opinion?

Gen. Stewart—The case in which he rendered a dissenting opinion, if the court pleases, this particular construction of this particular part of the constitution was invoked and this section of the constitution was invoked. But in this particular part it was a taxation question, a question of taxation.

Mr. Neal—May I ask the General does he know the date of this decision?

Gen. Stewart—Yes, sir. It is 1874, I believe.

The Court—Have you got the opinions here? Let us see it.

Gen. Stewart—No, sir. I have not that book. Only the annotation I have here.

Mr. Neal—Did you cite Humphreys?

Gen. Stewart—Green vs. Allen, 5 Humphreys, 215.

I find that opinion dissented from in a number of other cases. They can be found running through this brief.

The Court—You say the majority didn't pass upon that question?

Gen. Stewart—No. It was mere dictum. It is cited and recognized in several cases and annotated under this section of the constitution, and I read from the annotation, to cherish literature and science, which means to recognize, to protect, to aid, to comfort. Cherish means to protect, comfort, aid and so forth. So that it could not be any more than directory. It shall be the duty of the legislature to cherish literature and also to cherish science.

The Court—That would be a question of policy addressing itself to the legislature?

Gen. Stewart—If your Honor pleases, just as in the question where the question has been raised that the spirit of the constitution has been violated by a certain act they hold that this is a matter which addresses itself purely to the legislature. They have a right to say in their acts what is the spirit and what is not the spirit of the constitution. The question cannot be raised that the legislature violates the spirit of the constitution in any act. The spirit of the unwritten law or the unwritten part of he constitution.

As has been said only the express words of the constitution can be violated; but in determining that question the supreme court has said what the spirit of the constitution is, and in that addresses itself to the legislature. But, likewise this as Judge White says is merely a direction to the legislature. They are not bound by it, and it is left for them to interpret, and there is nothing binding about it at all.

Supposing then there should come within the minds of the people a conflict between literature and science? Then what would the legislature do? Wouldn't they have to interpret? It would go to the act, speaking to us through the statute book. Wouldn't they have to interpret their construction of this conflict which one should be recognized as higher or more in the public schools? Where there would be a conflict between literature and science? It is merely directory. And as he states, eloquently to me, that it merely expresses the policy or the feeling of the people at the time.

The Court—You say they cited Judge White approvingly in some other cases?

Gen. Stewart—Not stating it to be approved, your honor, but it is cited, and the presumption would be where it is cited in some of these cases that I can cite to your honor, in this brief, it would, of course, approve it. Now on that same proposition of cherishing science and literature, the case reported in 103 Tennessee, Page 209, which is to my mind the controling case, on the proposition, and we reach the last question—and the greatest question we might discuss on this, the case of Leeper versus the State of Tennessee, where the uniform textbook law was attacked and numerous questions raised, and in a very lengthy opinion by the supreme court they placed within the legislature the absolute power to control the public school system.

In this case, if your honor please, I want to read from it. In construeing Article 11, Section 12, the same article we are reading from here, cherishing literature and science they say: "We are of the opinion that the legislature under the constitutional provisions, may as well establish a uniform system of schools and taxation and a uniform system of criminal law and, of course, to execute them."

Now, I think this dictum announced in this dissenting opinion is to cherish literature and science. What else could it mean? What else could the constitution mean, if they had meant for the legislature to recognize literature and science, for instance, over and above the Bible in so many words? If they had intended that the legislature recognize science over literature, they would have said so. If they had intended that the legislature should pass laws recognizing science they would have said so affirmatively. They merely say it shall be the duty of the legislature to cherish literature and science. And who, who in the last analysis, if the court pleases, has the right to say whether they have or not? It is merely directory to the legislature.

The Court—Do you think that would be a question of public policy, addressing itself to the legislature?

Gen. Stewart—It might be.

The Court—Addressing itself to the legislature?

Gen. Stewart—It might be a question of public policy. But the point, the principal point I intend to make is that it is a matter that addresses itself to the legislature and its discretion.

Mr. Neal—May I ask a question?

Gen. Stewart—Go ahead.

Mr. Neal—I gather he admits it would be impossible to cherish science under this law?

Gen. Stewart—No, sir, I do not make any such admission; claiming that I do not come from a monkey, I cannot do it.

Mr. Malone—We do not think you did either, General.

Mr. Malone—Section 18, Article 2, of the constitution is the next, the question of the passage of bills, and since that relates to the house journal, the journal is not here, they did that—

The Court—I understood Judge Neal said that they threw that in, thinking they might find some irregularity.

Mr. Neal—Not exactly that, your honor; if any irregularity existed, we might take advantage of it.

The Court—You do not insist on that?

Mr. Neal—

Mr. Darrow—We have no contention on that.

The Court—Yes.

Gen. Stewart—The next one, and the one which Dr. Neal referred to as one of the most important ones, Section 3, Article 17, still of the constitution, the right of free worship:

Says Law Does Not Interfere With Worship.

"That all men have a natural and indefeasible right to worship Almighty God according to the dictates of their conscience, that no man can of right be compelled to attend, erect, or support any place of worship, that no human authority in no case whatever can control or interfere with the rights of conscience, that no preference shall ever be given by law to any religious establishment or mode of worship."

If your honor please, this law is as far removed from that interference with the provision in the constitution as it is from any other that is not even cited. This does not interfere with the religious worship—it does not even approach interference with religious worship. This addresses itself directly to the public school system of the state. This does not prevent any man from worshiping God as his conscience directs and dictates. A man can belong to the Baptist, the Methodist, the Lutheran, the Christian or any other church, but still this act would not interfere with any worship by any construction you might place on it. It is not a religious worship to every man who lives within the bounds of this sovereign jurisdiction, and this cannot interfere with it. How could it? How could it interfere in any particular with religious worship? You can attend the public schools of this state and go to any church you please. This does not require you to harbor within the four walls of your home any minister of any denomination, even. Or, what is there in this act that says you shall contribute to the maintenance of any particular religious sect or cult? There is nothing in the question, if your honor please, there is not an abridgement of the rights of religious freedom or worship.

Darrow Says Law Gives Preference to Bible.

Mr. Darrow—I suggest you eliminate that part you are on so far. The part we claim is that last clause, "no preference shall ever be given, by law, to any religious establishment or mode of worship."

Gen. Stewart—Yes, that "no preference shall ever be given, by law, to any religious establishment or mode of worship." Then, how could that interfere, Mr. Darrow?

Mr. Darrow—That is the part we claim is affected.

Gen. Stewart—In what wise?

Mr. Darrow—Giving preference to the Bible.

Gen. Stewart—To the Bible?

Mr. Darrow—Yes. Why not the Koran.

Gen. Stewart—Might as well give it to any other book?

Mr. Darrow—Certainly.

Gen. Stewart—And no preference shall ever be given by law to any religious establishment or mode of worship?

Mr. Darrow—Certainly.

Gen. Stewart—What is there in this that requires you to worship in any particular way?

Mr. Darrow—That is the part we claim.

Stewart Claims St. James Version Standard in Tennessee.

Gen. Stewart—I think so, too. There is as little in that as in any of the rest. If your honor please, the St. James Version of the Bible is the recognized one in this section of the country. The laws of the land recognize the Bible; the laws of the land recognize the law of God and Christianity as a part of the common law.

Mr. Malone—Mr. Attorney-General, may I ask a question?

Gen. Stewart—Certainly.

Mr. Malone—Does the law of the land or the law of the state of Tennessee recognize the Bible as a part of a course, in biology or science?

Gen. Stewart—I do not think the law of the land recognizes them as confusing one another in any particular.

Mr. Malone—Why does not this statute impose the duty of teaching the theory of creation, as taught in the Bible, and exclude under penalty of the law any other theory of creation; why does not that impose upon the course of science or specifically the course of biology in this state a particular religious opinion from a particular religious book?

Gen. Stewart—It is not a religious question.

Mr. Malone—I am asking why.

Gen. Stewart—You are getting right back to the proposition of the police power, where the legislature, through the exercise of police power, passes a law directing a particular curriculum in the schools.

Mr. Malone—I do not want to interrupt.

Gen. Stewart—All right, go ahead.

Mr. Malone—Not only do we maintain not only is the police power of the states not the power to direct any particular line of study, but it is not the law—

Gen. Stewart—This act could not turn his religious point of view or his religious purpose. The question involved here is, to my mind, the question of the exercise of the police power.

Mr. Neal—It does not mention the Bible?

Gen. Stewart—Yes, it mentions the Bible. The legislature, according to our laws, in my opinion, would have the right to preclude the teaching of geography. That is—

State Not Heathen.

Mr. Neal—Does not it prefer the Bible to the Koran?

Gen. Stewart—It does not mention the Koran.

Mr. Malone—Does not it prefer the Bible to the Koran?

Gen. Stewart—We are not living in a heathen country.

Mr. Malone—Will you answer my question? Does not it prefer the Bible to the Koran?

Gen. Stewart—We are not living in a heathen country, so how could it prefer the Bible to the Koran? You forced me then, in advance of the matter I am arguing now, to get down to the absolute basis of the proposition that it is the exercise of the police power; that is the question that is involved. That is what it must turn on.

Mr. Malone—The improper exercise—

Gen. Stewart—The improper exercise of the police power and dictation of what should be taught in the public schools?

Mr. Malone—Yes, sir.

Gen. Stewart—Do you say teaching the Bible in the public school is a religious matter?

Mr. Malone—No. I would say to base a theory set forth in any version of the Bible to be taught in the public school is an invasion of the rights of the citizen, whether exercised by the police power or by the legislature.

Gen. Stewart—Because it imposes a religious opinion?

Mr. Malone—Because it imposes a religious opinion, yes. What I mean is this: If there be in the state of Tennessee a single child or young man or young woman in your school who is a Jew, to impose any course of science a particular view of creation from the Bible is interfering, from our point of view, with his civil rights under our theory of the case. That is our contention.

Gen. Stewart—Mr. Malone, could not he go to school on Friday and study what is given him by the public school; then on Sunday study his Bible?

Mr. Malone—No, he should be given the same right in his views and his rights should not be interfered with by any other doctrine.

Gen. Stewart—It is not an invasion of a man's religious rights. He can go to church on Sunday or any other day that there might be a meeting, and worship according to the dictates of his conscience. It is not an invasion of a man's religious liberty or an invasion of a man's religious rights. That question cannot determine this act. It is a question of the exercise of the police power. That is what it is, and nothing else, and if they undertake to pass an act to state you shall not teach a certain Bible or theory of anything in your churches, an invasion of a private or civil act, then, according to my conception of this, it might interfere with this provision of the constitution. But this is the authority, on the part of the legislature of the state of Tennessee, to direct the expenditure of the school funds of the state, and through this act to require that the money shall not be spent in the teaching of the theories that conflict or contravene the Bible story of man's creation. It is an effort on the part of the legislature to control and direct the expenditure of state funds, which they have the right to do. It is an effort on the part of the legislature to control the public school system, which they have the right to do.

Insists it is Question of Police Power.

Gen. Stewart—That question cannot determine this act. It is a question of the exercise of the police powers of the state; that is what it is and nothing else, and if they undertake to pass an act saying you cannot teach the Bible or any certain book in any of your Bibles, that is an invasion of civil rights and that would interfere with their rights under the constitution. But this is a statement on the part of the legislature of the state of Tennessee, which directs the expenditure of the school funds of the state, and this is an act requiring that their money shall not be expended in teaching theories that contradict the Bible. It is an effort on the part of the legislature to control the expenditure of state funds, which it has the right to do. It is within the province of the legislature to control the public schools of the state. This is not an invasion of individual rights, nor of the right of worship in the different churches. If they taught there anything that conflicts with this act it would not prohibit attendance at such a church. That is not what it restricts, nor does it undertake to control one's conscience. I have gotten ahead of their assignment, however. Another question is as to the violation of Section 19, Article 1, of the constitution of Tennessee, as to the freedom of speech, the printing press, etc. From the formation of this Union, one of the inalienable rights of a citizen has been the right to speak freely on any subject. Being responsible, however, for the abuse of that privilege, or to prosecution, for the publication in papers investigating men in a public capacity, and by indictment for libel, where a jury shall have the right to determine under the law and the facts, under the direction of the court, as in any other criminal case. Now this assignment under freedom of speech, Dr. Neal insists upon. Under that question, I say, Mr. Scopes might have taken his stand on the street corners and expounded until he became hoarse, as a result of his effort and we could not interfere with him; but he cannot go into the public schools, or a school house, which is controlled by the legislature and supported by the public funds of the state and teach this theory. Under the exercise or the police power, we should have a right to object to it. The legislature has a right to control that. Now if your honor please, Mr. Hays said this morning, by way of injecting a little fun into this matter, I presume, what he conceived to be an act, the equal in viscious qualities to this, and prescribing the death penalty upon any man who might undertake to teach a certain theory or system—as to the earth being round I believe he said; I forget which it was.

Mr. Hays—Round. Round in our city.

Gen. Stewart — How is that? Round in your city? You must live on a hillside. Is it round in New York?

Mr. Hays—All round.

Gen. Stewart—The inference was that this act was absurd to him as an act carrying the death penalty for teaching a theory in contravention of what modern science claimed as a natural and well-known proposition. I presume that under this right to regulate liberty and freedom of thought and freedom of speech, that Mr. Hays would insist that the court should construe the act at bar in this manner—without reflecting, if Your Honor please, on Your Honor, or anybody—that the court in ruling on this would say. (Reading.)

Law of the Land.

"Law of the land and due process of law have been defined to mean one and the same thing. The law of the land as Daniel Webster has said, is the general law, which hears before it condemns, and proceeds upon inquiry before it renders judgment, and after hearing. The law of the land applies to all amendments, with certain restrictions." No property right is involved in the right of a man to teach in a public school. We come again to the proposition of the exercise of the police power of the state. A man has no vested right, he has no civil right, he has no inherent right, and no right that he can claim as a property right, as a teacher in a public school, except those which are subject to the control of the legislature. So there can be no serious contention there, if Your Honor please; that is a right that is subject to the constitution and subject to the acts of the legislature in the exercise of the police powers.

Darrow Says Statute is Void.

Mr. Darrow—No person shall be put to answer a criminal charge but by presentment or by indictment.

Gen. Stewart—What particular section do you mean there? Section 14, Article 1 of the constitution is as follows:

"Crimes punished by presentment, etc. That no person shall be put to answer any criminal charge, except—

Mr. Darrow—We mean indictment.

Gen. Stewart—Except by presentment, indictment or impeachment. The two things are void. The whole indictment?

Mr. Darrow—Yes, sir. It doesn't state any crime.

Gen. Stewart—It would void the statute, would it?

Mr. Darrow—We claim the statute is void; and that it is based on those two grounds.

The Court—That the statute is too meager, they claim, General. I think, and therefore, that the indictment, is too meager.

Mr. Neal—That under this law it is not possible to draw an indictment, and therefore this defendant was being tried without indictment.

Gen. Stewart—The wording of the indictment complies with the wording of the statute. In such a case it is generally held to be good.

The Court—As I understand, general, after disposing of the statute they say there is no indictment.

Mr. Darrow—On both grounds, Your Honor.

Gen. Stewart—There is no way of discussing them without discussing them together.

The Court—Of course the indictment could not be more comprehensive than the statute, and if the statute is too meager therefore, the indictment would be too meager.

Gen. Stewart—And if the statute is good the indictment is good.

Mr. Darrow—We claim that the indictment should set out what the offense was—what the doctrine was—what his version of the doctrine was.

Gen. Stewart—Undertake to set out the full and complete doctrine?

Mr. Darrow—Yes.

Gen. Stewart—I do not understand that to be the law. It would be impossible to frame an indictment properly under that, and no indictment can be presented. An indictment must state facts, and not conclusions of law. Of course there is no conclusions of law stated here. An indictment must charge the crime with certainty and show such facts and circumstances as constitute the crime; a mere statement of conclusion on the law, is sufficient. The law says it shall be a violation of the law for a man in our public schools to teach a theory that denies the divine theory of creation and that man descends from a lower order of animals. The indictment complies with the wording of the statute in toto. If the statute is good, then the indictment must be good. Now, if Your Honor please, they say it is too vague; he does not know what he is charged with. We must set out in our indictment that he taught Little Johnnie Jones that a man is descended from a monkey, a gorilla, or what not, and told him this in the following words, to-wit: It is not necessary that we state all that; it is sufficient under our law that he may know what he. is charged to answer. This indictment says that John Scopes, on such and such a date, taught a theory denying the divinity of Christ and that man is descended from a lower order of animals. He is notified sufficiently under this what he is here to defend. That is all that is necessary and all that is required under our law.

In Harris vs. State, in 71 Tennessee, Page 326—

Mr. Darrow—71 Tennessee?

Gen. Stewart—At page 326. In that case it is held that the words of the statute must be followed, or otherwise the defendant might be charged with one offense and convicted of another.

By our code, Section 5117, only such a degree of certainty is required as will enable the court who sits on it, to form judgment, and they comment, less strictness. As has always been held in this state—it has always been held in this state, that less strictness is required in indictments for misdeameanors than in felonies. That is from Section 5117, that is where they require that only such degree of certainty is required as will enable the court to pronounce judgment upon conviction. That the section is based upon that same section of the constitution.

All that is necessary under both of them is that the defendant may know what he is charged with and that the court may intelligently pronounce judgment upon conviction. That is all that is required, and that, in my opinion, makes it entirely sufficient. I see no reason why this indictment is too vague. If we had charged John Scopes with unlawfully teaching in the public schools of Rhea county and said no more, then, certainly, he would not be upon notice with what he has was charged to come here and defend. But we say that he has unlawfully taught a theory that denies the story of divine creation and has taught instead that man descended from a lower order of animals, and what could be plainer? What is there vague and indefinite and uncertain about that? You did not prepare a brief here to defend him on a charge of arson, did you? He is not here for transporting liquor, and he knows it. He is here for teaching a theory that denies the story of divine creation and that, if Your Honor please, is sufficient. The act is sufficient to notify him what he is charged with, and therefore the indictment is sufficient, and it complies with the requirements of the law. And when it meets that requirement, and the further requirement that it is sufficient for the court to know to be able to render judgment upon conviction. The next is Article 8, Section 11, general laws, only to be passed. "The legislature shall have no power to suspend any general law for the benefit of any particular individual inconsistent with the general laws of the land, nor to pass any law granting to any individual or individuals right, privileges, immunities or exemptions, other than such as may be by the same law extended to any member of the community who may be able to bring himself within the provision of such law. No corporation shall be created or its general powers increased or diminished by special law; but the general assembly shall provide by general law, for the organization of all corporations hereafter created which laws may, at any time, be altered or repealed; and no such alterations or repeal shall interfere with or divest rights which have become vested."

I don't see that there is anything in that assignment to discuss. One observation, however, I have discussed in discussing this sufficiency of the indictment—it was suggested in conversation between Mr. Darrow and myself that if a man is indicted for murder, he cannot simply be indicted for the unlawful murder of another—as Mr. Darrow says he must be told or he must be accused of murdering some particular man who must be named in the indictment. That is true as a matter of common sense. That is true as a matter of construction of our murder statute. It is true our murder statute says it shall be unlawful for any person to kill any reasonable creature in being. And, of course, you have to name who is killed.

(Reading) In that the act violates Section 2, Article 2 of the constitution of Tennessee. "No person to exercise power of more than one department."

Judge Chases Photographers.

The Court—Gentlemen, the jury will not be sworn this afternoon, and you photographers will have to move out.

Gen. Stewart—You might let the officers dismiss them for the day?

The Court—Yes. Let the jury go home, Mr. Officer?

Gen. Stewart—The next assignment, if the court please, is that no person or persons belonging to one of these departments shall exercise any of the powers properly belonging to either of the others, except in the cases herein directed or permitted.

Mr. Darrow—We are not going to argue that.

Gen. Stewart—We will just strike that then. They say they do not reply on the next assignment—Section 2, Article 2 of the constitution.

Mr. Neal—We do not insist on it.

Mr. Darrow—Oh, we don't care.

Mr. Stewart—Let's strike it then?

Mr. Neal—All right.

Gen. Stewart—They are willing that that be stricken. The next is, the indictment—

Mr. Darrow—Will you tell me what that is, to be sure?

Gen. Stewart—Under (j) Section 2, Article 2. The next is that the indictment is so vague as not to inform the defendant of the nature and cause of the accusation against him. I have already argued that. The next is that the statute is void. I have already argued that. And void for indefiniteness and uncertainly. And the next assignment is the only one, if Your Honor please—is the principal one, I think on which this case rests. It is the Fourteenth Amendment to the United States constitution, and that and the other—that and the constitution of Tennessee—raising the same questions are the ones that I think the case must terminate on. (Reading).

"All persons born or naturalized in the United States and subject to the jurisdiction thereof, as citizens of the United States and of the state wherein they reside. No state shall make or enforce any law which shall abridge the privileges or immunities of citizens of the United States: Nor shall any state deprive any person of life, liberty or property without due process of law, nor deny to any person within its jurisdiction the equal protection of the laws."

One Serious Contention.

Now, on that assignment, if the court please, comes the discussion of the exercise of police powers, and that assignment, I think, is the only one your honor which might be seriously considered. In the consideration of this assignment, I have made careful search of authorities, and while I have found much law in favor of the state's position, there are particularly two or three cases from which we shall quote, and largely, these are derterminative of the issues here. The case of Meyer vs. The State of Nebraska, which is reported in the supreme court reports, lawyers' edition, is a case recently decided by the supreme court of the United States, and in that we have an act of that state—Nebraska—which prohibited the teaching in any of the schools of that state—not just the public schools, but all schools—any language other than the English language to any pupil under the eighth grade. The supreme court held that act unconstitutional. They said that it contravened that it was an abridgement of the right—that it invaded the right of property, that it was unconstitutional on account of the Fourteenth Amendment. They hold in substance that the school teacher was deprived of the right to pursue his lawful occupation to teach German in the private and parochial schools of that state. And here is in part what they said.

The Court—Have you a copy of the opinion?

Gen. Stewart—Yes, sir; I have the book at the office. Further in deciding the case the court said, in part:

"The problem for our determination is whether the statute is construed to apply and unreasonably infringes the liberty granted by the Fourteenth Amendment." They pass directly upon this question. "While this court has not attempted to define with exactness the liberty thus guaranteed, the term has received much consideration, and some of the included things has been definitely stated. Without doubt it denotes not merely freedom from bodily restraint, but also the right of the individual to contract, to engage in any of the common occupations of life. Plaintiff in error taught this language in school as a part of his occupation —his right to thus teach and the right of parents to engage him, we think are within the liberty of the amendment."

Thus the line is drawn and in deciding the case, the supreme court held that this law was unconstitutional, but we call the court's especial attention that the court held it was unconstitutional because it affected all the schools—not only the public schools, but the private schools and in this connection we call the court's special attention to the comment of the supreme court in this opinion at the conclusion of the same, and just before decision. "The power of the state to compel attendance at some school and to make reasonable regulations for all schools, including a requirement that they shall give instruction in English is not questioned. Nor has challenge been made of the state's power to prescribe a curriculum for institutions which it supports. Those matters are not within the present controversy."

That is the very crux of this lawsuit. That is absolutely the question involved here, if Your Honor please. And the case of Leeper against the state of Tennessee—on this case, and the case of Leeper against the state of Tennessee we are willing to risk our rights.

The Court—That is the Nebraska case?

Nebraska Case Cited.

Gen. Stewart—Yes, sir.

The opinion in the Nebraska case says, "nor has challenge been made of the state's power to prescribe a curriculum for institutions which it supports." Here in Rhea county is a high school erected, supported and maintained by the public treasury, by the school fund that is taken from that treasury—by the money that is paid into the court from the pockets of the taxpapers of Rhea county and of the state of Tennessee. Isn't that a school that is supported by the state? And the supreme court of the United States says, "Nor has challenge been made of the state's power to prescribe a curriculum for institutions which it supports."

How much stronger could they make the language? How much more, Your Honor, would we have them say than to recognize the right of the state of Tennessee to direct and control the curriculum in the Rhea County High School. That is the question. I think that is settled; that is the highest tribunal of our nation speaking.

I want to cite 7 Mellory, 240, the Indiana case which holds, in substance, that the regulation of public schools is a set matter exclusively within the dominion of the legislature.

There are a great many authorities along this line shedding light over different angles. But, your honor, I think it is sufficient here for the state, insofar as anything else I might have to say here is concerned, to rightly, wholly and entirely in accord with what I have already said upon the case of the state of Tennessee vs. Leeper, that one from Blount county.

The Court—Have you the book?

Gen. Stewart—Yes, Your Honor. Your Honor, the case in Oregon, recently decided, in which Justice McReynolds also rendered the opinion, is at one with the Nebraska case.

The Court—Have you the opinion?

Gen. Stewart—Yes, sir.

The Court—I wish you would preserve that.

Gen. Stewart—It holds the same as the Indiana case which I just referred to, Your Honor.

Now, if Your Honor please, I prefere to read this Leeper case to the court.

The Court—I wish you would read the entire case if it is not too long.

Mr. Darrow—I guess he can state it in a minute. Take as long as you want, though.

Mr. Stewart—This is 103 Tennessee, 504. The defendant was convicted of violating the uniform textbook law and sentenced to pay a cost of $10. I will not read the indictment.

Mr. Darrow—Is that what you want to go into?

Gen. Stewart—You may read it if you care to. On this same question. (Reading from the State of Tennessee, 103, 504, Edward Leeper vs. State of Tennessee, the defendant, a public school teacher, beginning with the words, "did unlawfully use and permit to be used," etc., to "prescribed the terms upon which it may be done in the interest of the citizen.")

Gen. Stewart—Then they discuss the question of monopoly, and whether they have a right to make this restriction upon the publishers of these books. Then, going further into the question, the question of police power, they say: (Reading beginning with "It is said that the schools do not belong to the state," and ending with words "best interest of the citizens will be conserved.")

They come back upon that question, I thought I had gotten beyond that question.

(Reading beginning with words "We are of the opinion" and ending with words "prevent benefit from book dealers.")

Now, if Your Honor please, they wind up here with further remarks along that line, but they adopt the opinion there, as I just finished reading, and they say in State vs. Hawer, that the control of the public school system must be lodged somewhere, and where is a better place to lodge it than in the general assembly, composed of men from the different counties of the state, men who represent a certain standard in their legislative and senatorial district; men who are responsible to their constituency, to the citizens of Tennessee for the acts that they commit.

Here is a uniform system of public schools in the state of Tennessee. Who has the right to control it? If the legislature should not have the right to control them, then who ought to have the right to control them. Who may say what books shall be taught or what books shall not be taught; who has that right? The legislature has that right. If they don't have it, who could have it? Where could the power be lodged? Where in the state of Tennessee could you lodge a central power to control the uniform system, if the court please? I think the case of Leeper vs. State settled that question beyond peradventure of a doubt, and that it settles it definitely. I think it says that case construed with the case of the U. S.—I have forgotten the style of it—

The Court—Nebraska?

Gen. Stewart — Nebraska case. construed with that case, Your Honor. I think it is as plain as it can be possibly made that in the exercise of its police power the state legislature has the right to execute a uniform law regarding the uniform system of public schools. Who then has a right to control, who then has the right to control the management of these public schools, and they have a right to name the curriculum for each and every one of these public schools, because they have a right to control the system.

They do it in the exercise of their police power and the court will not refute this except as to where it is shown that there is an abuse of this power. It must be a reasonable use, and the reason is the one test, the only test that can be applied to it. And reason is the test we would want to apply to it, and we are willing that the test of reason be applied to it.

This, if the court please, the constitutionality of this act—the question is important that they have no right, that it is an abridgement of rights.

Your Honor, just a few more words.

The Police Officer—No, no talking in the courtroom.

Gen. Stewart—Your Honor, just a few more words and I am through.

Charges Attack on Legislature.

Attack is made upon the right of the legislature to pass such an act. The question has been made that it abridges the right of religious liberty; that it is an intervention of that section of the constitution. Much more might be said about it. I could make, in a very short time, a speech about it, but that is unnecessary and perhaps foolish; it would be sufficient to say that I believe, Your Honor, that this is important upon a construction of the constitution as to whether or not the state was, in the exercise of its police power, as to the right to control the curriculum in the public schools. The question on the invasion of religious liberty is not even raised in the case of the State vs. Marbury, the Nebraska case, where they passed a law you could not teach except in the English language. There is no question there in the violation of that part of the constitution. No question was made in that case. No question was made in the Leeper case it is an invasion if the court please, of any religious liberty, and they inject it into this case only because the Bible is mentioned.

Now, what is the difference? If the state has a right in the exercise of its police power to say you cannot teach Wentworth's arithmetic or Fry's geography, it has the same right to say you cannot teach any theory that denies the divine creation of man. This is true because the legislature is the judge of what shall be taught in the public schools and that is the reason it is true. Police power, the exercise of police power, the phrase which no man under God's shining sun has ever undertaken to define, what does it mean.

You might talk from now until doomsday and you could not define it; it passes down to the sound discretion of the legislature. They have a right to say and no one else has a right to say, and I say, Your Honor, that in the passage of this act the legislature abused no discretion, but used only the reasonable means at hand; they exercised a lawful and legal right that was given them by the constitution, the police power of the state, and I say that they were within their right, and I say that any effort to place any other construction upon this, or to invalidate any other part of the constitution, is an effort to becloud the true issues in the case.

Mr. Hayes—May I ask you a question?

Gen. Stewart—Yes, sir.

Hays Asks How Scopes Got Book.

Mr. Hayes—Did the state, under the power you have referred to, prescribe the book which Mr. Scopes taught in the schools?

Gen. Stewart—Did they do what?

Mr. Hays—Did the state, under the power you have referred to, prescribe the book which Mr. Scopes taught from, the manual that he was teaching from?

Gen. Stewart—There is no act on that, as I understand it.

Mr. Hayes—I thought you just stated that the state prescribed the school books; did they prescribe the school book that Mr Scopes was using?

Gen. Stewart—I said they had a right to.

Mr. Hayes—Did they exercise that right?

Mr. Malone—How did he get the book we mean, was it given to him by the state.

Gen. Stewart—That is a matter of proof; we are prepared to show that; do you want to put me on the witness stand?

Mr. Malone—No. I would like to—

(Laughter in the courtroom.)

The Court—We will take a few minutes recess.

(Thereupon a short recess was taken.)

Mr. Darrow—Shall I proceed?

The Court—I will hear you, Colonel.

Mr. Darrow—If the court please.

The Court—Have order in the courtroom. Get seats.

Mr. Darrow—I know my friend, McKenzie, whom I have learned not only to admire, but to love in our short acquaintance, didn't mean anything in referring to us lawyers who come from out of town. For myself, I have been treated with the greatest courtesy by the attorneys and the community.

The Court—No talking, please, in the courtroom.

Darrow Given Title.

Mr. Darrow—And I shall always remember that this court is the first one that ever gave me a great title of "Colonel" and I hope it will stick to me when I get back north.

The Court—I want you to take it back to your home with you, colonel.

Darrow's Speech—Holds Bryan Responsible.

Mr. Darrow—That is what I am trying to do.

But, so far as coming from other cities is concerned, why, Your Honor, it is easy here. I came from Chicago, and my friend, Malone, and friend Hays, came from New York, and on the other side we have a distinguished and very pleasant gentleman who came from California and another who is prosecuting this case, and who is responsible for this foolish mischievous and wicked act, who comes from Florida.

This case we have to argue is a case at law, and hard as it is for me to bring my mind to conceive it, almost impossible as it is to put my mind back into the sixteenth century, I am going to argue it as if it was serious, and as if it was a death struggle between two civilizations.

Let us see, now what there is

about it. We have been informed that the legislature has the right to prescribe the course of study in the public schools. Within reason, they no doubt have, no doubt. They could not prescribe it, I am inclined to think, under your constitution, if it omitted arithmetic and geography and writing, neither under the rest of the constitution if it shall remain in force in the state, could they prescribe it if the course of study was only to teach religion, because several hundred years ago, when our people believed in freedom, and when no men felt so sure of their own sophistry that they were willing to send a man to jail who did not believe them. The people of Tennessee adopted a constitution, and they made it broad and plain, and said that the people of Tennessee should always enjoy religious freedom in its broadest terms, so I assume, that no legislature could fix a course of study which violated that. For instance, suppose the legislature should say, we think the religious privileges and duties of the citizens of Tennessee are much more important than education, we agree with the distinguished governor of the state, if religion must go, or learning must go, why, let learning go. I do not know how much it would have to go, but let it go, and therefore we will establish a course in the publice schools of teaching that the Christian religion as unfolded in the Bible, is true, and that every other religion, or mode or system of ethics is false and to carry that out, no person in the public schools shall be permitted to read or hear anything except Genesis, Pilgrims Progress, Baxter's Saint Rest, and In His Image. Would that be constitutional? If it is, the constitution is a lie and a snare and the people have forgot what liberty means.

I remember, long ago, Mr. Bancroft wrote this sentence, which is true: "That it is all right to preserve freedom in constitutions, but when the spirit of freedom has fled, from the hearts of the people, then its matter is easily sacrificed under law." And so it is, unless there is left enough of the spirit of freedom in the state of Tennessee, and in the United States, there is not a single line of any constitution that can withstand bigotry and ignorance when it seeks to destroy the rights of the individual; and bigotry and ignorance are ever active. Here, we find today as brazen and as bold an attempt to destroy learning as was ever made in the middle ages, and the only difference is we have not provided that they shall be burned at the stake, but there is time for that, Your Honor, we have to approach these things gradually.

If This Law Holds—Reverts to Wicked Ancient Laws.

Now, let us see what we claim with reference to this law. If this proceeding both in form and substance, can prevail in this court, then Your Honor, no law—no matter how foolish, wicked, ambiguous, or ancient, but can come back to Tennessee. All the guarantees go for nothing. All of the past has gone, will be forgotten, if this can succeed.

I am going to begin with some of the simpler reasons why it is absolutely absurd to think that this statute, indictment, or any part of the proceedings in this case are legal, and I think the sooner we get rid of it in Tennessee the better for the peace of Tennessee, and the better for the pursuit of knowledge in the world, so let me begin at the beginning.

Let us take this statute as it is, the first point we made in this suit is that it is unconstitutional on account of the divergence and the difference between the statute and the caption, and because it contains more than one subject. Now, my distinguished friend was quite right, every constitution with which I am familiar has substantially this same provision, that the caption and the law must correspond. He is right in his reason. Why? Lots of things are put through in the night-time. Everybody does not read all of the statutes, even members of the legislature—I have been a member of the legislature myself, and I know how it is—they may vote for them without reading them, but the substance of the act is put in the caption, so it may be seen and read, and nothing can be in the act that is not contained in the caption. There is not any question about it, and only one subject shall be legislated on at once. Of course the caption may be broader than the act. My friend is entirely right about it. They may make a caption and the act may fall far short of it, but the substance of the act must be in the caption, and there can be no variance. Now, Your Honor, that is elementary, nobody need to brief on that, it is a sufficient brief to read the constitution, that one section, it is very short.

Now, let us see what they have done, there is not much dispute about the English language, I take it, here is the caption, "Public act, Chapter 37, 1925. An act prohibiting the teaching of the evolution theory in all the universities, normals and all the public schools of Tennessee which are supported in whole or in part by the public school funds of the state, and to prescribe penalties for the violation thereof."

Now what is it, an act to prohibit the teaching of the evolution theory in Tennessee? Well, is that the act? Is this statute to prevent the teaching of the evolution theory? There is not a word said in the statute about evolution, there is not a word said in the statute about preventing the teaching of the theory of evolution—not a word. This statute contains nothing whatever in reference to teaching the theory of evolution in the public schools of Tennessee. And, Your Honor, the caption contains nothing else—nothing else. Does the caption say anything about the Bible? Oh! no, does it say anything about the divine account contained in the Bible? Oh! no. If a man was interested in the peace and harmony and welfare of the citizens of Tennessee, if he was interested in intellectual freedom and religious freedom, if he was interested in the right to worship God as he saw fit, but he found out that chaos and disorder and riot could follow in the wake of this caption, and he found out that every religious prejudice inherent in the breast of man could be appealed to, by the law, the legislature was about to pass—there is not a single word in it. This caption says what follows is an act forbiding the teaching of evolution, and the Catholic could have gone home without any thought that his faith was about to be attacked, the Protestant could have gone home without any thought that his religion could be attacked, the intelligent scholarly Christian, who by the millions in the United States, find no inconsistency between evolution and religion, could have gone home without any fear that a narrow, ignorant, bigoted shrew of religion could have destroyed their religious freedom and their right to think and act and speak, and the nation and the state could have laid down peacefully to sleep that night without the slightest fear that religious hatred and bigotry was to be turned loose in a great state. Any question about it? Anything in this caption whatever about religion, or anything about measuring science and knowledge and learning by the book of Genesis, written when everybody thought the world was flat? Nothing. They went to bed in peace, probably, and they woke up to find this, which has not the slightest reference to it, which does not refer to evolution in any way, which is as claimed a religious statute, the growth of as plain religious ignorance and bigotry as any that justified the Spanish inquisition or the hanging of the witches in New England, or the countless iniquities under the name of what some people called religion, and persued the human race down to the last hundred years. That is what they found, and here is what it is: "Be it enacted by the general assembly of the state of Tennessee, that it shall be unlawful for any teacher in any of the universities, normals and all other public schools in the state, which are supported in whole or in part by the public school funds of the state, to teach" —what, teach evolution? Oh! no— "to teach the theory that denies the story of the divine creation of man, as taught in the Bible, and to teach instead that man has descended from a lower order of animals." That is what was foisted on the people of this state, under a caption which never meant it, and could give no hint of it, that it should be a crime in the state of Tennessee to teach any theory of the origin of man, except that contained in the divine account as recorded in the Bible. But the state of Tennessee under an honest and fair interpretation of the constitution has no more right to teach the Bible as the divine book than that the Koran is one, or the book of Mormons, or the book of Confucius, or the Budda, or the Essays of Emerson, or any one of the 10,000 books to which human souls have gone for consolation and aid in their troubles. Are they going to cut them out? They would have to pick the right caption at least, and they could not pick it out without violating the constitution, which is as old and as wise as Jefferson.

Certainly Violates Constitution.

Your Honor, there can be no sort of question, I submit, as a lawyer, I may be wrong, I have been wrong before—there is no more question that this violates the constitution in its provisions. The caption must state the substance and meaning of the act, and the act can contain nothing excepting the substance of the caption; and there be no more question about it than that two and two make four. They will have to arrange their cohorts and come back for another fight if the courts of Tennessee stand by their own constitution, and I presume they will.

It is binding on all the courts of Tennessee and on this court among the rest, and it would be a travesty that a caption such as this and a body such as this is woud be declared valid law in the state of Tennessee. So much for that. Now, as to the statute itself. It is full of weird, strange, impossible and imaginary provisions. Driven by bigotry and narrowness they come together and make this statute and bring this litigation. I cannot conceive anything greater.

What is this law? What does it mean? Help out the caption and read the law. "Be it enacted by the general assembly of the state of Tennessee that it shall be unlawful for any teacher in any of the universities, normals and all the public schools in the state which are supported in whole or in part by public school funds of the state, to teach any theory that denies the conception of the divine creation of man as put in the Bible and teach in its stead that man is descended from a lower order of animal."

The statute should be comprehensible. It should not be written in Chinese anyway. It should be in passing English. As you say, so that common, human beings would understand what it meant, and so a man would know whether he is liable to go to jail when he is teaching not so ambiguous as to be a snare or a trap to get someone who does not agree with you. It should be plain, simple and easy. Does this statute state what you shall teach and what you shall not? Oh, no! Oh, no! Not at all. Does it say you cannot teach the earth is round? Because Genesis says it is flat? No. Does it say you connot teach that the earth is millions of ages old, because the account in Genesis makes it less than six thousand years old? Oh, no. It doesn't state that. If it did you could understand it. It says you shan't teach any theory of the origin of man that is contrary to the divine theory contained in the Bible.

No Legislature Can Say What is Divine—Discusses Bible.

Now let us pass up the word "divine!" No legislature is strong enough in any state in the Union to characterize and pick any book as being divine. Let us take it as it is. What is the Bible? Your Honor,

I have read it myself. I might read it more or more wisely. Others may understand it better. Others may think they understand it better when they do not. But in a general way I know what it is. I know there are millions of people in the world who look on it as being a divine book, and I have not the slightest objection to it. I know there are millions of people in he world who derive consolation in their times of trouble and solace in times of distress from the Bible. I would be pretty near the last one in the world to do anything or take any action to take it away. I feel just exactly the same toward the religious creed of every human being who lives. If anybody finds anything in this life that brings them consolation and health and happiness I think they ought to have it whatever they get. I haven' any fault to find with them at all. But what is it? The Bible is not one book. The Bible is made up of sixty-six books written over a period of about one thousand years, some of them very early and some of them comparatively late. It is a book primarily of religion and morals. It is not a book of science. Never was and was never meant to be. Under it there is nothing prescribed that would tell you how to build a railroad or a steamboat or to make anything that would advance civilization. It is not a text-book or a text on chemistry. It is not big enough to be. It is not a book on geology; they knew nothing about geology. It is not a book on biology; they knew nothing about it. It is not a work on evolution; that is a mystery. It is not a work on astronomy. The man who looked out at the universe and studied the heavens had no thought but that the earth was the center of the universe. But we know better than that. We know that the sun is the center of the solar system. And that there are an infinity of other systems around about us. They thought the sun went around the earth and gave us light and gave us night. We know better. We know the earth turns on its axis to produce days and nights. They thought the earth was created 4,004 years before the Christian Era. We know better. I doubt if there is a person in Tennessee who does not know better. They told it the best they knew. And while suns may change all you may learn of chemistry, geometry and mathematics, there are no doubt certain primitive, elemental instincts in the organs of man that remain the same, he finds out what he can and yearns to know more and supplements his knowledge with hope and faith.

Bible is in Province of Religion—Accounts of Creation Conflict.

That is the province of religion and I haven't the slightest fault to find with it. Not the slightest in the world. One has one thought and one another, and instead of fighting each other as in the past, they should support and help each other. Let's see now. Can your Honor tell what is given as the origin of man as shown in the Bible? Is there any human being who can tell us? There are two conflicting accounts in the first two chapters. There are scattered all through it various acts and ideas, but to pass that up for the sake of argument no teacher in any school in the state of Tennessee can know that he is violating a law, but must test every one of its doctrines by the Bible, must he not? You cannot say two times two equals four or a man an educated man if evolution is forbidden. It does not specify what you cannot teach, but says you cannot teach anything that conflicts with the Bible. Then just imagine making it a criminal code that is so uncertain and impossible that every man must be sure that he has read everything in the Bible and not only read it but understands it, or he might violate the criminal code. Who is the chief mogul that can tell us what the Bible means? He or they should write a book and make it plain and distinct, so we would know. Let us look at it. There are in America at least five hundred different sects or churches, all of which quarrel with each other and the importance and nonimportance of certain things or the construction of certain passages. All along the line they do not agree among themselves and cannot agree among themselves. They never have and probably never will. There is a great division between the Catholics and the Protestants. There is such a disagreement that my client, who is a school-teacher, not only must know the subject he is teaching, but he must know everything about the Bible in reference to evolution. And he must be sure that he expresses his right or else some fellow will come along here, more ignorant perhaps than he and say, "You made a bad guess and I think you have committed a crime." No criminal statute can rest that way. There is not a chance for it, for this criminal statute and every criminal statute must be plain and simple. If Mr. Scopes is to be indicted and prosecuted because he taught a wrong theory of the origin of life why not tell him what he must teach. Why not say that you must teach that man was made of the dust; and still stranger not directly from the dust, without taking any chances on it, whatever, that Eve was made out of Adam's rib. You will know what I am talking about.

No Man Could Obey Law—No Court Could Enforce It

Now my client must be familiar with the whole book, and must know all about all of these warring sects of Christians and know which of them is right and which wrong, in order that he will not commit crime. Nothing was heard of all that until the fundamentalists got into Tennessee. I trust that when they prosecute their wildly made charge upon the intelligence of some other sect they may modify this mistake and state in simple language what was the account contained in the Bible that could not be taught. So, unless other sects have something to do with it, we must know just what we are charged with doing. This statute, I say, your Honor, is indefinite and uncertain. No man could obey it, no court could enforce it and it is bad for in-definiteness and uncertainty. Look at that indictment up there. If that is a good indictment I never saw a bad one. Now, I do not expect, your honor, my opinion to go because it is my opinion, because I am like all lawyers who practice law; I have made mistakes in my judgment of law. I will probably make more of them. I insist that you might just as well hand my client a piece of blank paper and then send the sheriff after him to jail him. Let me read this indictment.

Reads from Newspaper

I am reading from a newspaper. I forget what newspaper it was, but am sure it was right: "That John Thomas Scopes on April, 1925, did unlawfully and willfully teach in the public schools of Rhea County, Tennessee, which public schools are supported in part and in whole —" I don't know how that is possible, but we will pass that up—"In part or in whole by the public school funds of the state a certain theory and theories that deny the story of the divine creation of man as taught in the Bible and did teach instead thereof that man is descended from a lower order of animals." Now, then there is something that is very elementary. That is one of them and very elementary, because the constitutions of Tennessee provides and the constitution of pretty near every other state in the United States provide that an indictment must state in sufficient terms so that a man may be appraised of what is going to be the character of charge against him. Tennessee said that my friend the attorney-general says that John Scopes knows what he is here for. Yes, I know what he is here for, because the fundalmentalists are after everybody that thinks. I know why he is here. I know he is here because ignorance and bigotry are rampant, and it is a mighty strong combination, your Honor, it makes him fearful. But the state is bringing him here by indictment, and several things must be stated in the indictment; indictments must state facts, not law nor conclusions of law. It is all well enough to show that the indictment is good if it charges the offense in the language of the statute. In our state of Illinois, if one man kills another with malice aforethought, he would be guilty of murder, but an indictment would not be good that said John Jones killed another. It would not be good. It must tell more about it and how. It is not enough in this indictment to say that John Scopes taught something contrary to the divine account written by Moses—maybe—that is not enough. There are several reasons for it. First, it is good and right to know. Secondly, after the shooting is all over here and Scopes has paid his fine if he can raise his money, or has gone to jail if he cannot, somebody else will come along and indict him over again. But there is one thing I cannot account for, that is the hatred and the venom and feeling and the very strong religious combination. That I never could account for. There are a lot of things I cannot account for. Somebody may come along next week and indict him again, on the first indictment. It must be so plain that a second case will never occur. He can say to him, "I have cleared that off."

No Other Indictment Like This One

He can file a plea that he has already been put in jeopardy and convicted and paid the fine, so you cannot do it over again. There is no quesion about that, your Honor, in the slightest and the books are full of them. I have examined, I think all the criminal cases in Tennessee on this point. I don't like to speak with too much assurance, because sometimes you get held up on such a thing, but I assume that if they have got anything on the other side I would have heard from them, and I have, with the aid of my assistants and helpers, they doing most of the work, I have examined most all of them, and if there is another indictment in Tennessee like it I haven't found it, and plenty of indictments have been declared void in Tennessee because they did not tell us anything—plenty of them. I do not think there ever was another one like it in Tennessee, and I am not referring to the subject matter now because I know there never was, as far as the subject matter goes, but I am speaking of the form of it. Now, Mr. Scopes, on April 24 did unlawfully and wilfully teach in a public school of Rhea county, Tennessee, which public school is supported in whole or in part by the public school fund of the state, certain theories that deny the story of the divine creation of man. What did he teach? What did he teach? Who is it that can tell us that John Scopes taught certain theories that denied the story of the divine—the divine story of creation as recorded in the Bible. How did he know what text-books did he teach from? Who did he teach? Why did he teach? Not a word—all is silent. He taught, oh yes, the place mentioned is Rhea county. Well, that is some county—Maybe all over it, I don't know where he taught, he might have taught in a half a dozen schools in Rhea county on the one day and if he is indicted next year after this trial is over, if it is, for teaching in District No. 1, in Rhea county, he cannot plead that he has already been convicted, because this was over here in another district and at another place. What did he teach? What was the horrible thing he taught that was in conflict with Moses and what is it that is not in conflict with Moses? What shouldn't he have taught? What is the account contained in the Bible which he ignored, when he taught the doctrine of evolution which is taught by every—believed by every scientific man on earth. Joshua made the sun stand still. The fundamentalists will make the ages roll back. He should have been informed by the indictment what was the doctrine he should have taught and he should have been informed what he did teach so that he could prepare, without reading a whole book through, and without waiting for witnesses to testify—we should have been prepared to find out whether the thing he taught was in conflict with the Bible or what the Bible said about it. Let me call attention, your Honor, to one case they have heralded here—I don't know why. I will refer to it later. Let me show you a real indictment, gentlemen, in case you ever need to draw another one. You don't mind a little pleasantry, do you? Here is the case we have heard so much about.

Leeper Case Again

Leeper vs. State. My fellow is a leper, too, because he taught evolution. I am going to discuss this case a moment later to show that it has nothing to do with the subject. This man was indicted because under the school book law of this state the commission had decided certain books should be taught, and amongst the rest they decided that Frye's geography should be taught. That any teacher that did not follow the law and taught something else should be fined $25. Of course, it wasn't so bad as to teach evolution, although the statute doesn't say anything about evolution. Now they indicted him and this is what they said in the indictment. This is their leading case. "The grand jury for the State of Tennessee, upon their oaths present that Edward Leeper, heretofore, to-wit: On the 5th day of October, 1899, in the state and county aforesaid, being then and there a public schoolteacher and teaching the public school known as school No. 5, Sixth district, Blount county"—they pick that out all right—"did unlawfully use and permit to be used in said public school, after the state textbook commission had adopted and prescribed for use in the public schools of the state Frye's introductory geography as a uniform textbook another and different textbook on that branch than the one so adopted aforesaid, to-wit: Butler's geography and the new Eclectic elementary geography against the peace and dignity of the state. Now, your honor, would that have been a good indictment, if they had left all that out and said he taught some book not authorized by the board? He has got a right to know what he taught and where he taught it and all the necessary things to convict him of crime. Your Honor, he cannot be convicted in this case unless they prove what he taught and where he taught it, and we have got a right to know all that before we go into court—every word of it. The indictment isn't any more than so much blank paper. I insist, your Honor, that no such indictment was ever returned before on land or sea. Some men may pull one on me, but I don't think so—I don't think so. You might just as well indict a man for being no good—and we could find a lot of them down here probably and if we couldn't I could bring them down from Chicago—but only a man is held to answer for a specific thing and he must be told what that specific thing is before he gets into court. The statute is absolutely void, because they have violated the constitution in its caption and it is absolutely uncertain—the indictment is void because it is uncertain, and gives no fact or information and it seems to me the main thing they did in bringing this case was to try to violate as many provisions of the constitution as they could, to say nothing about all the spirit of freedom and independence that has cost the best blood in the world for ages, and it looks like it will cost some more. Let's see what else we have got. This legislation—this legislation and all simliar legislation that human ingenuity and malice can concoct, is void because it violates Section 13, Section 12 and Section 3. I want to call attention to that, your Honor, Section 12 is the section providing that the state should cherish science, literature and learning. Now, your Honor, I make it a rule to try not to argue anything that I do not believe in, unless I am caught in a pretty close corner and I want to say that the construction of the attorney-general given to that, I think, is correct and the court added a little to it, which I think makes your interpretation correct for what it is good for. It shows the policy of the state. It shows what the state is committed to. I do not believe that a statute could be set aside as unconstitutional simply because the legislature did not see fit to pass proper acts to enlighten and educate the yeomen of Tennessee.

Violates Right of Worship—Does Not Understand Religious Hatred

The state by constituion is committed to the doctrine of education, committed to schools. It is committed to teaching and I assume when it is committed to teaching it is committed to teaching the truth—ought to be anyhow—plenty of people to do the other. It is committed to teaching literaure and science. My friend has suggested that literature and science might conflict. I cannot quite see how, but that is another question. But that indicates the policy of the state of Tennessee and wherever it is used in construing the unconstitutionality of this act it can only be used as an indication of what the state meant and you could not pronounce a statute void on it, but we insist that this statute is absolutely void because it contravenes Section 3, which is headed "the right of worship free." Now, let's see, your Honor, there isn't any court in the world that can uphold the spirit of the law by simply upholding its letters. I read somewhere—I don't know where—that the letter killeth, but the spirit giveth life. I think I read it out of "The Prince of Peace." I don't know where I did, but I read it. If this section of the constitution which guarantees religious liberty in Tennessee cannot be sustained in the spirit it cannot be sustained in the letter. What does it mean? What does it mean? I know two intelligent people can agree only for a little distance, like a company walking along in a road. They may go together a few blocks and then one branches off. The remainder go together a few more blocks and another branches off and still further some one else branches off and the human minds are just that way, provided they are free, of course, the fundamentalists may be put in a trap so they cannot think differently if at all, probably not at all, but leave two free minds and they may go together a certain distance, but not all the way together. There are no two human machines alike and no two human beings have the same experiences and their ideas of life and philosophy grow out of their construction of the experiences that we meet on our journey through life. It is impossible, if you leave freedom in the world, to mold the opinions of one man upon the opinions of another—only tyranny can do it—and your constitutional provision, providing a freedom of religion, was meant to meet that emergency. I will go further—there is nothing else—since man—I don't know whether I dare say evolved—still, this isn't a school—since man was created out of the dust of the earth—out of hand—there is nothing else your Honor that has caused the difference of opinion, of bitterness, of hatred, of war, of cruelty, that religion has caused. With that, of course, it has given consolation to millions.

But it is one of those particular things that should be left solely between the individual and his Maker, or his God, or whatever takes expression with him, and it is no one else's concern.

500 Different Christian Creeds—Darrow Pseudo-Scientist

How many creeds and cults are there this whole world over? No man could enumerate them? At least as I have said, 500 different Christian creeds, all made up of differences, your honor, every one of them, and these subdivided into small differences, until they reach every member of every congregation. Because to think is to differ, and then there are any number of creeds older and any number of creeds younger, than the Christian creed, any number of them, the world has had them forever. They have come and they have gone, they have abided their time and have passed away, some of them are here still, some may be here forever, but there has been a multitude, due to the multitude and manifold differences in human beings, and it was meant by the constitutional convention of Tennessee to leave these questions of religion between man and whatever he worshiped, to leave him free. Has the Mohammedan any right to stay here and cherish his creed? Has the Buddist a right to live here and cherish his creed? Can the Chinaman who comes here to wash our clothes, can he bring his joss and worship it? Is there any man that holds a religious creed, no matter where he came from, or how old it is or how false it is, is there any man that can be prohibited by any act of the legislature of Tennessee? Impossible? The constitution of Tennessee, as I understand, was copied from the one that Jefferson wrote, so clear, simple, direct, to encourage the freedom of religious opinion, said in substance, that no act shall ever be passed to interfere with complete religious liberty. Now is this it or is not this it? What do you say? What does it do? We will say I am a scientist, no, I will take that back, I am a pseudo-scientist, because I believe in evolution, pseudo-scientist named by somebody, who neither knows or cares what science is, except to grab it by the throat and throttle it to death. I am a pseudo-scientist, and I believe in evolution. Can a legislative body say, "You cannot read a book or take a lesson, or make a talk on science until you first find out whether you are saying against Genesis. It can unless that constitutional provision protects me. It can. Can it say to the astronomer, you cannot turn your telescope upon the infinite planets and suns and stars that fill space, lest you find that the earth is not the center of the universe and there is not any firmament between us and the heaven. Can it? It could—except for the work of Thomas Jefferson, which has been woven into every state constitution of the Union, and has stayed there like the flaming sword to protect the rights of man against ignorance and bigotry, and when it is permitted to overwhelm them, then we are taken in a sea of blood and ruin that all the miseries and tortures and carion of the middle ages would be as nothing. They would need to call back these men once more. But are the provisions of the constitutions that they left, are they enough to protect you and me, and every one else in a land which we thought was free? Now, let us see what it says: "All men have a natural and indefeasible right to worship Almighty God according to the dictates of their own conscience."

That takes care, even of the despised modernist, who dares to be intelligent. "That no man can of right be compelled to attend, erect or support any place of worship, or to maintain any minister against his consent; that no human authority can in any case whatever control or interfere with the rights of conscience in any case whatever"—that does not mean whatever, that means, "barring fundamentalist propaganda. It does not mean whatever at all times, sometimes may be—and that "no preference shall be given by law to any religious establishment or mode of worship." Does it? Could you get any more preference, your honor, by law? Let us see. Here is the state of Tennessee, living peacefully, surrounded by its beautiful mountains, each one of which contains evidence that the earth is millions of years old,—people quiet, not all agreeing upon any one subject, and not necessary. If I could not live in peace with people I did not agree with, why, what? I could not live. Here is the state of Tennessee going along in its own business, teaching evolution for years, state boards handing out books on evolution, professors in colleges, teachers in schools, lawyers at the bar, physicians, ministers, a great percentage of the intelligent citizens of the state of Tennessee evolutionists, have not even thought it was necessary to leave their church. They believed that they could appreciate and understand and make their own simple and human doctrine of the Nazarine, to love their neighbor, be kindly with them, not to place a fine on and not try to send to jail some man who did not believe as they believed, and got along all right with it, too, until something happened. They have not thought it necessary to give up their church, because they believed that all that was here was not made on the first six days of creation, or that it had come by a slow process of developments extending over the ages, that one thing grew out of another. There are people who believed that organic life and the plants and the animals and man and the mind of man, and the religion of man are the subjects of evolution, and they have not got through, and that the God in which they believed did not finish creation on the first day, but that he is still working to make something better and higher still out of human beings, who are next to God, and that evolution has been working forever and will work forever—they believe it.

A Crime in the State to Get Learning

And along comes somebody who says we have got to believe it as I believe it. It is a crime to know more than I know. And they publish a law to inhibit learning. Now, what is in the way of it? First, what does the law say? This law says that it shall be a criminal offense to teach in the public schools any account of the origin of man that is in conflict with the divine account in the Bible. It makes the Bible the yard stick to measure every man's intellect, to measure every man's intelligence and to measure every man's learning. Are your mathematics good? Turn to I Elijah ii, is your philosophy good? See II Samuel iii, is your astronomy good? See Genesis, Chapter 2, Verse 7, is your chemistry good? See—well, chemistry, see Deuteronomy iii-6, or anything that tells about brimstone. Every bit of knowledge that the mind has, must be submitted to a religious test. Now, let us see, it is a travesty upon language, it is a travesty upon justice, it is a travesty upon the constitution to say that any citizen of Tennessee can be deprived of his rights by a legislative body in the face of the constitution. Tell me, your honor, if this is not good, then what? Then, where are we coming out? I want to argue that in connection with another question here which is equally plain. Of course, I used to hear when I was a boy you could lead a horse to water, but you could not make him drink—water. I could lead a man to water, but I could not make him drink, either. And you can close your eyes and you won't see, cannot see, refuse to open your eyes—stick your fingers in your ears and you cannot hear—if you want to. But your life and my life and the life of every American citizen depends after all upon the tolerance and forebearance of his fellowman. If men are not tolerant, if men cannot respect each other's opinions, if men cannot live and let live, then no man's life is safe, no man's life is safe.

Here is a country made up of Englishmen, Irishmen, Scotch, German, Europeans, Asiatics, Africans, men of every sort and men of every creed and men of every scientific belief; who is going to begin this sorting out and say, "I shall measure you; I know you are a fool, or worse; I know and I have read a creed telling what I know and I will make people go to Heaven even if they don't want to go with me, I will make them do it." Where is the man that is wise enough to do it?

Statute Under Police Power

This statute is passed under the police power of this state. Is there any kind of question about that? Counsel have argued that the legislature has the right to say what shall be taught in the public school. Yes, within limits, they have. We do not doubt it, but they probably cannot say writing and arithmetic could not be taught, and certainly they cannot say nothing can be taught unless it is first ascertained that it agrees with the Scriptures; certainly they cannot say that.

But this is passed under the police power. Let me call your honor's attention to this. This is a criminal statute, nothing else. It is not any amendment to the school law of the state. It makes it a crime in the caption to teach evolution and in the body of the act to teach something else, purely and simply a criminal statute.

There is no doubt about the law in this state. Show me that Barber's case will you? (Taking book from counsel.)

There isn't the slightest doubt about it, or in any other state. Your honor, I have got a case there, but I have not got my glasses.

Associate Counsel—Here they are.

Mr. Darrow—Thank you.

There isn't the slightest doubt about it. Can you pass a law under the police powers of the state; that a thing cannot be done in Dayton, but they can do it down in Chattanooga? Oh, no. What is good for Chattanooga is good for Dayton; I would not be sure that what is good for Dayton is good for Chattanooga, but I will put it the other way.

Any law passed under the police power must be uniform in its application; must be uniform. What do you mean by a police law? Well, your honor, that calls up visions of policemen and grand juries and jails and penitentiaries and electrocutionary establishments, and all that, and wickedness of heart; that is police power. True, it may extend to public health and public morals, and a few other things. I do not imagine evolution hurts the health of anyone, and probably not the morals, excepting as all enlightment may and the ignorant think, of course, that it does, but it is not passed for them, your honor, oh, no. It is not passed because it is best for the public morals, that they shall not know anything about evolution, but because it is contrary to the divine account contained in Genesis, that is all, that is the basis of it.

Now let me see about that. Any police statute must rest directly upon crime, or what is analagous to it; it has that smack, anyhow. Talk about the police power and the policemen and all the rest of them with their clubs and so on, you shudder and wonder what you have been doing, and that is the police power.

Now, any such law must be uniform in its application, there cannot be any doubt about that, not the slightest. Here, for instance, the good people of—well, I guess these are good people, Nashville, wasn't it? Whether the common people down there—

Mr. Neal—That is a Tennessee case.

Is Bath on Sunday Wicked?

Mr. Darrow—Anyhow, it is a Tennessee case. Good people stirred up the community, by somebody, I don't know who, passed a law which said it was a misdemeanor to carry on barbering on Sunday, and that it should be a misdemeanor for anyone engaged in the business of barbering to shave, shampoo and cut hair or to keep open the bath rooms on Sunday.

(Laughter in courtroom.)

Mr. Darrow—Well, of course, I suppose it would be wicked to take a bath on Sunday, I don't know, but that was not the trouble with this statute. It would have been all right to forbid the good people of Tennessee from taking a bath on Sunday, but that was not the trouble. A barber could not give a bath on Sunday, anybody else could. No barber shall be permitted to give a bath on Sunday, and the supreme court seemed to take judicial notice of the fact that people take a bath on Sunday just the same as any other day. Foreigners come in there in the habit of bathing on Sundays just as any other time, and they could keep shops open, but a barber shop, no. The supreme court said that would not do, you could not let a hotel get away with what a barber shop can't. (Laughter.)

And so they held that this law was unconstitutional, under the provision of the constitution which says laws must be uniform. There is no question about the theory of it. If there were not, why, they would be passing laws against—the fundamentalists would be passing laws against the Congregationalists and Unitarians—I cannot remember all the names—Universalists—they might graduate the law according to how orthodox or unorthodox the church was. You cannot do it; they have to be general. The supreme court of this state has decided it and it does not admit of a doubt.

Now, I will just read one section of the opinion: The act is for the benefit of all individuals, barbers excepted; we know that all of the best hotels have bathrooms for the use of guests, that they accept pay for baths and permit them on Sunday.

Charges Class Legislation

(Reading from Barbers case, 2 Pickle, beginning with "that in many cases the barber has bathroom" to "for this and other things the act is held void.")

That in the case in 2 Pickle that I read from. Why they named this Pickle I have not found out yet.

But there is another in 16 Cates, page 12. This is a case, your honor, where they passed a law:

(Reading from above book beginning with words "that it shall be unlawful for any jobbing," to "It shall be unlawful.")

If it is unlawful for these corporations to discharge an individual because they didn't vote a certain ticket, this must have been passed against the wicked democrats up here. Up in our state it is the republicans who do all that, but still, it shall be unlawful to discharge any man if he don't vote a certain way or buy at a certain place if he did buy at a certain place, that only applied to corporations; if John Smith had a little ranch upon the mountain or had hired a man he could discharge him all right if he didn't vote the right ticket or go to the right church or any old reason. And the supreme court of the state said, "Oh, no, you cannot pass that sort of a law." What is sauce for the goose must be sauce for the gander. You cannot pass a law making it a crime for a corporation to discharge a man because he voted differently and leave private individuals to do it. And they passed this law.

Let us look at this act, your honor. Here is a law which makes it a crime to teach evolution in the caption. I don't know whether we have discussed that or not, but it makes it a crime in the body of the act to teach any theory of the origin of man excepting that contained in the divine account, which we find in the Bible. All right. Now that act applies to what? Teachers in the public schools. Now I have seen somewhere a statement of Mr. Bryan's that the fellow that made the pay check had a right to regulate the teachers. All right, let us see. I do not question the right of the legislature to fix the courses of study, but the state of Tennessee has no right under the police power of the state to carve out a law which applies to schoolteachers, a law which is a crimnal statute and nothing else; which makes no effort to prescribe the school law or course of study. It says that John Smith who teaches evolution is a crimnal if he teaches it in the public schools. There is no question about this act; there is no question where it belongs; there is no question of its origin. Nobody would claim that the act could be passed for a minute excepting that teaching evolution was in the nature of a criminal act; that it smacked of policemen and criminals and jails and grand juries; that it was in the nature of something that was criminal and, therefore, the state should forbid it.

It cannot stand a minute in this court on any theory than that it is a criminal act, simply because they say it contravenes the teaching of Moses without telling us what those teachings are. Now, if this is the subject of a criminal act, then it cannot make a criminal out of a teacher in the public schools and leave a man free to teach it in a private school. It cannot make it criminal for a teacher in the public schools to teach evolution, and for the same man to stand among the hustings and teach it. It cannot make it a criminal act for this teacher to teach evolution and permit books upon evolution to be sold in every store in the state of Tennessee and to permit the newspapers from foreign cities to bring into your peaceful community the horrible utterances of evolution. Oh, no, nothing like that. If the state of Tennessee has any force in this day of fundamentalism, in this day when religious bigotry and hatred is being kindled all over our land, see what can be done?

Now, your honor, there is an old saying that nits make lice. I don't know whether you know what it makes possible down here in Tennessee? I know, I was raised in Ohio. It is a good idea to clear the nits, safer and easier.

To Strangle Puppies Is Good When They Grow Into Mad Dogs, Maybe

To strangle puppies is good when they grow up into mad dogs, maybe. I will tell you what is going to happen, and I do not pretend to be a prophet, but I do not need to be a prophet to know. Your honor knows the fires that have been lighted in America to kindle religious bigotry and hate. You can take judicial notice of them if you cannot of anything else. You know that there is no suspicion which possesses the minds of men like bigotry and ignorance and hatred.

If today—

The Court—Sorry to interrupt your argument, but it is adjourning time.

Mr. Darrow—If I may I can close in five minutes. I can close in five minutes in the morning, only a few.

If today, your honor—give me five minutes more, I will not talk five minutes.

The Court—Proceed tomorrow.

Mr. Darrow—I shall not talk long, your honor, I will tell you that.

If today you can take a thing like evolution and make it a crime to teach it in the public school, tomorrow you can make it a crime to teach it in the private schools, and the next year you can make it a a crime to teach it to the hustings or in the church. At the next session you may ban books and the newspapers. Soon you may set Catholic against Protestant and Protestant against Protestant, and try to foist your own religion upon the minds of men. If you can do one you can do the other. Ignorance and fanaticism is ever busy and needs feeding. Always it is feeding and gloating for more. Today it is the public school teachers, tomorrow the private. The next day the preachers and the lecturers, the magazines, the books, the newspapers. After while, your honor, it is the setting of man against man and creed against creed until with flying banners and beating drums we are marching backward to the glorious ages of the sixteenth century when bigots lighted fagots to burn the men who dared to bring any intelligence and enlightment and culture to the human mind.

Tomorrow I will say a few words.

The Court—You gentlemen send down your authorities to my room at the hotel, on both sides, and your briefs, if you have such.

Court is adjourned to 9:00 o'clock tomorrow morning.

CHAPTER III.

THIRD DAY'S PROCEEDINGS—TUESDAY, JULY 14, 1925.

Darrow Objects to Prayer

Court met pursuant to recess.

Present as before.

Whereupon:

Immediately upon the rapping of the bailiff for order in the courtroom, and before the regular session was opened, the following proceedings occurred:

The Court—Rev. Stribling will you open with prayer?

Mr. Darrow—Your honor, I want to make an objection before the jury comes in.

The Court—What is it, Mr. Darrow?

Mr. Darrow—I object to prayer and I object to the jury being present when the court rules on the objection.

Gen. Stewart—What is it?

The Court—He objects to the court being opened with prayer, especially in the presence of the jury.

Mr. Stewart—The jury is not here.

The Court—Are any of the jury in the courtroom?

(No response.)

The Court—No, I do not want to be unreasonable about anything, but I believe I have a right, I am responsible for the conduct of the court, it has been my custom since I have been judge to have prayers in the courtroom when it was convenient and I know of no reason why I should not follow up this custom, so I will overrule the objection.

Mr. Darrow—May we ask if there are any members of the jury in the courtroom?

The Court—Yes, everyone stand up.

Mr. Darrow—May I make the record?

The Court—Yes.

(The bailiff raps for order.)

Mr. Darrow—Just a minute.

The Court—Yes.

Mr. Darrow—I understand from the court himself that he has sometimes opened the court with prayer and sometimes not, and we took no exceptions on the first day, but seeing this is persisted in every session, and the nature of this case being one where it is claimed by the state that there is a conflict between science and religion, above all other cases there should be no part taken outside of the evidence in this case and no attempt by means of prayer or in any other way to influence the deliberation and consideration of the jury of the facts in this case.

For that reason we object to the opening of the court with prayer and I am going to ask the reporters to take down the prayer and make specific objections again to any such parts as we think are especially obnoxious to our case.

The Court—Do you want to say anything.

Gen. Stewart—Go ahead, Gen. McKenzie.

Mr. McKenzie—That matter has been passed upon by our supreme court. Judge Shepherd took a case from the court, when the jury, after retiring to consider their verdict, at the suggestion of one of them to bow in prayer, asked divine guidance, afterwards delivering a verdict not excepted to, and afterwards taken to the supreme court: It was commendable to the jury to ask divine guidance.

No Objection to Secret Prayer

Mr. Darrow—I do not object to the jury or anyone else praying in secret or in private, but I do object to the turning of this courtroom into a meeting house in the trial of this case. You have no right to do it.

The Court—You have a right to

put your exceptions in the record.

Gen. Stewart—In order that the record may show the state's position, the state makes no contention, as stated by counsel for the defense, that this is a conflict between science and religion insofar as the merits are concerned, it is a case involving the fact as to whether or not a school-teacher has taught a doctrine prohibited by statute, and we, for the state, think it is quite proper to open the court with prayer if the court sees fit to do it, and such an idea extended by the agnostic counsel for the defense is foreign to the thoughts and ideas of the people who do not know anything about infidelity and care less.

Mr. Hays—May I ask to enter an exception to the statement "agnostic counsel for the defense."

Mr. Malone—I would like to reply to this remark of the attorney-general. Whereas I respect my colleagues, Mr. Darrow's right to believe or not to believe as long as he is as honest in his unbelief as I am in my belief. As one of the members of counsel who is not an agnostic, I would like to state the objection from my point of view. Your honor has the discretion to have a prayer or not to have a prayer. There was no exception offered and I can assure the court when we talked it over among ourselves as colleagues, there was no exception felt to the opening of these proceedings by prayer the first day, but I would like to ask your honor whether in all the trials over which your honor has presided, this court has had a clergyman every morning of every day of every trial to open the court with prayer?

Our objection goes to the fact that we believe that this daily opening of the court with prayers, those prayers we have already heard, having been duly argumentative that they help to increase the atmosphere of hositility to our point of view, which already exists in this community by widespread propaganda.

"A God Fearing Country"—Stewart

Gen. Stewart—In reply to that there is still no question involved in this lawsuit as to whether or not Scopes taught a doctrine prohibited by the statute, that is that man descended from a lower order of animals. So far as creating an atmosphere of hostility is concerned, I would advise Mr. Malone that this is a God fearing country.

Mr. Malone—And it is no more God fearing country than that from which I came.

The Court—Gentlemen, do not turn this into an argument.

Mr. Darrow—I would like to reply to counsel, that this statute says no doctrine shall be taught which is contrary to the divine account contained in the Bible. So there is no question about the religious character of these proceedings.

The Court—This court has no purpose except to find the truth and do justice to all the issues involved in this case.

In answer to counsel for the defendant, as to my custom, I will say the several years I have been on the bench I have used my discretion in opening the court with prayer, at times when there was a minister present and it was convenient to do so; other times when there was no large assemblage of people and no minister present, I have not always followed this custom, but I think it is a matter wholly within the discretion of the court.

I have instructed the ministers who have been invited to my rostrum to open the court with prayer, to make no reference to the issues involved in this case. I see nothing that might influence the court or jury as to the issues. I believe in prayer myself; I constantly invoke divine guidance myself, when I am on the bench and off the bench; I see no reason why I should not continue to do this. It is not the purpose of this court to bias or prejudice the mind of any individual, but to do right in all matters under investigation.

Arguing about Prayer in the Court

STEWART MALONE DARROW

Judge Overrules Objection

Therefore, I am pleased to overrule the objection of counsel and invite Dr. Stribling to open the court with prayer.

Mr. Darrow—I note an exception, your honor.

Thereupon Dr. Stribling proceeded to offer the following prayer:

Dr. Stribling—Our Father, to Thee we give all the praise for every good thing in life and we invoke Thy blessings upon us this morning, as accountable beings to Thee as we enter into the duties of this day. It matters not what our relation to man may be. We have a responsibility to fulfill, righteously the tasks that are ours to do and we would ask Thee this morning, oh God, to make us fully conscious of Thy presence and to give unto us minds that are willing to be directed in the way Thou wouldst have us do. We pray, our Father, to bless the proceedings of this court, bless the court, the judge, as he presides, and may there be in every heart and in every mind a reverence to the Great Creator of the world.

We ask Thee, our Father, to help us, every one to find our place in our relation to every other man, so that we can best serve, can best know human interests and can best sympathize with the needs of every heart.

To this end we ask that Thou wilt enlighten our minds and lead us to understand and know truth in all its every phase, we ask it in the name of our Blessed Redeemer, Jesus Christ, amen.

The Court—Open court, Mr. Sheriff, pursuant to adjournment. Be seated.

Thereupon court was regularly opened.

The Court—Any motions this morning, gentlemen? Any further motions? Col. Darrow, did I understand you to finish your argument or not?

Mr. Darrow—Your honor, I only reserved the right this morning in looking over my points to see whether I had forgotten something. I find that I covered everything that I wished to cover and submit the argument now.

The Court—Anything further from the state? Of course not.

Gen. Stewart—No, sir; not on that.

The Court—Well, of course, these are profound questions, gentlemen, which you present to me. I worked late last night; but the lights were out until 8:30 and I couldn't do anything until that time. As I said yesterday it is not my disposition to guess if I can avoid doing so, when deciding issues involved in a lawsuit and therefore I will have to ask the indulgence of the court this morning, until I finish my investigation, and the preparation of my ruling upon these questions. I don't know how long that will be—possibly two hours.

Neal Presents Demurrer

Mr. Neal—We have a demurrer that we wish to file and consider for decision nunc pro tunc?

The Court—Involving the same questions?

Mr. Neal—Yes, sir.

The Court—You might consider it as filed now, and let me act on both together if you desire? Is that agreeable to you, Gen. Stewart?

Gen. Stewart—I would like to see the demurrer?

The Court—Have you furnished him a copy?

Mr. Neal—No, sir; but we will.

The Court—If you want to make any pictures, boys, make them now. I will have to excuse you from the stage.

Mr. Neal—The same questions are raised.

The Court—You raised the same questions by a different route.

Mr. Neal—Yes, sir; that is right.

The Court—You are not sure as to the method.

Mr. Darrow—That is all there is to it.

The Court—If you want to make any pictures, I will give you fifteen minutes.

Gen. Stewart—Your honor, just adjourn until this afternoon. It is 10:00 o'clock now?

Mr. Darrow—May we let the record show, just to save bringing up this question again, of prayer—may the record show, without any further objection, on each and every morning that the motion is made and the same ruling is made?

The Court—Yes, sir. Let the record show it.

Mr. Darrow—I don't care to emphasize it at all. I just want to save it.

The Court—Let the record show that it will be treated as made and overruled every morning.

Mr. Neal—Just a moment, I will hand you a copy of the demurrer. Do you have the other motion, judge?

The Court—Yes, sir; I have it. It is in the hands of the court stenographer. I haven't it with me here. Is Mr. Buchanan here?

Mr. Fain—Yes, sir; he is in the transcribing room.

Gen. Stewart—Did I understand, your honor, we would just adjourn until noon?

The Court—No; I haven't said that. We have been adjourning at 11:30.

Gen. Stewart—If your honor wants that much time, I want you to have it. We just want to know definitely if the court wants it.

The Court—Let's see if they want to make any pictures, and then I will make the announcements.

(After photographers completed the taking of pictures.)

The Court—The court will recess until 1:00 o'clock.

AFTERNOON SESSION

1:00 p. m.

Tuesday, July 14, 1925

Judge Warns Reporters

Whereupon the policeman rapped for order and announced that court would reconvene at 2:30 o'clock p. m.

2:15 o'clock p. m.

Whereupon the court announced as follows:

The Court—I want to announce that I gave strict instructions to the stenographer that my opinion was not to be released to any person or to give any information out. If any member of the press has any intimation as to what my opinion is—no person knows except myself and the stenographer—and sends it out before I begin to read it, I will deal with them for contempt of court.

3:45 o'clock p. m.,

Present as before.

Whereupon:

The court was called to order.

Mr. Hayes—Before your honor presents a decision or the proceedings go further, may I present a petition to the court, addressed to the Hon. John T. Raulston, presiding judge, Rhea county court. We, the following named representatives of various well-known religious organizations, churches, synagogues, do hereby petition your honor that if you continue your custom of opening the daily sessions of the court of Rhea county with prayer—

Gen. Stewart—Your honor, just a minute, I submit that is absolutely out of order.

Mr. Hayes—Mr. Stewart—

Gen. Stewart—This is not an assembly met for any purpose of hearing a motion of that sort, or any thing of that sort. Your honor has passed upon the motion.

Mr. Hayes—I insist upon making this motion.

Gen. Stewart—I am making my exception to the court, will you please keep your mouth shut.

Mr. Hayes—Will your honor hear my motion?

Gen. Stewart—I am making my exception to the court.

The Court—I will hear it.

Gen. Stewart—It is entirely out of order. And I except to it with all the vehemence of my nature.

The Court—I will hear it, proceed, Mr. Hayes.

Petition from Unitarians, Jews and Congregationalists

Mr. Hayes—(Reading the petition.)

To the Hon. John T. Raulston, Presiding Judge, Rhea County Court:

We, the following representatives of various well-known religious organizations, churches and synagogues, do hereby petition your honor that, if you continue your custom of opening the daily sessions of the court of Rhea county with prayer, you select the officiating clergymen from among other than fundamentalist churches in alteration with fundamentalist clergymen.

We beg you to consider the fact that among the persons intimately connected with, and actively participating in this trial of Mr. John T. Scopes there are many to whom the prayers of the fundamentalists are not spiritually uplifting and are occasionally offensive. Inasmuch as by your own ruling all the people in the courtroom are required to participate in the prayers by rising, it seems to us only just and right that we should occasionally hear a prayer which requires no mental reservations on our part and in which we can conscientiously participate.

Signed:

REV. CHARLES POTTER,
Minister, West Side Unitarian church, New York.

RABBI JEROME MARK,
Temple Beth-El, Knoxville, Tenn.

REV. FRED W. HAGAN,
First Congregational church, Huntington, W. Va.

REV. D. M. WELCH,
Minister, Knoxville Unitarian church.

Mr. Hayes—My motion, your honor, is, without, of course, giving up our exception to your honor's ruling, that if the court denies that, this petition be granted and that we have an opportunity to hear prayer by men who think that God has shown His divinity in the wonders of the world, in the book of nature, quite as much as in the book of the revealed word.

Court Refers Petition to Pastors' Association

The Court—I shall refer that petition to the pastors' association of this town, and I shall ask them—

(Laughter and loud applause, and rapping for order by the policeman.)

The Court—I shall ask the pastors' association from now on to name the man who is to conduct prayer. I shall have no voice, make no suggestions as to who they name, but I will invite the men named by the association to conduct the prayer each morning.

Now, I have an announcement to make.

Mr. Hayes—May I ask your honor if this is a decision on my motion?

The Court—Yes, sir.

Mr. Hayes—So that I may except, so that I may save the record.

Mr. Neal—Your honor knows that the men your honor refers this motion to, are not among the class of men that signed the petition.

The Court—I see by the press one minister has resigned his post recently because Dr. Potter was not allowed to preach in his church and I take it he is in sympathy with Dr. Potter and his doctrine, the others are perhaps fundamentalists, I don't know.

Scoop of Judges Opinion

Now, I have a very serious matter to speak of, I dictated my opinion in this case, which is lengthy. I have been about some four hours in the preparation of the opinion. I gave it to the court stenographer, a reputable court stenographer in secret, with the instruction that no living person know anything as to the conclusions I had reached until I had begun to read my opinion from the bench. I have not intimated to any living soul what my opinion was, except to the stenographer who took the decision.

I am now informed that the newspapers in the large cities are being now sold, which undertake to state what my opinion is. Now any person that sent out any such information as that, sent it out without the authority of this court and if I find that they have corruptly secured said information I shall deal with them as the law directs. Now on account of this improper conduct, apparently at least improper conduct, of some person or persons, the court has decided to withhold his decision until tomorrow morning and tomorrow morning, after the opening of the court, the decision will be read. Now I want, when the crowd is gone from the court room—I want all the members of the press to meet me in this court room. I want to talk with them about this matter. If I find that some representative here has used in stratagem or used any corrupt means or has in any manner secured my opinion, or as to the result of it, and sent it out, I shall promptly deal with him and of course excuse him from any further presence in this court room, so when the crowd is gone I desire that the newspaper men stay with me.

Mr. Stewart—Does Your Honor want the attorneys on either side?

The Court—Yes, sir, the attorneys on both sides.

Mr. Malone—Would it not be possible for us to dispose of the motion or the business which has accumulated quite naturally, because witnesses are on their way here and some are here and we would like to get along with the greatest possible expedition? We regret sincerely that this difficuly has arisen to disturb the court.

The Court—Col. Malone, we cannot go any further until I decide these questions that are before me and I think I have announced satisfactory reasons for not doing so this afternoon. I regret myself very much to have this delay. Of course I don't mind so much personally, because I am absolutely exhausted in the preparation of this opinion.

Mr. Neal—Could you go into the question of the competency of the witnesses?

The Court—No, sir, not until the proof is offered. Court will adjourn until 9 o'clock in the morning.

Whereupon an adjournment was taken until 9:00 a. m., July 15, 1925.

Judge Meets with Newspapermen

The Court—I have information gentlemen, that the newspapers are being sold in the eastern cities now, which undertake to state what my action was on the motion that is pending before me.

Richard J. Beemish—Was that a deduction?

The Court—I understand it purports to be information. Did you see the wire, Mr. Stewart?

Mr. Stewart—I saw the wire, Mr. Bell, who had that?

The Court—Let's see that.

Mr. Stewart—Mr. Losh, just read it to him.

At this point Mr. William J. Losh handed the message to the court.

A Voice—Your Honor—

The Court—Let me hear this telegram.

Mr. Beemish—Won't we be allowed, won't it be read out, please, so that we can all hear it?

The Court—St. Louis Star out final, carrying story law been held constitutional by judge.

Appoints Committee to Investigate

Now if this is a deduction, gentlemen, of course, they have a right to guess, so I think it is proper that I appoint a committee of pressmen to ascertain what these papers are carrying and ascertain if they are carrying this as a true story.

Mr. Beemish—That would be very fair.

The Court—I will appoint on this committee, on my own motion, because this matter is more important to me than to anyone else, Mr. Earl Shaub, Richard Beemish, Bert Kinser, Forrest Davis and Tony Muto. I wish you gentlemen would be prepared to report to me as soon as you can. I will hear you at any time. You may be excused.

CHAPTER IV.

FOURTH DAY'S PROCEEDINGS—WEDNESDAY, JULY 15, 1925.

Court met pursuant to recess.

Rev. Dr. Potter Chosen for Prayer.

The Bailiff—Is Preacher Stribling in the house?

The Court—Will everyone stand up? Mr. Chairman of the ministers' association, have you—who did you appoint as the minister to open court with prayer?

The Rev. Stribling—The Rev. Dr. Potter.

The Court—Dr. Potter, come forth to the judge's rostrum and open court with prayer.

Mr. Potter—Oh, Thou to Whom all pray and for Whom are many names, lift up our hearts this morning that we may seek Thy truth. May we in all things uphold the ends of justice and seek that those things may be done which will most redound in honor to Thy glory and to the progress of mankind toward Thy truth. Amen.

The Court—Everybody rise, please.

The Court—Open court, Mr. Sheriff.

The Bailiff—Oyez, oyez, this honorable circuit court is now open pursuant to adjournment. Sit down, please.

Neal Renews Objection to Prayer

Dr. Neal—I want to renew our objection to the prayer and I want the courtesy of the court just a moment to explain my particular attitude. I join with counsel on this side in their objection, but I think that it is such an important matter that I would like the courtesy of the court just a moment to explain my individual reaction or attitude toward this particular exception.

The Court—I will hear you, Judge Neal.

Dr. Neal—Being very breif, indeed. First, may it please your honor, I would like that you read from a case a very well-known principle of law, and I think you will agree with me when I read it. "The courts will take judicial notice that the religious world is divided into numerous sects and of the general doctrines—" this is quoting from the case of State vs. District Board, 76 Wis., 177—"the courts will take judicial knowledge that the religious world is divided into numerous sects and of the general doctrines maintained by each sect; for these things pertain to general history, and may fairly be presumed to be subjects of common knowledge. Thus they will take cognizance, without averment, of the facts that there are numerous religious sects called Christian, respectively maintaining different and conflicting doctrines; that some of these believe the doctrine of predestination, while others do not; some the doctrine of eternal punishment of the wicked while others repudiate it; some the doctrines of the apostolic succession and the authority of the priesthood, while others reject both; some that the Holy Scriptures are the only sufficient rules of faith and practice, while others believe that the only safe guide to human thought, opinion and action is the illuminating power of the divine spirit upon the humble and devout heart; some in the necessity and efficacy of the sacraments of the church, while others reject them entirely; and some in the literal truth of the Scriptures, while others believe them to be allegorical, teaching spiritual truths alone, or chiefly."

Now, may it please your honor, we differ, of course, very widely with the attorney-general in his opening statement that this is not a religious case. We differ very widely with him in his interpretation of this act—in his effort to simply split the act in two and take the latter clause as the whole of the act. Therefore, believing as we do firmly that certain great religious questions are involved in this case and appealing to the general knowledge of the court, that any religious atmosphere injected in the proceedings must necessarily be of one particular faith—not that we are religious or irreligious, but simply because this is a religious question—that the whole atmosphere of the court in every respect should be neutral; that the court should receive its sole information in this case from the facts presented by witnesses to the jury and the law presented by the lawyers. That, may it please your honor, is my reason for joining in this objection to the daily prayer.

Sue Hicks Replys to Neal

Sue K. Hicks—I have set over here and remained quiet these three days while the defense counsel have been constantly bringing up objections to these prayers. I want to make a statement in behalf of the court. I have been in this court for about five years and I know that every time that a minister has been in this court room when court was opened that the court was opened with prayer, and I think that their objections, your honor, should be put on the record, if they want them on the record, but this constant heckling every morning should be avoided. We are trying to avoid any religious controversy and we maintain that there is no religious controversy in this case. Their very opposition contradicts their own-selves. They say, your honor, that evolution is not—does not contradict the Bible—does not contradict Christianity. Why are they objecting to prayers if it doesn't contradict the Bible—doesn't contradict Christianity? Now, his case there that he reads dealt only with the sects of the church. This morning's prayer has been opened by a Unitarian. It has been opened by a Baptist and by a Methodist on the different mornings, and other denominations, and I think that the case that he cited is entirely out of order. It has no bearing on the controversy and we think that, your honor from now on should stop any such arguments as this arising and ask the defense to put their objections on the record and stop this here heckling in court in opening court every morning.

The Court—The court in selecting ministers to open the court with prayer has had no regard to denominational lines and no concern about sects. The court believes that any religious society that is worthy of the name should believe in God and believe in divine guidance. The court has no purpose by opening the court with prayer to influence anybody wrongfully, but hopes that such may influence somebody rightfully. It has been my custom at times when there has been no minister in the court, I have called on some good old pious man whom I knew was good, who believed in God, to open the court with prayer. I don't think it hurts anybody and I think it may help somebody. So I overrule the objection.

Darrow Takes Execption to Remarks of Court—Stewart Apologizes

Mr. Darrow—Your honor, I want to take exception to the remarks of the court.

The Court—Let the exceptions go in the record.

Gen. Stewart—Your honor, on yesterday Mr. Darrow and I had an agreement that the record would show each morning that they excepted to the prayer. Perhaps the other attorneys did not understand that, but hereafter it will just show that without any statement being made in open court.

Your honor, I want, before the court proceeds with business— I want to make just a statement of explanation. On yesterday afternoon, if the court please, near the hour of adjournment, I said a thing which upon reflection and deliberation I feel sorry for. Sometimes under the stress of circumstances, perhaps we all do things that we should not do and that is about the only consolation I have to get out of it. Mr. Hays was presenting a matter to the court to which I desired to object and did object and when I interposed my objection, feeling that Mr. Hays did not give me an opportunity to address myself to the court, I expressed myself toward him in a rather discourteous manner I feel. I meant at the time to be emphatic, but I did not mean to be discourteous. The least that one lawyer can do toward another that is in his attitude toward another lawyer, in the trial of the case, is to be courteous to him and I feel very much ashamed when I feel that I have not been courteous to anybody. Mr. Hays has treated me with much courtesy and I am sure he did not mean on yesterday to try to drown me out with his voice. I know that as soon as I said it I knew I had said the wrong thing and I want to say to him this morning, and the court publicly, that there was nothing back of what I said at all, except a temporarily ruffled temper. I am sorry for it and I apologize for it.

Mr. Hays—If your honor please.

The Court—I recognize Mr. Hays.

Hays Accepts Apology

Mr. Hays—I am happy to accept the apology of the attorney-general, with the knowledge that Mr. Stewart realizes that when he speaks he is speaking in the name of the sovereign state of Tennessee and I would like to condition that upon the suggestion that there be no further reference or allusions that are disrespectful to the state from which counsel for the defense come and no reference or allusions to the economic, political, social or religious views of counsel for the defense and I wish to warn counsel for the prosecution that if statements of that sort are made in the presence of the jury that we should regard them as prejudical and take exception to them. Permit me to say personally that there are two qualities I much admire in a man. One is that he is human and the other that he is courteous. The outburst on yesterday proves that the attorney-general was human, and the apology proves that he has the courtesy of a southern gentleman.

Neal Demands Further Apology—Stewart Stands Pat

Dr. Neal—I submit as the local counsel in this case, I am not at all satisfied with the apology of Gen. Stewart and he knows why. In that discourteous action yesterday was included another very grave discourtesy to one of my colleagues. I have given him every opportunity to apologize privately for his remark and he has refused, and now I ask him in public to erase from the record the slurring, discourteous remark that he made in regard to another colleague of mine in this case and he knows very well what I refer to.

Gen. Stewart—The very thing that Mr. Hays and I were trying to avoid is being injected again into the case by Mr. Neal. The offense has already been committed and Mr. Neal is attempting to inject into this record the very thing that Mr. Hays and I were trying to avoid. It is very obvious that Dr. Neal is not familiar with court procedure. Even lawyers say things and do things that they should explain and sometimes apologize for. When I do a thing that I feel badly about I apologize. So long as I speak what I conceive to be the truth, I apologize to no man.

(The officer calls the audience to order.)

Mr. Neal—I still think the attorney-general's remarks were extremely discourteous and uncalled for.

Beamish Reports on News Leak

The Court—Is the chairman of the press committee present? If so, I will hear the report from the chairman.

Mr. Richard Beamish (chairman of the press committee) (Reading)—The committee appointed by your honorable court, consisting of Richard J. Beamish, chairman; Phillip Kingsley, Earl L. Shaub, Forrest Davis, and Tony Muto, to investigate a reported news leak of the substance of your honor's opinion upon the motion to quash the indictment of John T. Scopes, respectfully reports: That it has been ascertained that the brief bulletin to the effect that the decision would uphold the indictment was based upon information which the sender of the bulletin believed to be correct and truthful; that the sender did not obtain this information from your honor's stenographer, nor in any improper or unethical manner; that no good ground exists for further investigation and the committee recommends that the sender of the bulletin be not disturbed in his relations with the court. Signed: Richard J. Beamish, chairman; Earl L. Shaub, Forrest Davis, and Tony Muto. One member of the committee, Mr. Phillip Kingsley, desired to have another meeting, and asked that his name be not included. The other members signed, and joined with the chairman in submitting this report. If the court desires any further information, or any additional details, we will be glad to submit them.

Court—I think the court is entitled to know how this information was had, if you can furnish me that information.

Information Came from Court

Mr. Beamish—Upon investigation, we find that the information came from the court.

The Court—Well—

Mr. Beamish—The circumstances are that the young man who sent the message, met the judge upon his way to the hotel. The judge, I am informed, had a bundle of papers under his arm. The young man asked him if that was his decision. The court replied, No, that the decision was being copied by a stenographer. The next question was, will you read that decision this afternoon? The reply was, that is my intention. The next question was, will you adjourn until tomorrow? To which the reply was, yes, I think so. The inference was that if the motion to quash the indictment was refused, there would be an adjournment. If the motion to quash was affirmed the trial would be ended. It was pure deduction. The young man then sent the message.

The Court—Who is he?

Mr. Beamish—Mr. Hutchinson.

Court—Come around, Mr. Hutchinson.

Hutchinson Before the Bar

(Mr. Hutchinson comes before the bar.)

Court—I have endeavored since this trial began to be extremely courteous, and do anything I could do for you gentlemen. I do not believe any pressman has a right to ask the court a question except for direct information which the question indicates that he wants. I do not think you had any right to inquire if the court would adjourn until tomorrow.

The Court—Young man, do you want to make any statement at all?

Mr. Hutchinson—I would be very glad to talk to the judge in chambers; I don't think I ought to do so here.

Mr. Beamish—I would ask that any other question be taken up in chambers.

The Court—Anything that I have said, that goes into the press—I have had an honest purpose in making this inquiry; it is no reflection upon you and the court does not mean to reflect upon you at all.

Mr. Beemish—I will say, your honor, for the purpose of the record, that Mr. Hutchinson is an upright, conscientious and thoroughly honest newspaper man and has the approval of the entire corps of journalists.

The Court—He comes to me from Senator Keller, of this state, recommending him very highly. I am sure he had no sinister motive.

Mr. Beemish—I think he had not.

The Court—I want to be fair to all the press and to put you all on the same basis; I think it is proper for me to suggest that you be as courteous to me as I try to be to you.

Mr. Beemish—We want to, your honor.

The Court—And if you want information ask me directly and I will give you a direct answer; if I want to give you information, I will; if it is not proper, I will not, but I prefer that if you want to ask me a question not to put me on notice as to the information you want, and then take advantage of the answer I may give.

So, you may be excused.

The Policeman—We will have to have order.

Mr. Darrow—May I say a word, and then be through in a very short time?

The Court—Yes.

Darrow is Agnostic—Says Infidel Means Nothing

Mr. Darrow—I don't want the court to think I take any exceptions to Mr. Stewart's statement,—of course, the weather is warm, and we may all go a little further at times than we ought, but he is perfectly justified in saying that I am an agnostic, for I am, and I do not consider it an insult, but rather a compliment to be called an agnostic. I do not pretend to know where many ignorant men are sure; that is all agnosticism means. He did, however, use a word, "infidel," although Mr. Stewart says he thinks I am wrong, but I am quite certain I am not. Of course, the word "in fidel" has no meaning whatever. Everybody is an infidel that does not believe in the prevailing religion, among the Saracens, everybody is an infidel that does not believe with them, and in a Mohammedan country, everybody that is not a Mohammedan is an infidel, and among the Christians, everybody is an infidel that is not a Christian, or professes to be. It has no generic meaning, and I don't think I am fairly classified under it. But, I do say this, and I have no doubt the attorney-general will agree with me; I don't know what their particular brand of religion may be; I presume amongst the six or seven there are six or seven different brands, if you analyze it closely enough.

But, while I take no offense for anybody to say in any way that I am an agnostic, for I am, I think everybody's religious rights and religious liberties are protected under the consitution of Tennessee, and if not, they would be protected under the fellowship that we owe to each other, and I do not think that anybody's religious creed should be used for the purpose of prejudicing or influencing any action in this case.

That is all I shall insist on through this case.

The fact that I am an agnostic ought not to weigh in the balance as to whether Mr. Scopes is innocent or guilty. And, all I ask for is that if counsel thinks it is wise to refer again to it that it shall not be done in such a way in the presence of the jury as to in any manner influence anybody, and I think I am right on that. I do not take any offense whatever in his having said I was an agnostic, although I hate to be accused of such a foolish thing as infidelity because everybody in the world can be accused of that.

The Court—What do you say, Gen. Stewart?

Gen. Stewart—I think we are wasting a lot of valuable time, your honor, in felicitation, and I am ready if these gentlemen will join me, in trying this law suit as lawyers. I would like to get done with this thing.

The Court—I think Col. Darrow is correct when he suggests no reference be made to the religious belief of any counsel in the presence of the jury; that it might prejudice the jury in the trial, and I shall expect that no such references will be made during the trial of this case.

Now, the court is about to read his opinion on the motion to quash the indictment, but I shall expect absolute order in the courtroom because people are entitled to hear this opinion.

Let us have order. No talking, now; let us have order in the courtroom.

If you gentlemen want to make my picture, make it now. (Laughter in the courtroom.)

Then I will proceed to read.

Court Officer—Order in the courtroom. No talking. (Rapping with gavel.)

(Following the photographing of the court.)

Raulston Reads Opinion on Overruling Motion to Quash

State of Tennessee vs. John T. Scopes.

Court—This case is now before me on a motion to quash the indictment on the following grounds:

"First — (a) Because the act, which is the basis of the indictment, and which the defendant is charged with violating, is unconstitutional and void, in that it violates Section 17, Article 11 of the constitution of Tennessee, which reads as follows:

"Section 17. Origin and frame of bills. Bills may originate in either house; but may be amended, altered or rejected by the other. No bill shall become a law which embraces more than one subject, that subject to be expressed in the title. All acts which repeal, revive or amend former laws, shall recite in their caption, or otherwise, the title or substance of the law repealed, revived or amended."

It is insisted by the defendant, through his counsel, that the body of the act involved in this case is not germane to the caption, and that the caption is too general in its terms, and that, therefore the act is unconstitutional and void.

In passing upon this provision of our constitution, our supreme court has said:

"Any provision of a statute germane to the subject expressed in the title directly or indirectly relating to that subject, and having a natural connection therewith, and not foreign thereto, is embraced in the title.

"It is not necessary that the title should express fully what is contained in the body of the act, for it was not intended that the title should express everything contained in the act. So long as the subject matter of the body of the act is germane to that expressed in the title, there is an obedience to the mandates of the constitution."

The general title to the act is one which is broad and comprehensive and covers all legislation germane to the general subject stated. The title may cover more than the body, but it must not cover less. It need not index the details of the act, nor give a synopsis thereof.

It is further said:

"The title of a legislative bill may be broader and more comprehensive than the subject of legislation contained in the body of the act, so that the one real subject of legslation is expressed in the title, and not obscured by foreign matters."

In the case at bar the caption of the act involved provides, among other things, that the purpose of the act is to prohibit the teaching of the evolution theory in the public schools, etc., of the state of Tennessee. It is true that this provision is rather general in its nature and in my conception of the terms employed in the caption and the body those used in the caption are broader and more comprehensive than those employed in the body of the act; but in my opinion the caption covers all the legislation provided for in the body, and is germane thereto, and in no way obscures the legislation provided for.

The purpose of this provision in our present constitution was to remedy an existing evil, and prevent laws on other subjects from being tacked on to a bill upon a wholly different subject, which tacked on laws this way sometimes eluded the attention of the legislature and were passed without sufficient consideration, and when passed, often remained for a time undiscovered, for the reason that the title of the act failed to call attention to the same, and to prevent smuggling through the legislature important measures without due notice to the members of the legislature as to the nature and purport of the matter under consideration.

In my judgment, the caption of this act is sufficient to put any member of the legislature on notice as to what the nature of the proposed legislation is, and that really the caption is more comprehensive than the body of the act. Therefore, I am content to overule this ground.

(b) In that it violates Section 12, Article 11 of the constitution of Tennessee which reads as follows:

"Section 12—Education to be cherished; common school fund; poll tax; whites and negroes; colleges, etc., rights of—knowledge, learning and virtue, being essential to the preservation of republican institutions, and the diffusion of the opportunities and advantages of education throughout the different portions of the state, being highly conducive to the promotion of this end, it shall be the duty of the general assembly in all future periods of this government to cherish literature and science. And the funds called the common school fund and all the lands and proceeds thereof, dividends, stocks and other property of every description whatever, heretofore by law appropriated by the general assembly of this state for the use of common schools, and all such as shall hereafter be appropriated, shall remain a perpetual fund, the principal of which shall never be diminished by legislative appropriations; and the interest thereof shall be inviolably appropriated to the support and encouragement of common schools throughout the state, and for the equal benefit of all the people thereof; and no law shall be made authorizing said fund or any part thereof to be diverted to any other use than the support and encouragement of common schools. The state taxes, derived hereafter from polls shall be appropriated to educational purposes, in such manner as the general assembly shall from time to time direct by law. No school established or aided under this section shall allow white and negro children to be received as scholars together in the same school. The above provisions shall not prevent the legislature from carrying into effect any laws that have been passed in favor of the colleges, universities or academies, or from authorizing heirs or distributees to receive and enjoy escheated property under such laws as may be passed from time to time."

It is not seriously insisted by the defendant in this case that the indictment should be quashed on this ground. But that there may be no doubt as to the defendant's rights under this section, I will briefly state the law relative thereto.

This section of the constitution makes it the express duty of every general assembly, at all times, to foster, and cherish literature and science. As one of the chief means of accomplishing this most important purpose, the constitution contemplated the establishment of a common school system, and provided the common school fund. But this provision of the constitution is merely directory to the legislature and indicates the popular feeling and the public policy of the people of the state on this great question.

The courts are not concerned in questions of public policy or the motive that prompts the passage or enactment of any particular legislation. The policy, motive or wisdom of the statutes address themselves to the legislative department of the state, and not the judicial department. Therefore, this court has no concern and no jurisdiction to pass upon this question, and is contented to overrule on this ground.

(c) In that it violates Section 18, Article 2 of the constitution of the state of Tennessee, which reads as follows:

"Sec. 18. Of the Passage of Bills —Every bill shall be read once, on three different days, and be passed each time in the house where it originated, before transmission to the other. No bill shall become a law until it shall have been read and passed, on three different days in each house, and shall have received on its final passage in each house, the assent of a majority of all the members to which that house shall be entitled under this constitution; and shall have been signed by the respective speakers in open session, the fact of such signing to be noted on the journal; and shall have received the approval of the governor, or shall have been otherwise passed under the provisions of this constitution."

As I understand the position of defendant's counsel at bar, there is no insistence that this ground is good, and no evidence before the court that would indicate the invalidity of this act, because of any violation of this section of the constitution. Therefore, the same is overruled.

(d) In that it violates Section 3, Article 1 of the constitution of Tennessess, which reads as follows:

"Sec. 3. Right of Worship Free.—That all men have a natural and indefeasible right to worship Almighty God according to the dictates of their own conscience; that no man can of right, be compelled to attend, erect, or support any place of worship, or to maintain any minister against his consent; that no human authority can, in any case whatever, control or interfere with the rights of conscience; and that no preference shall ever be given, by law, to any religious establishment or mode of worship."

And also:

(e) In that it violates Section 19, Article 1 of the constitution of Tennessee, which reads as follows:

"Sec. 19. Printing Presses Free; Freedom of Speech, etc., Secured.—That the printing presses shall be free to every person to examine the proceedings of the legislature or of any branch or officer of the government, and no law shall ever be made to restrain the right thereof."

"The free communication of thoughts and opinions, is one of the invaluable rights of man, and every citizen may freely speak, write and print on any subject, being responsible for the abuse of that liberty. But in the prosecutions for the publications of papers investigating the official conduct of officers, or men in public capacity, the truth thereof may be given in evidence; and in all indictments for libel, the jury shall have the right to determine the law and the facts, under the direction of the court as in other criminal cases."

Act Does not Interfere with Worship

It will be observed that the first provision in this section of our constitution provides that all men shall have the natural and indefeasible right to worship Almighty God according to the dictates of their own consciences. I fail to see how this act in any wise interferes or in the least restrains any person from worshiping God in the manner that best pleaseth him. It gives no preference to any particular religion or mode of worship. Our public schools are not maintained as places of worship, but, on the contrary, were designed, instituted, and are maintained for the purpose of mental and moral development and discipline.

This section fully provides that: "No man can of right be compelled to attend, erect, or support any place of worship, or to maintain any minister against his consent; that no human authority can in any case whatever control or interfere with the right of conscience; that no preference shall be given by law to any religion or established mode of worship."

I cannot conceive how the teachers' rights under this provision of the constitution would be violated by the act in issue. There is no law in the state of Tennessee that undertakes to compel this defendant, or any other citizen, to accept employment in the public schools. The relations between the teacher and his employer are purely contractual and if his conscience constrains him to teach the evolution theory, he can find opportunities elsewhere in other schools in the state, to follow the dictates of his conscience, and give full expression to his beliefs and convictions upon this and other subjects without any interferenc from the state of Tennessee or its authorities, so far as this act is concerned. Neither do I see how the act lays any restraint on his right to worship according to the dictates of his conscience. Under the provisions of this act this defendant, or any other person, can entertain any religious belief which most appeals to their conscience. He can attend any church or connect himself with any denomination or contribute to the erection of buildings to be used for public worship, as he sees fit. (The court is pleased to overrule these grounds.)

(f) In that it violates Section 8, Article 1 of the constitution of Tennessee, which reads as follows:

"Sec. 8. No Man to Be Disturbed but by Law.—That no man shall be taken or imprisoned or disseized of his freehold, liberties, or privileges, or outlawed, or exiled, or in any manner destroyed or deprived of his life, liberty or property, but by the judgment of his peers or the law of the land."

As the court understands, the defendant insists that this section of the constitution is the foundation for what is generally termed the law of the land.

"The law of the land means the law which embraces all persons who are in, or who may come into like situation and circumstances. It may be made to extend to all citizens, or to be consigned, under proper limitations, to particular classes. If the class be a proper one it matters not how few the persons are who may be included in it, if all who are in, or who may come into the like situation and circumstances, be embraced in the class, the law is general, and not partial."

Law of the Land

The law of the land hears before it condemns; it proceeds upon inquiry, and renders judgment only after trial.

"Legislation general in its operation upon the subjects to which it relates, and enforceable in the usual mode established in the administration of government with respect to kindred matters, that is, by process or proceedings adapted, to the nature of the case, is the law of the land."

As the court understands the pro visions of the statute involved in the case at bar, it applies alike to all persons coming into the like situation and circumstances, so far as public schools are concerned. That is, it applies alike to all those who see proper to engage themselves as teachers in the public schools of the state of Tennessee. Therefore, I am of the opinion that this statute is not violative of this section of the constitution and that it does not unlawfully deprive this defendant of any of his liberties, privileges, or property, and for this reason the court is pleased to overrule this ground.

(g) In that the act and the indictment and the proceedings herein are violative of Section 9, Article 1 of the constitution of Tennessee, which reads as follows:

"Sec. 9—Rights of the accused in criminal prosecutions. That in all criminal prosecutions, the accused hath the right to be heard by himself and his counsel; to demand the nature and cause of the accusation against him, and have a copy thereof, to meet the witnesses face to face, to have compulsory process for obtaining witnesses in his favor, and in prosecutions by indictment or presentment, a speedy public trial, by an impartial jury of the county in which the crime shall have been committed, and shall not be compelled to give evidence against himself."

And also:

(h) In that the act, prosecution and proceedings herein violate Section 14, Article 1, of the constitution of Tennessee, which reads as follows:

"Sec. 14—Crimes punished by presentment, etc.—That no person shall be put to answer any criminal charge but by presentment. indictment or impeachment."

As the court conceives, both of these grounds are predicated upon the same objection to the statute and the indictment, therefore, they will be considered together. One objection, as the court understands, that is insisted upon is that both the statute and the indictment are too vague and uncertain to put him on notice of the nature of the accusation brought against him. The requirement, as the court, understands, is this: The description of the offense charged in the indictment must be sufficient in definiteness, certainty and precision to enable the accused to know what offense he is charged with and to understand the special nature of the charge he is called upon to answer; to enable the court to see from the indictment a definite offense, so that the court may apply its judgment and determine the penalty or punishment prescribed by way, and also to enable the accused to protect himself from a second prosecution for the same offense.

"A description distinguishing the offense from all other similar offenses is not required. That degree or precision in the descriptions of the particular offense cannot be given in the indictment so as to distinguish it per se from all other cases of a similar nature. Such discrimination amounting to identification must rest in averment by plea and in the proof; and its absence in description in indictment can be no test of the certainty required either for defense against the present prosecution or for protection against a future prosecution for the same matter."

"The description of a statutory offense in the words of the statute is sufficient, and renders the indictment sufficiently certain if it gives the defendant notice of the nature of the charge against him."

The statute involved in this case, in part, reads as follows:

"Sec. 1—Be it enacted by the general assembly of the state of Tennessee, that it shall be unlawful for any teacher in any of the universities, normals, and all other public schools of the state which are supported in whole or in part by the public school fund of the state, to teach any theory that denies the story of the divine creation of man as taught in the Bible and teach instead that man has descended from a lower order of animals."

The indictment, in part, reads:

"That John Thomas Scopes, heretofore on the 24th day of April, 1925, in the county aforesaid, then and there unlawfully did wilfully teach in the public schools of Rhea county, Tennessee, which said public schools are supported in part, or in whole by the public school fund of the state, a certain theory and theories that denied the story of the divine creation of man as taught in the Bible, but did teach instead thereof, that man is descended from a lower order of animals, he, the said John Thomas Scopes, being at the time, and prior thereto, a teacher in the public schools of Rhea county, Tennessee, as aforesaid, against the peace and dignity of the state."

(i) In that the act violates Section 8, Article XI of the Constitution of Tennessee which reads as follows:

"Sec. 8—General laws only to be passed; corporations only to be provided for by general laws—The legislature shall have no power to suspend any general law for the benefit of any particular individual nor to pass any law for the benefit of individuals, inconsistent with the general laws of the land; nor to pass any law granting to any individual or individuals rights, privileges, immunities or exemptions other than such as may be by the same law, extended to any member of the community who may be able to bring himself within the provisions of such law. No corporation shall be created, or its powers increased or diminished by special laws, but the General Assembly shall provide by general laws, for the organization of all corporations hereafter created, which laws may, at any time, be altered or repealed; and no such alteration or repeal shall interfere with or divest rights which have become vested."

The court is of the opinion that what has been said in discussing Section 8 of the first article of the constitution of Tennessee above, would also be applicable to the objection made under this ground. In the defining and construing individual rights under this section, our supreme court said:

"If the classification is made under this section, everyone who is in, or may come into the situation and circumstances which constitute the reasons for and the basis of the classification, must be entitled to the rights, privileges, immunities and exemptions conferred by the statute or it would be partial and void. If the classification is made under Section 8 of the first article of the constitution, everyone who is in or may come into the situation and circumstances which constitute the reasons for the basis of the classification, must be subjected to the disabilities, duties, obligations and burdens imposed by the statute, or it would be partial and void. It follows that the cases that have been decided upon either of the subsections are of equal value in arriving at the meaning of the expression and requirement that all class legislation must be so framed as to extend to and embrace equally all persons who are in or may come into the like situation and circumstances, constituting the reasons for and basis of the classification. Class legislation which has applied equally to all that are in or that may come into the like situation and circumstances and which makes a reasonable and natural classification, is valid and constitutional."

Therefore, the court is pleased to overrule this ground.

(j) In that the act violates Section 2, Article 2 of the constitution of Tennessee; which reads as follows:

Sec. 2. No person or persons belonging to one of these departments shall exercise any of the powers properly belonging to either of the others, except in the cases herein directed or permitted."

So far as the court can recall there is no insistence by the defendant that this ground should be sustained by the court, and for that reason it is passed and overruled.

Second (a) That the indictment is so vague as not to inform the defendant of the nature and cause of the accusation against him.

And also.

(b) That the statute upon which the indictment is based is void for indefiniteness and lack of certainty.

The questions raised by these sections, have been discussed in another part of this opinion, fully, and the grounds stated upon which the same questions have been overruled. Therefore, these are overruled without further comment.

Third (a) In that the act and the indictment violate Section 1 of the Fourteenth amendment of the constitution of the United States, which reads as follows:

Does Not Violate U. S. Fourteenth Amendment

"Sec. 1, Art. XIV. All persons born or naturalized in the United States, and subject to the jurisdiction thereof, are citizens of the United States and of the state wherein they reside. No state shall make or enforce any law which shall abridge the privileges or immunities of citizens of the United States; nor shall any state deprive any person of life, liberty or property, without due process of law, nor deny to any person within its jurisdiction the equal protection of the laws."

As the court conceives, the defendants raised the same question under this assignment of this ground as they did under Section 8 of Article 1 of the constitution of Tennessee, except, they insist that the act involved in the case at bar, not only violates Section 8 of Article 1 of the constitution of Tennessee, but in like particular violates Article 1 of the Fourteenth Amendment to the constitution of the United States.

In the case of Meyer vs. State of Nebraska, decided by Justice McReynolds, and quoted in 67 Law Ed., United States. Reports on page 390, a case wherein the plaintiff in error was tried and convicted upon an indictment in Hamilton county, Nebraska, under a charge that on May 25, 1920, while an instructor in Zion parochial school, he unlawfully taught the subject of reading in the German language to Raymond Parpart, a child of ten years, who had not attended and successfully passed the eighth grade, the opinion was based upon an act relating to the teaching of foreign languages in the state, approved April 9, 1919, which was as follows:

"Sec. 1. No person, individually or as a teacher, shall, in any private, denominational, parochial or public school, teach any subject to any person in any language other than the English language.

"Sec. 2. Languages other than the English langauge may be taught as languages only after a pupil shall have attained and successfully passed the eighth grade, as evidenced by a certificate of graduation issued by the county superintendent of the county in which the child resides.

"Sec. 3. Any person who violates any of the provisions of this act shall be deemed guilty of a misdemeanor and upon conviction shall be subject to a fine of not less than twenty-five ($25) dollars, nor more than one hundred ($100) dollars, or be confined in the county jail for any period not exceeding thirty days for each offense.

Sec. 4. Whereas, an emergency exists this act shall be enforced from and after its passage and approval."

The supreme court of the state affirmed a judgment of conviction. It declared the offense charged and established was the direct and intentional teaching of the German language as a distinct subject to a child who had not passed the eighth grade in the parochial school maintained by the Zion Evangelical Lutheran congregation, a collection of Biblical stories being used therefor, and it held that the statute forbidding this did not conflict with the Fourteenth Amendment, but was a valid exercise of the people's power.

In deciding this case, Justice McReynolds said:

"The problem for our determination is, whether the statute, as construed and applied, unreasonably infringes the liberty guaranteed to the plaintiff in error by the Fourteenth Amendment, 'No state . . . shall deprive any person of life, liberty or property without due process of law.'

"While this court has not attempted to define with exactness the liberty thus guaranteed, the term has received much consideration and some of the included things have been definitely stated. Without it denotes, not merely freedom from bodily restraint, but also the right of the individual to contract, to engage in any of the common occupations of life, to acquire useful knowledge, to marry, establish a home and to bring up children, to worship God according to the dictates of his own conscience, and generally, to enjoy these privileges long recognized at common law as essential to the orderly pursuit of happiness by free men."

"That the state may do much, go very far indeed, in order to improve the quality of its citizens, physically, mentally and morally, is clear; but the individual has certain fundamental rights which must be respected. The protection of the constitution extends to all—to those who speak other languages as well as to those born with the English on the tongue. Perhaps it would be highly advantageous if all had ready understanding of our ordinary speech, but this cannot be coerced by methods which conflict with the constitution—a desirable end cannot be prompted by prohibited means."

"The desire of the legislature to foster (a) homogeneous people with American ideals, prepared readily to understand current discussions of civic matters, is easy to appreciate. Unfortunate experiences during the late war, and aversion toward every characteristic of truculent adversaries, was certainly enough to quicken that aspiration. But the means adopted, we think, exceed the limitations upon the power of the state and conflict with rights assured to plaintiff in error. The interference is plain enough, and no adequate reason therefor in time of peace and domestic tranquility has been shown."

"But the power of the state to compel attendance at some school and to make reasonable regulation for all schools, including a requirement that they shall give instructions in English, is not questioned, nor has challenge been made of the state's power to prescribe a curriculum for institutions which it supports. Those matters are not in the present controversy. Our concern is with the prohibition approved by the supreme court."

Court Presents Law

In the case of Pierce et als. vs. Society of the Sisters of the Holy Names of Jesus and Mary, decided about June 1, 1925, also by Justice McReynolds, coming up from the state of Oregon. This case also involved the right of a citizen as guaranteed by the Fourteenth Amendment of the constitution of the United States. The court, in commenting, said:

"No question is raised concerning the power of a state reasonablyto regulate all schools, to inspect, supervise, and examine them, their teachers and pupils, to require that all children of proper age attend some school, that teachers shall be of good moral character, and of patriotic disposition, that certain studies, plainly essential to good citizenship, must be taught, and that nothing be taught which is inimical to the public welfare."

"Under the doctrine of Meyer vs. Nebraska, 262 U S., Page 390, we think it is entirely plain that the act of 1922 unreasonably interferes with the liberty of parents and guardians to direct the upbringing and education of children under their control. As often heretofore pointed out, rights granted by the constitution may not be breached by legislation, which has no reasonable relation to some other purpose than the competency of the statutes. The fundamental theory of liberty upon which all governments in the United States repose, precludes the general power of the state to standardize its children by forcing them to accept instruction from public teachers only. The child is not the mere creature of the state; those who nurture him and direct his destiny have the right, coupled with the high duty to recognize and prepare him for additional obligations."

In the Meyer case the statute, in part, provided:

"No person individually, or as a teacher, shall in any private, denominational or parochial or public school, teach any subject to any person in any language other than the English language."

In passing on the constitutionality of this statute, the court held it unconstitutional under the Fourteenth Amendment to the constitution of the United States. But this act is, as is apparent from its reading, applied to all schools in the state of Oregon, and an obedience to its provisions would have made it impossible for any child, regardless of its nationality, ancestry and purposes in life, to have been taught by any teacher any subject, except in the English language, and I think the court properly held that this was an infringement upon the rights of individuals living in that state; but, as above indicated, it will be observed that the court in passing upon this act, observed, "the power of the state to compel attendance at some school and to make reasonable regulations for schools, including a requirement that they shall give instruction in English," is not questioned. Nor has challenge been made of the "state's power to prescribe a curriculum for institutions which it supports."

It is true that the last quotation above referred to would be classed, in legal parlance, as dictum.

In the case of Pierce vs. Society of Sisters, etc., the act required the children of the state of Oregon to attend public schools, in which the court said that "the child is not the mere creature of the state, and those who nurture him and direct his destiny have a right, coupled with high duty, to recognize and prepare him for additional obligations."

In the Oregon case, the Nebraska case is referred to without any suggestion or intimation that the dictum therein is not good law.

Leeper Again

In the case of Leeper vs. The State, reported in 19th Pickle, page 500, wherein it was insisted that the act involved therein was unconstitutional under Section 8 of Article 1 of the constitution of the state of Tennessee, the supreme court of Tennessee said that "The state may establish a uniform series of books to be taught in the school which it provides and controls seems to be a proposition as evident as that it may provide a uniform system of schools, which we take it is not now an open question."

In deciding the Leeper case the court referred to, with approval, the case of the State vs. Hayworth, 122 Indiana, 462, thusly: The reasoning of the court in the case of State vs. Hayworth is so satisfactory and conclusive that we cannot, perhaps, do better than give a synopsis of it. It was held that such an act does not infringe in the slightest degree upon the right of local self-government; that essentially and inferentially the schools in which are educated and trained the children that are to become the rulers of the commonwealth are matters of state, and not local, jurisdiction; that in such matters the state is a unit and the legislature the source of power; that the establishment and control of public schools is a function of the general assembly, both under the constitution, and because it is a matter of state concern. Being a matter of legislative control, the legislature may abandon one plan and try another, if it sees proper, and the courts cannot interfere. It is further pertinently said, that it is impossible to conceive the existence of a uniform system of public schools without powers lodged somewhere to make them uniform, and in the absence of express constitutional provisions the power must necessarily reside in the legislature, and hence it has the power to prescribe a course of study as well as the books to be used, and how they shall be obtained and distributed, and its discretion as to methods cannot be controlled by the courts. We find neither reason nor authority that suggests a doubt as to the power of the legislature to require a designated series of books to be used in school.

The rule prevailing in Tennessee by which the courts are governed in passing upon the constitutionality of statutes is this: The rule of construction that every intendment and presumption is in favor of the constitutionality of the statute and that every doubt must be solved so as to sustain it; and where it is subject to two constructions, that which will sustain its constitutionality must be adopted.

Under the holdings in the Oregon case and in the Nebraska case, and in the Leeper-Tennessee case, the court is satisfied that the act involved in the case at bar does not violate the Fourteenth amendment to the constitution of the United States, and is, therefore, pleased to overrule this ground.

The court, having passed on each ground chronologically, and given the reasons therefor, is now pleased to overrule the whole motion, and require the defendant to plead further.

(Following the reading by the court of the opinion on motion to quash the indictment).

Defense Excepts to Court's Ruling

Mr. McElwee—Your honor, we desire to enter an exception to your honor's ruling in overruling our motion to quash the indictment and in holding the act under which Mr. Scopes is being prosecuted meets the requirements and is not in conflict with the constitution of Tennessee, or of the constitution of the United States. We do this out of abundance of precaution and to keep the record straight in event that a record may be made in this case ultimately.

Mr. Neal—May it please your honor, I would like to remind your honor that at this moment we would like to have considered filed our demurrer, which is absolutely the same as the motion to quash, and I assume that your honor will probably take the same action.

The Court—To be frank, Judge Neal, you handed me a copy of the demurrer, but I have had such great responsibilities that I have not seen it .

Mr. Neal—Well, we assure your honor that it is simply for the purpose of procedure and the record; the demurrer is exactly the same as the motion to quash.

The Court—You mean raises the same questions?

Mr. Neal—It does, yes.

The Court—In different form?

Gen. Stewart—I would like to see the demurrer.

The Court—Did you give me the original?

Mr. Neal—Yes, sir; I gave you the original.

The Court—I have not it. Hand it to the attorney-general.

Mr. Neal—We will advise the attorney-general that the motion to quash, if he would substitute the word demurrer, is the same.

Gen. Stewart—Yes, I understand that, but we want to see the instrument filed as the demurrer.

The Court—Well, just let it be filed and then let the attorney-general see it.

Gen. Stewart—Is that the one the court has filed; I am asking, is that the one?

Mr. Neal—The original I handed to your honor.

The Court—I do not know. I have had so many papers, telegrams and letters, I may have laid it aside.

Mr. Neal—We will file this, may it please your honor, to satisfy the attorney-general. It will take only a moment.

The Court—All right, file that.

Mr. Neal—We file that.

The Court—Hand it to the clerk and let him mark it filed.

Gen. Stewart—I did not see it. I do not know just what objections we may want to interpose.

(At this point, Mr. Hays walked over to Gen. Stewart, standing in front of the judge's stand, whereupon)

Mr. Hays—I don't suppose you object to shaking hands, after this is all over? (Extending hand.)

Gen. Stewart (shaking hands with Mr. Hays)—That is all right.

The Court—The court will take a ten-minute recess.

(Court thereupon recessed for ten minutes.)

Mr. Hicks—If the court please, before you recess, we would like to call our witnesses.

The Court—Not just now.

The Court—I desire to announce to the press that my copy of the opinion fails to show my action on grounds "D" and "E" on page 9, just before the letter "F" on that page. Just after the word "fit" there should be written in, "The court is pleased to overrule these grounds," there being two incorporated and discussed together.

Dr. Neal—In regard to the demurrer, we have not been able—this copy was simply nothing but a memorandum and not complete, and if it so happens that the copy I gave your honor was the one that was filed—I did not find it there—we would like the record to show we filed the demurrer, and we will file it in the exact terms of the motion to dismiss.

The Court—This was a very—what time is it?

Gen. Stewart—11:13, your honor.

Mr. Neal—Let's dispose of this.

The Court—Oh, yes. This has required quite a bit of energy, as you must know, for the court to read the opinion that has just been delivered in the atmosphere by which he was surrounded, and I am inclined to adjourn the court and give you gentlemen also an opportunity to get your demurrer together, or get from me the copy, if I can find it. I have so many papers I will do my best—that you might have your demurrer ready to file at 1 o'clock.

Mr. Neal—May I make a suggestion?

The Court—Yes.

Mr. Malone—I make it, as the court knows, with the greatest respect for your wishes, and I know you are worn out and you are tired, and yet I hope that it will be possible for all sides so to co-operate, so that we can move at a greater speed. I do not like to speak of personal matters, but we are lawyers with clients and the importunities are very great for us to speed up and return to our practice, and I hope we will be able to take as few adjournments as possible.

Gen. Stewart—There was a thing that occurred this morning in the absence of Mr. Malone and I heartily agree with his views. We are all lawyers, and I hope we can co-operate, and I am sure we will to expedite the trial of this case as rapidly as possible.

Mr. Hays—Would it not be possible to continue court for a later hour than the usual hour for adjournment?

The Court—I will take that up. My custom in life is never to cross my bridge until I get to it, and when I get to that I will determine it. The court will adjourn until 1 o'clock, instead of 1:30. Let the court stand adjourned.

AFTERNOON SESSION

Court Thanks Little Girl for Flowers

The Court—Everyone stand up. Open court, Mr. Sheriff.

The Bailiff—Oyez, oyez, this honorable court is now open pursuant to adjournment. Be seated, please.

Mr. Neal—Your honor, please, I would like to straighten out this.

The Court—I desire to thank the lady, little girl or whoever it may be that is so mindful of the court as to send up this beautiful bouquet. (Applause.)

Mr. Neal—May it please your honor.

The Court—I will hear you, judge.

Mr. Neal—We wish to straighten out this question of the demurrer that have—both the motion to quash and the demurrer, both having been filed, and I think that I have satisfied the attorny-general that they are identical and I presume your honor will rule on them.

The Court—Are you satisfied, Gen. Stewart, that they are identical?

Gen. Stewart—Yes, sir, they are identical, but, of course, we except to the filing of the demurrer because their motion to quash has previously been filed and I want to preserve any exception to the filing of it for that reason.

Mr. Neal—We filed it to be considered filed as of before.

The Court—Well, that exception, of course, is purely technical and I will overrule it and let the demurrer be filed and then I will overrule the demurrer?

Mr. Neal—And then we except to your ruling, may it please your honor.

The Court—Yes, sir. Are you ready to proceed now, Mr. Attorney-General?

Gen. Stewart—I think so, your honor, we prefer to call our witnesses and I take it first that the jury would be brought in. First let me call our witnesses, your honor.

State's Witnesses Called

The Court—Call your witnesses and see if you are ready to proceed.

Sue K. Hicks—I want all these witnesses to meet me outside in front of the door of Mr. McKenzie's office right immediately after your names are called. Answer to your names as they are called. Harry Shelton—

The Court—Mr. Sheriff, take these names and call them outside, please. If there is an officer at the door let him repeat the call.

Sue K. Hicks—Orville Gannoway, Morris Stout, Howard Morgan, F. E. Robinson, Jack Hudson, Fraser Hutchinson, James N. Benson.

Gen. Stewart—Is Walter White, the prosecutor here? Walter White? Col. Darrow, we may have to get you to agree to what we can prove if we cannot find the witnesses.

Mr. Darrow—We might round them up later in the day. (Laughter in the courtroom.)

Gen. Stewart—These witnesses have already been subpoenaed, I am informed and we expected to get to them on yesterday, but there has been this delay and we will go out for a conference, if your honor will give us about five minutes.

The Court—Do you want a conference?

Gen. Stewart—Just about five minutes, I think, is all we require.

The Court—Col. Darrow, will you want any time?

Mr. Darrow—Time for what?

The Court—Time for a conference? If you do we might make the conferences simultaneous.

Mr. Darrow—We are all ready.

The Court—The court will be at ease for a few minutes and let you talk and laugh a little while if you want to.

Thereupon after the recess the following proceedings occurred:

Mr. Malone—If the court please—

The Court—(Rapping for order.)

Mr. Malone—If the court, please, we are informed, we do not know from how reliable a source, that the witnesses for the state are in the building, and if they are in the building, we know of no reason why there should be any further delay.

The Court—Mr. Sheriff, notify the counsel to come in, if they have finished their conference. They were having a little conference.

Mr. Malone—Apparently.

(Laughter in the courtroom.)

Thereupon a policeman returned to the bench and announced to the court that the attorneys would be in in a few minutes.

The Court—All right.

Thereafter the following occurred after the lapse of a few moments:

The Court—Tell the attorney-general to come in.

Thereupon the policeman rapped for order.

Jury Called

The Court—Let the clerk call the jury. Call the jury, please. When your names are called, gentlemen, come in and have seats in the jury-box.

Thereupon the clerk called the names of the jurors and the policeman repeated them as follows:

W. F. Roberson, J. W. Dagley, Jim Riley.

The Court—Are they responding, any of them? Call outside. We excluded them from the courtroom and I judge they are still excluded. Call the jury from the outside, you will have to begin all over again.

Thereupon the names were called as follows:

W. F. Roberson, J. W. Dagley, Jim Riley, W. G. Taylor, R. L. Gentry, J. R. Thompson, W. G. Smith, J. R. Goodrich, J. H. Bowman, Bill Day, R. F. West, J. S. Wright.

The Court—Have your seats in the jury box, gentlemen.

A Newspaper Reporter—Can we have chairs, judge?

The Court—Gentlemen, I do not believe the whole courtroom should expect the judge to look after chairs. Let the sheriff do that, appeal to the sheriff.

Gentlemen, let me see the jury.

I wish you would call the jury again, Mr. Clerk, and if your names are not correct, stop the clerk and correct them. Answer to your names, and if not correct, indicate it.

Whereupon the names were called again, as follows:

W. F. Roberson, J. W. Dagley, Jim Riley.

A Juror—J. W. Riley.

The Court—J. W. Riley, he prefers.

Thereupon the calling of the names was continued.

W. G. Taylor, R. L. Gentry, J. R. Thompson, W. G. Smith, J. R. Goodrich, J. H. Bowman, Bill Day.

A Juror—W. G. Day.

The Court—W. G. Day.

Thereupon the calling continued.

R. F. West, J. S. Wright.

The Court—All present.

Mr. McKenzie—As a matter of suggestion, I wish at this time to ask the court to make the announcement to the people, and ask them that they not carry off the chairs of the attorneys. We are a necessary evil in the courtroom, supposed to be a part of it.

Thereupon the policeman announced that no chairs should be carried off from the attorneys, from either the state or the defense or the press.

The Court—Are you ready to read the indictment?

Gen. Stewart—The indictment has been read, your honor, but we can read it again.

Foreman Requests Electric Fans

Juror Thompson—If it ain't out of order, I would like to make the request, the unanimous request of the jury to take up the matter of some electric fans here. This heat is fearful. While I think I could stand my part of it—

The Court—The county judge is

the man you would have to appeal to on that.

The Juror—He is a mighty nice man and some intimation from you would do some good.

Mr. McKenzie—Nothing would give me greater pleasure than to have them installed, but on account of the depleted state of the treasury I do not believe the county can do it.

Mr. Malone—I will buy some fans.

The Court—Col. Thompson, I will divide my fan. Perhaps we can borrow some small fans, and place them on the table, Mr. County Judge. Maybe we can place some small fans on the table.

Are there any further preliminary matters, before the jury is sworn, or before the plea, I mean?

Gen. Stewart—The state is ready.

The Court—What is your plea, gentlemen?

Defendant Pleads Not Guilty

Mr. Neal—Not guilty, may it please your honor.

The Court—Not guilty. Now gentlemen, I shall ask the counsel for both sides to make an opening statement, please, in which you will please briefly outline what your theory is in the case, before I swear the jury.

Gen. Stewart—It is the insistence of the state in this case, that the defendant, John Thomas Scopes, has violated the antievolution law, what is known as the antievolution law, by teaching in the public schools of Rhea county the theory tending to show that man and mankind is descended from a lower order of animals. Therefore, he has taught a theory which denies the story of divine creation of man as taught by the Bible.

Mr. Hays—If the court pleases, may I for the purpose of the record, on the opening statement of the attorney-general move to dismiss the case of the prosecution?

The Court—Yes, and I overrule the motion.

Mr. Hays—And I take an exception.

The Court—Yes, sir.

Now, I will have your statement, gentlemen? Order in the courtroom. I will swear the jury later, when I get these issues made up.

Col. Malone, I will hear from you, sir?

Mr. Malone—If the court please, for the purpose of brevity, though it is impossible to be as brief as the present conception of the prosecution's case, and for the purpose of accuracy, I will stick to my notes, with regard to the statement of the defense. It is going to take a long while, so I do not want to keep your honor standing.

The Court—Col. Malone, I don't want any argumentative statement made. I just want a brief statement of your theory.

Mr. Malone—I understand that, your honor.

The Court—Yes.

Mr. Malone—But we have more than one theory.

The Court—Yes; your theories, then. Put it in the plural.

Malone's Statement of Defense Theory of Case

The defense believes that "God is a spirit and they that worship Him must worship Him in spirit and in truth."

The defendant, John T. Scopes, has been indicted for the alleged violation of an act passed by the Tennessee legislature, which prohibits the teaching of the evolution theory in all the universities, normal schools and public schools of Tennessee, which may be supported in whole or in part by the public school funds of the state.

Section 1 of the act provides:

"Be it enacted by the general assembly of the state of Tennessee that it shall be unlawful for any teacher in any of the universities, normals and all other public schools in the state, which are supported in whole or in part by the public school fund of the state, to teach any theory that denies the story of the divine creation of man as taught in the Bible, and to teach instead that man has descended from a lower order of animals."

Section 2 provides:

"Be it further enacted that any teacher found guilty of a violation of this act shall be guilty of a misdemeanor and upon conviction shall be fined not less than $100, nor more than $500, for each offense."

In contradiction of the opinion of the legal leader of the prosecution, the attorney-general, the defense contends that before you, gentlemen of the jury, can convict the defendant, Scopes, of a violation of this act, the prosecution must prove two things:

First — That Scopes taught a theory that denies the story of the divine creation of man as taught in the Bible, and

Second—That instead and in the place of this theory he taught that man is descended from a lower order of animals.

Scopes Must Have Taught Evolution and Also Denied Bible Story

The defense contends that to convict Scopes the prosecution must prove that Scopes not only taught the theory of evolution, but that he also, and at the same time, denied the theory of creation as set forth in the Bible.

The defense contends the prosecution must prove that the defendant, Scopes, did these two things and that what he taught was a violation of the statute.

We will prove that whether this statute be constitutional or unconstitutional the Defendant Scopes did not and could not violate it. We maintain that since the Defendant Scopes has been indicted under a statute which prohibits the teaching of the evolutionary theory, the prosecution must prove as part of its case what evolution is.

So that there shall be no misunderstanding and that no one shall be able to misinterpret or misrepresent our position we wish to state at the beginning of the case that the defense believes there is a direct conflict between the theory of evolution and the theories of creation as set forth in the Book of Genesis.

Bible Story Not Scientifically Correct

Neither do we believe that the stories of creation as set forth in the Bible are reconciliable or scientifically correct. The defense will also prove by credible testimony that there is more than one theory of creation set forth in the Bible and that they are conflicting. But we shall make it perfectly clear that while this is the view of the defense we shall show by the testimony of men learned in science and theology that there are millions of people who believe in evolution and in the stories of creation as set forth in the Bible and who find no conflict between the two. The defense maintains that this is a matter of faith and interpretation, which each individual must determine for himself, and if you, men of the jury, are able to reconcile the theory of evolution and the theories of creation as set forth in the Bible, you are not only entitled to your view, but you will be supported in it by millions of your citizens who are of high culture, learning and deep religious faith.

The defense will prove these facts to you and you will determine the question for yourself.

No Conflict Between Evolution and Christianity

While the defense thinks there is a conflict between evolution and the Old Testament, we believe there is no conflict between evolution and Christianity. There may be a conflict between evolution and the peculiar ideas of Christianity, which are held by Mr. Bryan as the evangelical leader of the prosecution, but we deny that the evangelical leader of the prosecution is an authorized spokesman for the Christians of the United States. The defense maintains that there is a clear distinction between God the church, the Bible, Christianity and Mr. Bryan. * * * (Here Mr. Malone referred to Mr. Bryan's introduction to Jefferson's "Statute of Religious Freedom").

The great political leader in commenting on Jefferson's principles said:

Reads from Bryan's Writings

"The conciseness of Jefferson's style is well illustrated in this statute. Read it over. There is not a superfluous word, and yet there is enough to guard religious liberty. It is not strange that this doctrine so well set forth by Jefferson more than a century ago is now a part of the constitution and bill of rights of every state of this Union. Not only is that today the law of this land, but it is spreading throughout the world. It was only a short time ago that the Czar of Russia issued a decree in which he acknowledged the right of all the subjects of his empire to worship God according to the dictates of their own conscience, and I believe that when we come to measure the relative importance of things, the importance of an act like that, the very foundation upon which we build religious liberty—the importance of an act like that, which, gradually spreading, has become the creed of 80,000,000 people, and is ultimately to become the creed of all the world—when we come to consider the vast importance of a thing like that, how can we compare lands or earthly possessions with it?

"In the preamble to this statute, Jefferson set forth the main reasons urged by those who believed in religious freedom. Let me call attention to some of the more important ones. He said, in the first place, that to attempt to compel people to accept a religious doctrine by act of law was to make not Christians but hypocrites. That was one of the reasons, and it was a strong one. He said, too, that there was no earthly judge who was competent to sit in a case and try a man for his religious opinions, for the judgment of the court, he said, would not be a judgment of law, but would be the personal opinion of the judge. What could be more true. No man who has religious convictions himself bears them so lightly that he can lay them aside and act as a judge when another man's religious convictions are involved. Then he suggested—and I think that I am justified in elaborating upon this suggestion a moment—that religion does not need the support of the government to enable it to overcome error. Let me give the exact words of his report, for I cannot change them without doing injury to them:

"'And finally, that truth is great and will prevail if left to herself; that she is the proper and sufficient antagonist of error and has nothing to fear from the conflict unless, by human interposition, disarmed of her natural weapons—free argument and debate; errors cease to be dangerous when it is permitted freely to contradict them.'

"Tell me that Jefferson lacked reverence for religion He rather lacks reverence who believes that religion is unable to defend herself in a contest with error. He places a low estimate upon the strength of religion who thinks that the wisdom of God must be supplemented by the force of man's puny arm."

Jefferson paid a tribute to the power of truth when he said that truth was able to overcome error in the open field; and that it was this sublime confidence in the triumph of truth that distinguished him from many of the other great men of his time. In fact, of all the men who have lived upon this earth. I know of no man who surpassed Jefferson in his confidence in the ultimate triumph of truth; and upon what can people build if not upon faith in truth? Take from man his belief in the triumph of that which is right and he builds upon the sand. Give to man an abiding faith in the triumph of that which is true and you give him the foundation of a moral character that can withstand all reverses.

In the preamble to the statute for religious freedom.

Bryan Said Religion Not Subject to Legislation

Jefferson put first that which I want to speak of last. It was that the regulation of the opinions of men on religious questions by law is contrary to the laws of God and to the plans of God. He pointed out that God had it in His power to control man's mind and body, but that He did not see fit to coerce the mind or the body into obedience to even the Divine Will; and that if God Himself was not willing to use coercion to force man to accept certain religious views, man, uninspired and liable to error, ought not to use the means that Jehovah would not employ. Jefferson realized that our religion was a religion of love and not a religion of force."

No Science Can Be Taught Without Recognizing Evolution

These words were written by William Jennings Bryan and the defense appeals from the fundamentalist, Bryan of today, to the modernist, Bryan, of yesterday.

We maintain and we shall prove that Christianity is bound up with no scientific theory, that it has survived 2,000 years in the face of all the discoveries of science and that Christianity will continue to grow in respect and influence if the people recognize that there is no conflict with science and Christianity. We will show that science occupies a field of learning separate and apart from the learning of theology which the clergy expound. We will show that throughout the ages, every scientific discovery or new invention has been met by the opposition of people like those behind this prosecution who have pretended that man's inventive genius was contrary to Christianity. We shall prove by experts and scientists in every field of scientific knowledge that there is no branch of science which can be taught today without teaching the theory of evolution, and that this applies to geology, biology, botany, astronomy, medicine, chemistry, bacteriology, embryology, zoology, sanitation, forestry and agriculture. We will show that it will have been impossible for men like Luther Burbank and others without knowledge and faith in the theory of evolution to produce their invaluable experiments and results.

Do Not Contend Man Came from Monkeys

The prosecution has twice since the beginning of the trial referred to man as descended from monkeys. This may be the understanding of the theory of evolution of the prosecution. It is not the view, opinion or knowledge of evolution held by the defense. No scientist of any preeminent standing today holds such a view. The most that science says today is that there is an order of men like mamals which are more capable of walking erect than other animals, and more capable than other animals in the use of the forefeet as hands.

There are indications that not 6,000 years ago, but through the long course of the ages from this order came man in one direction, and monkeys in the other. All that science says is that probably some time not 6,000 years ago, but in the course of the ages, and all that science says today is that there are tendencies which indicate the validity of this opinion.

Human Embryo Has Tail

For the purpose of illustration, we hope to show you from embryology about the development of a child from a single cell to its birth. In the course of this development the cell divides repeatedly as growth proceeds and the mass grows. The parts begin to appear at first without resemblance to those of a human being. The arms and legs, for example, first appear as little rounded knobs without fingers or toes. Gradually they elongate and toes and fingers appear. At the end of four months the work of development is practically completed except for proportion. At an early stage, perhaps at the end of one month, the embryo has a tail about one-fourth as long as the rest of the body. This, of course, is not the tail of a monkey, but the tail in formation which is part of the embryo. It also has gill slits; not the gill slits of a fish, but the gill slits of an embryo baby. One of these later plays an important part in connection with hearing. At six months the body is covered with a complete coating of hair which it loses before birth.

All these stages of development can be seen, preserved and are used in the course of instruction in any of the great medical schools of the country. The embryo becomes a human being when it is born.

Evolution never stops from the beginning of the one cell until the human being returns in death to lifeless dust. We wish to set before you evidence of this character in order to stress the importance of the theory of evolution. If the teaching of the theory of evolution in this field is to be excluded by law you will have to find adequate training for your doctors in medical schools outside of your state or you will have to import physicians from Chicago and New York, as the defendant Scopes had to import Mr. Darrow and myself.

Evolution Theory Vital to Agriculture

We expect to show you how vital to agriculture is the theory of evolution in connection with the development of important varieties of crops, plants, strawberries, peaches and other products essential to the life and prosperity of the people.

We expect to show you how vital is the theory of evolution to geology. We expect to offer you testimony as to the gradual building of the earth, its age and how its age is determined. We expect to show you how by the evolution of the earth's crust it is possible to tell where earthquakes are most likely to occur, so that mankind, for its safety, may have warning.

Moses No Edison

Much of this learning we hope to set before you will not be found in the Bible, but we maintain that all scientific truth cannot be contained in the Bible since so many truths that we all know about have been discovered since the Bible was written. Moses never heard about steam, electricity, the telegraph, the telephone, the radio, the aeroplane, farming machinery, and Moses knew nothing about scientific thought and principles from which these vast accomplishments of the inventive genius of mankind have been produced.

The purpose of the defense will be to set before you all available facts and information from every branch of science to aid you in forming an opinion of what evolution is, and of what value to progress and comfort is the theory of evolution, for you are the judges of the law and the facts, and the defense wishes to aid you in every way to intelligent opinion.

Denies Attempt to Destroy Christianity

The defense denies that it is part of any movement or conspiracy on the part of scientists to destroy the authority of Christianity or the Bible. The defense denies that any such conspiracy exists except in the mind and purposes of the evangelical leader of the prosecution. The defense maintains that the book of Genesis is in part a hymn, in part an allegory and a work of religious interpretations written by men who believed that the earth was flat and whose authority cannot be accepted to control the teachings of science in our schools.

The narrow purpose of the defense is to establish the innocence of the defendant Scopes. The broad purpose of the defense will be to prove that the Bible is a work of religious aspiration and rules of conduct which must be kept in the field of theology.

The defense maintains that there is no more justification for imposing the conflicting views of the Bible on courses of biology than there would be for imposing the views of biologsts on courses of comparative religion. We maintain that science and religion embrace two separate and distinct fields of thought and learning.

We remember that Jesus said: "Render unto Caesar the things that are Caesar's and unto God the things that are God's."

Stewart Objects to Mention of Bryan

Gen. Stewart—Your honor, I except to that part of the statement that has brought in Mr. Bryan's name.

Court—Have you finished your statement, Mr. Malone?

Mr. Malone—No, sir, I have not.

Gen. Stewart—And that you strike his name out.

Court—I hardly think that Col. Bryan's name should be injected into your statement, Col. Malone. I will just exclude it—eliminate it.

Mr. Malone—Will your honor hear me first?

Court—I will hear you.

Mr. Malone—I suppose this court, at any rate, will take judicial notice of the fact that Mr. Bryan is a most important member of this prosecution, in the court's mind, and in my mind. I suppose the court will take judicial notice of the fact that Col. Bryan is a recognized leader of his day and Col. Bryan's name is used in this connection in the same way that any other great leader's name would be used in that connection. My relations with Mr. Bryan have been such for so many years, he would be the last one to think anything I have to say here would have any personality in it. There is no reflection upon him in presenting our views, where we are representing conflicting ideas. I maintain that I have a right to use Mr. Bryan's name as representative of the views conflicting with our own.

Court Sustains Objection

Court—I do not think Mr. Bryan's personal views are involved in this case, so I think it is not proper in connection with this statement to mention him, and sustain the motion to eliminate his name.

Mr. Malone—Your honor , will you give me an exception?

Court—Yes.

Mr. Malone—Shall I continue?

Court—Yes.

(Mr. Malone resumes reading on the fourth page of his written statement.)

Insert Mr. Malone's statement, page 8, between immediately following first paragraph thereon, after the word 'Force,' insert:)

These words, your honor, were written twenty years ago by a member of the prosecution in this case, whom I have described as the evangelical spokesman of the prosecution, and we of the defense appeal from his fundamentalist views of today to his philosophical views of yesterday, when he was a modernist to our point of view.

Gen. Stewart—Your honor, I want to interpose an objection again. He is treading upon the soil your honor directed him not to tread upon.

The Court—Yes, Col. Malone, I would like that you not make further reference to Col. Bryan. Let that be excluded.

Mr. Malone—Yes, your honor, I do not think Mr. Bryan is the least sensitive about it.

Bryan Speaks

Mr. Bryan—Not a bit.

The Court—It is not a question of whether it gives offense, it is a question of your legal rights.

Mr. Malone—I believe I am acting in my legal rights and if your honor excludes that, I will take an exception.

Mr. Bryan—The court can do as it pleases in carrying out its rules; but I ask no protection from the court, and when the proper time comes I shall be able to show the gentlemen that I stand today just where I did, but that this has nothing to do with the case at bar.

Mr. Malone—One of the reasons for the defense was—

(Loud applause in the court room.)

The Court—I will have to exclude you, gentlemen, the jury is present now, and I cannot tolerate any expression of feeling on the issues in this case at all in the presence of the jury.

Mr. Malone—Your honor, I have been granted an exception?

The Court—Proceed. Yes.

Mr. Malone—We maintain and we shall prove that Christianity is bound up with no scientific theory—

Gen. Stewart—Your honor, at this juncture, while Mr. Malone is on his feet, I think it is improper for him to argue that it is a religious question. Your honor has excluded in overruling the motion to quash this morning any such arguments as that, because your honor held that it did not violate that clause of the constitution that guaranteed religious liberty, and I think the statement is entirely out of order.

The Court—Have you finished your statement, Col. Malone?

Mr. Malone—I have not, sir. You know, your honor, we have been waiting for a long while to get busy here, so I must have a little bit of time.

The Court—You may proceed. It is difficult for me to cut out parts of his statement while he is reading You may proceed and I will rule on these questions when they are presented later in proof.

Mr. Malone (Continued reading) —"We maintain, and we shall prove that Christianity is bound up with no scientific theory, that it has survived 2,000 years in the face of all of the discoveries of science, and —I would like a little quiet. I have got a loud voice, but I cannot talk during a lot of debating."

(Order was restored.)

(Mr. Malone continues reading down through pages 8, 9, down to and including, "The development of a child from a single cell to its birth," whereupon:

State Objects to Any Theory of Evo lution Being Read

Mr. B. G. McKenzie—If the court please, we desire to enter an exception to Mr. Malone reading any theories in regard to evolution. In other words, may it please this honorable court, the question will present itself to your honor as to whether or not these scientific witnesses are competent. He is undertaking now, under his statement to influence the jury by reading a statement to them.

Mr. Malone—Just a moment, your honor will hear me on that statement.

Mr. McKenzie—Wait until I get through.

Mr. Malone—Yes, but I do not want his honor to rule until you hear me.

Mr. McKenzie—We do not want him to read that. Naturally if it is read, and your honor rules these scientific witnesses are incompetent for the defense, then Mr. Malone has no right to read a theory or theories on evolution in the presence of this jury, in order to prejudice them one way or the other, and present an argument in support of it.

The Court—I think it is proper for the court to withhold his rulings upon these questions until the evidence is offered. I will instruct the jury that this attorney, gentlemen, is merely making a statement that is not proof in the case. He is merely outlining what he hopes to prove, what his theory is. While it is your duty to hear the statement, but keep in mind that it is not evidence, and that you are not to consider this statement in determining the issues, but the purpose of the statement is to get before your minds in the beginning what they hope or propose to prove. Now, the court may later allow them to prove these theories, or the court may not allow them to be proven. So, it is difficult for me to chop his statement up, so I will just let you proceed.

Mr. Malone—Your honor, I entirely agree with the court, and I could not have stated it better myself. The defense is not pretending to give testimony, the defense is merely explaining its theory, and if when we offer this testimony, your honor does not want it, he can reject it.

The Court—Yes.

Mr. Malone (continuing reading) —"All these stages in development, etc.

(Following conclusion of Mr. Malone's statement.)

Gen. McKenzie—If your honor, please, we again renew our motion to strike the argument and instruct the jury that it is unprecedented and unknown to the forms of law, for a lawyer to attempt to discuss his case before the jury before the issues are made up. Your honor asked both sides to present the issues not to the jury, but to your honor. Then, your honor submits the issues to the jury, the testimony is given by the witnesses, and your honor gives them the law.

Mr. Malone—That is the procedure.

Gen. McKenzie—This is wholly improper, argumentative. It is not a statement as to what the issues are. Your honor has already held that this act is constitutional, it being the law of the land, there is but one issue before this court and jury, and that is, did the defendant violate the statute. That statute interpretes itself, and says that whenever a man teaches that man descended from a lower order of animals as contradistinguished from the record of the creation of man as given by the word of God, that he is guilty. Does the proof show that he did that, that is the only issue, if it please the honorable court, before this jury. My friend is talking about a theory of evolution that it took him two years to write, that speech. (laughter.) That is not proper, if your honor please, if it is proper, it would be like a couple of gentlemen over in my country, where they were engaged and were trying a lawsuit before a justice of the peace, and they had a large number of witnesses. Finally one lawyer said, "let us have a conference," and they went out to confer, and they came back in and said, "if your honor please, the witnesses in this case, some of them are not very well, others are awfully ignorant, and we have just agreed among ourselves to dispense with the evidence and argue the case." That is what my good friend Malone wants to do. (Loud laughter and officer rapping for order.) And that is exactly what he has done, and hence I make that motion to instruct the jury that they must not consider Col. Malone's argument for the present, but to give him a chance after a while to shoot again.

Mr. Malone—Your honor if my brother is the spokesman momentarily of the defense, I am very happy that the judge has explained his theory. We are willing that the prosecution should state all of the theories they have got about this, and if they have got any more, we would like to hear them. We do not want to shut them off from stating anything in their minds. And so far as I am concerned, I believe your honor correctly instructed the jury that what I have stated to the court and the jury that is our theory of the case. We are prepared to back it up by the evidence and by the evidence to the jury. The jury, we believe, is an intelligent body of citizens that know the difference between testimony taken from the witnesses, and oratorical flights of the judge and myself.

Gen. McKenzie—If your honor please, we understand, and for the present there could not have been anything in the minds of the lawyers —we are not mediocre as lawyers—

Mr. Malone—That is not what I meant.

Gen. McKenzie—The only mistake the good Lord made is that he did not withhold the completion of the job until he could have got a conference with you.

Mr. Malone—I rather think you are right. (Laughter in court room.)

The Court—Any further statement from the state's side?

Gen. Stewart—None whatever.

(Jury was thereupon sworn by the court in due form.)

Examination of White

(Direct examination of Mr. Walter White continued by Attorney-General A. T. Stewart):

Gen. Stewart—Col. Darrow has very kindly consented not to be captious in objecting. I may use a few leading questions in order to get the evidence out.

Q—Mr. White, do you know what particular books, or what particular subjects, Mr. Scopes taught in the high school?

A—He was a science teacher; he taught chemistry, biology and other subjects in the science course.

Q—Did he teach this book, Hunter's biology?
A—Yes, sir.
Q—Will you file that book as Exhibit 1 to your testimony?
A—Yes, sir.
Q—What school did he teach in, Mr. White?
A—The Rhea County Central High school, here in Dayton.
Q—Is that school supported by state and county funds?
A—Yes, sir.
Q—You say it was in this county, Rhea county?
A—Yes, sir.
Q—How long have you been superintendent of public instruction?
A—A few days more than six and a half years.
Q—Has Mr. Scopes been teaching in the high school here for more than a year?
A—No, sir, he taught last year only.
Q—Do you know when this last term of school that he taught was out?
A—May 1, 1925.
Q—Do you remember when the prosecution in this case was first begun, Mr. White?
A—May 5th—May 5th, 1925.
Q—Some three or four days after adjournment of the school?
A—Four days after the school completed its term.
Q—Did you have any conversation with him concerning this teaching of Hunter's biology, after the passage of this law or at any time?
A—I talked with him about it on the afternoon of May 4th, 1925, the day before this—
Q—Trial?
A—This trial was started.
Q—He had already been arrested then?
A—No, sir, he had not been arrested.
Mr. Darrow—It was before the preliminary hearing?
Q—About the 5th of May, the warrant was sworn out. The 5th of May?
A—Yes, sir.
Q—When was the preliminary trial?
A—On Saturday, May 9.
Q—You talked with him then after school had adjourned?
A—Yes, sir.
Q—That was on the fourth of May?
A—Yes, sir.
Q—School adjourned on May 2?
A—The first of May, 1925.
Q—What was the conversation between you and the defendant Scopes as to the teaching of Hunter's biology?
Mr. Darrow—Was that after the defendant had had his trial?
Gen. Stewart—That is, of course, with reference to his admission that he committed the offense, prior to the trial.
Court—The admission that he taught it?
Gen. Stewart—Yes, sir. We think it is competent as an admission.
Mr. Darrow—It is.
Court—Proceed.
A—Mr. Scopes said that he taught this biology, and that he had reviewed the entire book.
Mr. Darrow—What is the last part of that statement.
Q—How is that?
A—That he had reviewed the entire book during certain days in April, somewhere, after having taught it to the boys. That he had taught this book, and had reviewed the entire subject, as it is customary for the teacher to do, and among other things he said he could not teach that book without teaching evolution. And I defended the evolution statute, and he said—
Mr. Darrow—We except to that.
Court—Not what you defended, but what you said.
A—The substance of what I said about this? I told Prof. Scopes that he had violated the Tennessee statutes.
Q—Were you at that time discussing this new law that was passed?
A—Yes, sir.
Q—That law was passed on the twenty-first of March, was it?
A—The twenty-first of March, of this year, and he said he couldn't teach biology without violating this law.
Q—And he said that in teaching

biology, he was teaching evolution and that would be in violation of the law.

A—Yes, sir.

Q—What was the date of this conversation?

A—He came up while Mr. Rappleyea and I were discussing this, and then all three of us discussed it for some little time after the crowd scattered. It grew out of a conversation between Mr. Rappleyea and myself in regard to the law.

Q—He said he had taught it here in Rhea county?

A—Yes, sir.

Q—And said he had reviewed it somewhere about the twenty-first of April?

A—Somewhere along there in April, 1925.

Q—You say it was customary to review the books there at the defendant's school?

A—Yes, sir.

Q—Did he say to you in reference to this book that he had taught that part that pertained to evolution?

A—Yes, sir.

Q—What did he say?

A—He admitted that he had taught that. He said that he couldn't teach the book without teaching that and he could not teach that without violating the statute.

Q—Did he say that it was unconstitutional?

A—He defended his course by saying that the statute was unconstitutional.

Q—He taught that in the high school here in Dayton?

A—Yes, sir.

Q—In Rhea county?

Q—Mr. White, I will ask you if this is the King James version of the Bible, and to file it as an exhibit to your testimony?

Mr. Hays—Do you mean to file that in evidence?

Gen. Stewart—We offer this in evidence, yes, sir, as explanatory of what the act relates to when it says "Bible."

What is the Bible?

Mr. Hays—What is the Bible? Different secas of Christians disagree in their answers to this question. They agree that the Bible is the inspired word of God, that the Creator of the universe is its Author, and that it is a book of divine instruction as to the creation of man, his relation to, dependence and accountability to, God. The historical and literary features of the Bible are of the greatest value, but its distinctive feature is its claim to teach a system of religion revealed by direct inspiration from God. It bases its demand for the reverence and allegiance of mankind upon the direct authority of God Himself. The various Protestant sects of Christians use the King James version, published in London in 1611, while Catholics use the Douay version, of which the Old Testament was published by the English college at Douay, in France, in 1609, and the New Testament by the English college at Rheims in 1582, and these two versions are often called, respectively, the Protestant Bible and the Catholic Bible. The original manuscripts containing the inspired word of God, written in Hebrew, in Aremaic and in Greek, have all been lost for many hundreds of years, and each of the Bibles mentioned is a translation, not of those manuscrips, but of translations thereof into the Greek and Latin. The earliest copy of the Old Testament in Hebrew now in existence was made as late as the eleventh century, though there are partial copies made in the ninth and ten centuries. The oldest know Greek manuscripts of the Bible, except a few fragments, belong to the fourth and fifth centuries. Each party claims for its own version the most accurate presentation of the inspired word as delivered to mankind and contained in the original scriptures." Which version does the Tennessee legislature call for? Does it intend to distinguish between the different religious sects in passing this law? Does it mean the Protestant, the St. James version, rather than the Catholic or Douay Bible? They could be required to call some witness here to testify what the Bible is. The court says further: "The versions differ in many particulars. There are differences of translation, many of which seem unimportant, though Catholics claim that there are cases of wilful perversion of the Scriptures in King James' translation, from which erroneous doctrines and inferences have been drawn. The Lord's prayer is differently translated in the two versions. Of the different translations of the Lord's prayer in later versions of the Bible, the following language of a Protestant has been quoted with approval by a Catholic author: 'Even the Lord's prayer has been tampered with and a discord thrown into the daily devotions. The inspired text is changed and unsettled, the faith of the people in God's Holy Word is undetermined, and aid and comfort given the enemy of all religion.' The Douay version also contains six whole books and portions of other books which are not included in King James' version. The Catholic church regards these as a part of the inspired Scriptures, entitled to the same faith and reverence as the other portions of the Bible, while the Protestant churches do not recognize them as a part of the Scriptures." "There are many sects of Christians and their differences grow out of their different constructions of various parts of the Scriptures — the different conclusions drawn as to the effect of the same words."

Should Designate Violation

In other words, your honor, they should be required to designate the violation of the law. The court may take judicial notice of the Bible, but the court does not take judicial notice of a fact that is at issue between the parties. It can only take judicial notice of matters that are of common knowledge. That is a matter to be proven.

Now, your honor, in the Encyclopedia of Evidence I find this: "It has been held that when a fact of history is in issue before a jury, it must be proved." And, again, "Generally speaking, however, courts are bound to take notice only of the public laws and the facts established thereby, and the official capacity and seals of some officers."

This is a criminal statute and should be strictly construed. There is nothing in the statute that shows they should be controlled in their teaching by the St. James version. The statute might have said that, but it did not. And yet, with an unaccountable confidence they have presented a book to your honor, and attempt to put that book in evidence with the confidence of a man not learned in religion, because any man learned in religion knows it is no more the version of the Bible than a dozen or half a dozen other books. Therefore, your honor, we object to the Bible going in evidence, or that book going in evidence, but insist that the prosecution prove what the Bible is before they put it in evidence.

Court—Mr. Hays, would you raise the same objection if they attempted to file any other Bible?

Mr. Hays—Not if they put it in evidence, and someone testifies that the King James version is the Bible; and then the jury could believe or disbelieve the statement.

Court—Let your objection be overruled. Let it be introduced as the Bible.

Mr. Hays—We except to that.

Mr. Darrow—What parts of the Bible are you going to introduce, anyway? Is it the whole book, and each of us to read from it such passages as we want? I am asking for this purpose—I don't know what your practice is here, but as a rule in our courts you have to have a certified copy or you cover it by stipulation.

Gen. Stewart—Just make such excerpts as you care to take out of it. We file the whole book, and the judge can order it attached to the supreme court record in the original form.

Court—I can order the book itself sent up.

Mr. Darrow—Yes, and save that much work. We just want to know what particular edition it is.

Court—Well, you may inspect it and cross-examine the witness about it.

Mr. Darrow—No—Just a minute. Not just any Bible.

Gen. Stewart—Of course we were going into the story of the creation.

Mr. Darrow—We want it so we can get a copy of the same book; that is all.

Gen. Stewart—This is the—

Mr. Darrow—Scofield.

Gen. Stewart — Holman's Pronouncing Edition of the Holy Bible, containing the Old and New Testaments. Translating—Text comformable to that of 1611, known as the authorized or King James version.

Bible as an Exhibit

Mr. Hays—Now, if the purpose is to offer this book in evidence, we take exception to it. The act provides that one shall not teach a certain theory, different from what is taught in the Bible, and now he undertakes to provide that he shall not teach a theory contrary to the St. James' version of the Bible. If the court should take judicial notice of this exhibit as the Bible, you must likewise take judicial notice that there are various Bibles. And the King James' version is not necessarily the Bible and when they introduce one book in evidence, we are saying there are several different books called the Bible. It is not relevant unless those books are the same. You know there is a Hebrew Bible, of some thirty-nine books; and there is a Protestant Bible, and a Catholic Bible—the Protestant of sixty-six and the Catholic of eighty books; and you have the King James' version, and a revised version and there are 30,000 differences between the King James' version and it. You have the King James' version and it. You have the King James version here; there are thousands of Bibles. Who is to say that the King James version is the Bible? The prosecution will have to prove what Bible it is, and they will have to state the theory as taught in the Bible, and I presume the prosecution will be able to point out which theory of the creation as taught in the Bible they relied upon in prosecuting Mr. Scopes. We will insist upon an answer to this question. In People vs. Ring, an Illinois case, the court says:

Mr. Malone—What publishers?

Gen. Stewart—A. J. Holman & Co., Philadelphia, publishers.

Mr. Darrow—I didn't know the edition exactly. I am sure we can get that on sale here, can't we.

Gen. Stewart—How is that?

Mr. Darrow — Do you know whether we can find that on sale here?

Gen. Stewart—Yes, sir, you can find that same edition on sale, I think, at Robinson's drug store.

The Court—Of course, you can certify it. It is a Bible in common use. If we can't find it, we will have to get an extra one in the case, when we—I take it that another copy of this same Bible can be secured without difficulty, surely, at Mr. Robinson's.

Mr. Darrow—In this small town, I don't know.

Court—If you can't get that here, you can get it in some other town.

Teaching by Scopes

Direct examination of Mr. White continues by Gen. Stewart:

Q—On Pages 194 and 195 of this book, (biology) where the doctrine of evolution and the evolutionary tree is shown by a drawing. Did Mr. Scopes say that he reviewed that about the 20th of April, with the rest of the book?

A—It is my understanding that he reviewed the important parts of the book and that he reviewed that part, that refers to Charles Darwin's theory of evolution.

Q—And the same thing applies on Pages 252 and 253 and Pages 254 and 255?

A—Yes, sir.

Q—Which would refer to evolution?

Mr. Darrow—I turned down the leaves of that.

Gen. Stewart—They are marked right now. I want to call attention to the particular parts of that book.

Mr. Darrow—They are marked here.

Gen. Stewart—That is the particular part of the book that bears upon the theory of evolution.

Mr. Darrow—Then, we make the same reservation as we did to the Bible.

Gen. Stewart—Yes, you can certify the book.

Mr. Darrow—Hadn't you better mark this?

The Court—The stenographer will mark the books, colonel. He will mark them as exhibits. Better mark them as we go along.

Mr. Darrow—Your honor, one of the court reporters has called my attention to the fact they were terribly crowded coming in the courtroom over there.

The Court—Well, Mr. Sheriff, do the best you can; you are right there.

Mr. Darrow—I don't like to disturb anybody, but at the same time—

A Voice—One of them has quit.

The Court—Can you get along all right?

Mr. Darrow—What I was getting at—What I was getting at, it is pretty hard to get out for the reporter.

The Court—If they are crowded there, let them speak up.

Mr. Darrow—We are interested in them, anyhow. I have not asked him to read the first two chapters of Genesis of the Bible, nor any of the chapters of that book.

Gen. Stewart—We might ask you to read it before argument.

Mr. Darrow—We can do that on any argument. But, I don't care to burden the record with all of that; of course, the first two chapters of the Bible—

Gen. Stewart—Better mark it?

Mr. Darrow—Your honor, may we have it indicated?

Gen. Stewart—I mean as an exhibit.

Mr. Darrow—Oh, yes.

Gen. Stewart—This is Exhibit 1, will you mark it?

Bible as Exhibit

Thereupon said book was marked Exhibit 1 and the Bible was marked Exhibit 2.

Gen. Stewart—You may cross-examine, if you care to.

Cross-examination by Mr. Darrow:

Q—Mr. Witness, will you please speak loud?

A—Yes, sir.

Q—For the reporters. This book of Hunter's, what is the name of that book?

Gen. Stewart—Biology. I thought you were asking me about the book.

Mr. Darrow—I am asking the witnesses.

A—George William Hunter's Civic Biology.

Q—Where did Mr. Scopes get it?

A—In the course of study, Mr. Robinson, the book man for this section handled the books.

Q—That was the official book adopted by the board, was it not?

A—In Tennessee, the board of education does not adopt books.

Q—Who does?

A—The Tennessee textbook commission adopts the book.

Q—Official book adopted by the Tennessee textbook commission?

A—That was the official book adopted by the Tennessee textbook commission in 1919, but the contract expired August 31, 1924, a five-year contract.

Q—Had any other book been adopted in the meantime?

A—No, sir.

Q—And these books were to be purchased at certain places, were they?

A—Certain depositories in Tennessee.

Q—The Robinson store was one of those depositories, was it?

A—Yes, sir.

Q—So, he taught this, which was the official book at that time?

A—Yes, sir.

Q—And did you ever have any talk with him before the time it was charged he taught it?

A—I did not.

Q—You are charging he taught it on the fourth day of May?

A—Yes, sir.

Gen. Stewart—How is that?

Mr. Darrow: Q—You never said anything to him about it or to any other teacher about not teaching it?

A—No, sir; I did not for these reasons—

Q—I don't care anything. about the reason, but you may give it.

A—Under the Tennessee law, I have not—

Q—Nobody ever said anything to you about it, did they?

A—No, sir.

Q—You never complained of Mr. Scopes as a teacher?

Scopes' Work Satisfactory

A—I had no complaint against his work in general.

Q—That is what I am speaking of.

A—No complaint against his work in general.

Q—That's all, do you know how long this book has been used?

A—It has been used since 1909, the school year of 1909.

Mr. Darrow—That is all.

Gen. Stewart—That is all.

The Witness—All right.

Gen. Stewart—Step down.

(Witness excused.)

Howard Morgan's Testimony

Howard Morgan, a witness in behalf of the prosecution, having been first duly sworn, testified as follows:

Direct examination by General Stewart:

Q—Your name is Howard Morgan?

A—Yes, sir.

Q—You are Mr. Luke Morgan's son?

A—Yes, sir.

Mr. Darrow—Will you speak a little louder? Some of these reporters say they cannot hear.

Gen. Stewart—You both will have to speak a little louder.

Q—You are Mr. Luke Morgan's son?

A—Yes, sir.

Q—Your father is in the bank here, Dayton Bank and Trust company?

A—Yes, sir.

Q—How old are you?

A—14 years.

Q—Did you attend school here at Dayton last year?

A—Yes, sir.

Q—What school?

A—High School.

Q—Central High school.

A—Yes, sir.

Q—Did you study anything under Prof. Scopes?

A—Yes, sir.

Q—Did you study this book, General Science?

A—Yes, sir.

Q—Do you want to see it?

Mr. Darrow—Will you mark the number?

The Court—Let the stenographer mark it.

Mr. Darrow—Is that the one you just showed me?

Gen. Stewart—No, it is another book, General Science, by Lewis Elhuff.

The Court—Let it be marked first.

Gen. Stewart—Were you studying that book in April of this year, Howard?

A—Yes, sir.

Gen. Stewart—Mark this 3.

Whereupon said book was marked exhibit 3.

Q—Did Prof. Scopes teach it to you?

A—Yes, sir.

Q—Who did you study it under?

A—Prof. Scopes.

Q—When did you complete the book?

A—Latter part of April.

Q—When was school out?

A—First or second of May.

Q—You studied it then up to a week or so before school was out?

A—Yes, sir.

Q—Now, you say you were studying this book in April; how did Prof. Scopes teach that book to you? I mean by that did he ask you questions and you answered them or did he give you lectures, or both? Just explain to the jury here now, these gentlemen here in front of you, how he taught the books to you.

A—Well, sometimes he would ask us questions and then he would lecture to us on different subjects in the book.

Q—Sometimes he asked you questions and sometimes lectured to you on different subjects in the book?

A—Yes, sir.

Q—Did he ever undertake to teach you anything about evolution?

A—Yes, sir.

Q—Did he undertake to teach you anything about any theory—

Mr. Darrow—I think, your honor, I will object to that. Ask him what it is.

Gen. Stewart—Q—What did he teach you in reference to?

The Court—What is the difference?

Mr. Darrow—Why—

The Court—All right.

Gen. Stewart—Q—About any evolutionary theory as to where man came from.

(Laughter in courtroom.)

Gen. Stewart—Just state in your own words, Howard, what he taught you and when it was.

A—It was along about the 2d of April.

Q—Of this year?

As Boy Heard Story

A—Yes, sir; of this year. He said that the earth was once a hot molten mass, too hot for plant or animal life to exist upon it; in the sea the earth cooled off; there was a little germ of one cell organism formed, and this organism kept evolving until it got to be a pretty good-sized animal, and then came on to be a land animal, and it kept on evolving, and from this was man.

Q—Let me repeat that; perhaps a little stronger than you. If I don't get it right, you correct me.

Mr. Hayes—Go to the head of the class.

Gen. Stewart—He said that in the beginning, the earth was a crystaline mass, too hot for any life to exist upon it; that it cooled off and finally the soil formed and the sea formed, plant life was on the earth, and that in the sea animal life began with a little one-celled animal.

A—Yes, sir.

Q—Which evolved and evolved and finally got bigger and became a land animal?

A—Yes, sir.

Q—And the culmination of which was man?

A—Yes, sir.

Q—Is that right?

A—Yes, sir.

Q—Did he say—

Mr. Darrow—Would you mind asking what he said?

Gen. Stewart—Yes, sir; I will do that.

Q—Now, when was it he taught you this that we have just repeated?

A—Well, it was in April.

Q—During class?

A—Yes, sir.

Q—What were you studying, what subject, when he said that?

A—We were studying General Science book.

Q—This General Science book?

A—Yes, sir.

Q—That is the theory that he taught you about a man being a little germ and sprouting in the sea, and so forth, and finally culminating and coming out on dry land; is that in this book? Could you find it in this book?

A—No, sir; I couldn't find it.

Q—Did you look for it?

A—No, sir.

Q—If it is in there, you could not find it?

A—No, sir.

Gen. Stewart—I hand it to you, gentlemen, to find it. (Handing book to counsel.)

Gen. Stewart—I ask you further, Howard, how did he classify man with reference to other animals; what did he say about them?

A—Well, the book and he both classified man along with cats and dogs, cows, horses, monkeys, lions, horses and all that.

Q—What did he say they were?

A—Mammals.

Q—Classified them along with dogs, cats, horses, monkeys and cows?

A—Yes, sir.

Q—You say this was along about the 2d or 3d of April of this year?

A—Yes, sir.

Q—In high school of Rhea county.

A—Yes, sir.

Q—At Dayton?

A—Yes, sir.

Gen. Stewart—Cross-examine.

Cross-examination by Mr. Darrow:

Q—Let's see, your name is what?

A—Howard Morgan.

Q—Now, Howard, what do you mean by classify?

A—Well, it means classify these animals we mentioned, that men were just the same as them, in other words—

Q—He didn't say a cat was the same as a man?

A—No, sir; he said man had a reasoning power; that these animals did not.

Q—There is some doubt about that, but that is what he said, is it?

(Laughter in the courtroom.)

The Court—Order.

Gen. Stewart—With some men.

Mr. Darrow—A great many.

Q—Now, Howard, he said they were all mammals, didn't he?

A—Yes, sir.

Q—Did he tell you what a mammal was, or don't you remember?

A—Well, he just said these animals were mammals and man was a mammal.

Q—No; but did he tell you what distinguished mammals from other animals?

A—I don't remember.

Q—If he did, you have forgotten it? Didn't he say that mammals were those beings which suckled their young?

A—I don't remember about that.

Q—You don't remember?

A—No.

Q—Do you remember what he said that made any animal a mammal, what it was or don't you remember?

A—I don't remember.

Q—But he said that all of them were mammals?

A—All what?

Q—Dogs and horses, monkeys, cows, man, whales, I cannot state all of them, but he said all of those were mammals?

Whale Stumps Him

A—Yes, sir; but I don't know about the whales; he said all these other ones.

(Laughter in the courtroom.)

The Court—Order.

Mr. Darrow:

Q—You might never have seen a whale suckling its young?

A—I did not.

Q—But the others were all mammals?

A—Yes, sir.

Q—You don't know whether he told you why they were mammals or not, did you?

A—No, sir.

Q—And you don't know whether they were mammals or not, only what he told you?

A—I just know what he said; he said they were mammals.

Q—And you didn't know that the definition of a mammal was a species that suckled its young, did you?

A—No, sir.

Q—Well, did he tell you anything else that was wicked?

A—No, not that I remember of.

Q—Will you please step down here; I cannot come down there or I would.

(Witness steps down to counsel's table.)

Q—Is this one of the books he taught you from?

(Handing book to witness.)

A—Yes, sir.

Q—Now, read that and see whether you remember that after you read it.

A—Examples of mammals, lions, monkey, lion, cat, dog, horse, cow, monkey and man.

Q—Isn't there some more that you remember when you look it over?

A—I don't remember.

Q—And I will read over this, and then see whether you can remember. Heading is Mammals. Mammals compose a group of animals which are the most highly developed of all; the egg produced by the female is microscopic in size and fertilized within the body of the mother.

Q—Do you remember that? Anyway, you studied it, didn't you?

A—Yes, sir.

Q—And you are like the rest of us, you don't remember all you study, I suppose. Well, we are all that way.

(Reading.) And there grows into the young animal all the parts of an adult.

That is the grown being, adult, I suppose. I don't suppose I dare read this.

(Reading.) After birth the young are nourished for a time by milk secreted by the mammary glands of the mother.

Q—Do you remember this is in the book?

A—It is in the book, but I don't remember him saying anything about it.

Q—Well, you read this anyhow?

A—Yes, sir.

Q—Examples of mammals are the elephant, lion, mink, cat, dog, horse, cow, monkey and man.

A—Yes, sir.

Q—Now, he said the earth was once a molten mass of liquid, didn't he?

A—Yes, sir.

Q—By molten, you understand melted?

A—Yes, sir.

Q—Running molten mass of liquid, and that it slowly cooled until a crust was formed on it?

A—Yes, sir.

Q—After that, after it got cooled enough, and the soil came, that plants grew; is that right?

A—Yes, sir; yes, sir.

Q—And that the first life was in the sea.

Q—And that it developed into life on the land?

A—Yes, sir.

Q—And finally into the highest organism which is known as man?

A—Yes, sir.

Q—Now, that is about what he taught you?

Q—It has not hurt you any, has it?

A—No, sir.

Mr. Darrow—That's all.

(Laughter in courtroom.)

Mr. Hays—Q—Is there anything in this book that says man is descended from a monkey, you have read the book?

A—Yes, sir.

Q—That man descended from monkey?

A—No, sir; not that I know of.

Gen. Stewart—It is not in the book about man coming from the same cell that the monkey came from, either, Col. Darrow.

A—I could not find it, Mr. Darrow.

Mr. Darrow—Well, it doesn't.

Mr. Malone—Not even by what he said it descended.

Gen. Stewart—Come down. (Witness excused.)

Another Pupil's Story

Harry Shelton, a witness in behalf of the prosecution, having been first duly sworn, testified as follows:

Direct examination—By Gen. Stewart.

Q—Your name is Harry Shelton?

A—Yes, sir.

Q—Did you go to the high school up here?

A—Yes, sir.

Q—Study under Prof. Scopes?

A—Yes, sir.

Q—When was school out, Harry?

A—May first.

Q—This year?

Q—What class were you in?

A—Biology.

Q—Among others, did you study this Civic Biology?

A—Yes, sir.

Q—Prof. Scopes teach it to you?

A—Yes, sir.

Q—When did you have a review of it?

A—Along in April some time; I don't remember what day.

Q—Around the middle part of April. How long before school was out?

A—About three weeks, I guess.

Q—About three weeks, that would be about the middle of April, then?

A—Yes, sir.

Q—Did you study—did Prof. Scopes teach you anything about evolution during that time?

A—He taught that all forms of life begin with the cell.

Q—Begin with the cell?

A—Yes, sir.

Q—Did he teach you during that time, during that review, did he teach you these pages, 194 and 195? Did you review this with the other?

A—Yes, sir; reviewed the whole book.

Q—Reviewed the whole book; that was along about the middle of April, and he taught you this particular book at that time?

Gen. Stewart—That is all I want to ask you.

Cross-examination—By Mr. Darrow.

Q—How old are you?

A—Seventeen.

Q—Prof. Scopes said that all forms of life came from a single cell, didn't he?

A—Yes, sir.

Q—Did anybody every tell you before?

A—No, sir.

Q—That is all you remember that he told you about biology, wasn't it?

A—Yes, sir.

Q—Are you a church member?

A—Sir?

Q—Are you a church member?

A—Yes, sir.

Q—Do you still belong?

A—Yes, sir.

Q—You didn't leave church when he told you all forms of life began with a single cell?

A—No, sir.

Mr. Darrow—That is all.

The Court—No talking in the courtroom. Who do you want next?

Mr. Darrow—That is all.

Gen. Stewart—That is all.

(Witness excused.)

F. E. Robinson, a witness in behalf of the prosecution, having been first duly sworn, testified as follows:

Direct examination—By Mr. Stewart.

Q—You are Robinson, known as Robinson's drug store?

A—Yes, sir.

Q—Where all this thing started?

A—Yes, sir.

Q—Did you have any conversation with Scopes along about the time that this trial started with reference to his teaching the theory of evolution?

A—Yes, sir.

Q—Just state what that was, if you remember it.

A—Well—

Mr. Darrow—Just a minute; what is the question?

(Question read.)

Gen. Stewart—About the time this trial started.

Mr. Darrow—Get the date of it.

Gen. Stewart—Q—That was along—about May 4 or 5?

Admitted Violating Law

A—I don't remember what date; it was the next week after school was out. Scopes said that any teacher in the state who was teaching Hunter's Biology was violating the law; that science teachers could not teach Hunter's Biology without violating the law.

Q—That Hunter's Biology—

A—That is the adopted book—

Mr. Darrow—We will admit it was accepted.

Gen. Stewart—And you except only to Walter White's testimony?

A—That was a state adopted book, and Dr. Rappleyea said you have been teaching this book? And he said yes. He said if you got the book out of stock, and ask if he had taught this in regard to evolution, since this law was passed. He said: yes, I reviewed the book. And he said: Well, you have been violating the law. He said so has every other man violated the law. He said when it was passed Prof. Ferguson discussed the law that a man could not teach science from any of the books published now without violating the law.

Gen. Stewart—On evolution?

Mr. Darrow—He said biology.

The Witness—Biology.

Gen. Stewart—I didn't mean to prompt him, but he was speaking about that.

Mr. Darrow—Oh, I know you did not.

Gen. Stewart—Q—You say Dr. Rappleyea got the book out. Did he open it and examine it?

A—Yes, sir.

Q—Where the evolutionary tree is?

A—Yes, sir.

Q—Did you say he taught this along with the rest of this?

A—He said he had reviewed that the last two or three weeks of school.

Q—Page 194 of the biology?

Mr. Darrow—Will you read that?

Gen. Stewart—On page 194, where the evolutionary tree is. He said he discussed this with Prof. Ferguson?

A—Yes, sir.

Q—And that was about the time it was passed that he discussed it?

A—When it was published in the papers; yes, sir. When it was being passed.

Q—That was before he reviewed the book?

A—Yes, sir.

Q—Who is Prof. Ferguson?

A—He is principal of the Rhea County High school, Central High school, where Scopes taught.

Q—In this high school?

A—Yes, sir; under Prof. Ferguson.

Q—Under Prof. Ferguson?

A—Yes, sir.

Q—And Scopes said he taught this book in Rhea county?

A—Yes, sir.

Q—And that was along—well, about the middle of April, you say then?

A—Well, he said that he had reviewed it the last two or three weeks.

Q—And the school was out the first of May?

A—Yes, sir; the first of May.

Q—And you are the chairman of the School Board of this county?

A—Yes, sir.

Q—And Scopes told you that he knew of the law?

A—Yes, sir.

Q—And you discussed it with him?

A—Yes, sir.

Mr. Stewart—I think that is all. You may cross-examine.

Cross-examination by Mr. Darrow:

The Court—If counsel for the state will stop; they are talking too loud.

Gen. Stewart—Beg pardon, I didn't get that?

The Court—You are talking loud; the lawyers were.

Gen. Stewart—Just conferring with each other.

Mr. Darrow—I will be very careful while he is looking through it. I will wait.

Q—He showed you a book which has been marked "a civic biology," or entitled "A Civic Biology," which I hold in my hand?

A—Yes, sir.

Q—You were selling them, were you not?

A—Yes, sir.

Q—And you were a member of the school board?

A—Yes, sir.

(Laughter in the courtroom.)

Mr. Darrow—I think someone ought to advise you that you are not bound to answer these questions.

Gen. Stewart — The law says teach, not sell.

(Laughter in the courtroom.)

Mr. Darrow—And this part on page 194 was read, was it?

A—Yes, sir.

Q—That was opened and read.

A—Yes, sir. That was opened and read.

Q—And any part of 195?

A—Yes, sir.

Q—And then I think another page, another I think some place. I don't remember.

Q—Well, now read it. It has not been read?

Doctrine of Evolution

"The Doctrine of Evolution—We have now learned that animal forms may be arranged so as to begin with very simple one-celled forms and culminate with a group which contains man himself. This arrangeuent is called the evolutionary series. Evolution means change, and these groups are believed by scientists to represent stages in complexity of development of life on the earth. Geology teaches that millions of years ago, life upon the earth was very simple, and that gradually more and more complex forms of life appeared, as the rocks formed latest in time show the most highly developed forms of animal life. The great English scientist, Charles Darwin, from this and other evidence, explained the theory of evolution. This is the belief that simple forms of life on earth slowly and gradually gave rise to those more complex and that thus ultimately the most complex forms came into existence."

Q—Did you examine this evolutionary tree?

A—Yes, sir.

Q—You don't know whether man is in there, do you?

A—Yes, sir; man is in here.

Q—I am afraid they left him out. You put him in with the mammals, but nothing in there—the word man is not written in there, is it?

A—I don't believe it is; the word man is not, but it says in the books

Q—I am going to read the rest on the other page. This table down here:

"The number of Animal Species. Over 500,000 species of animals are known to exist today."

That wasn't read was it?

A—I think the whole book was read.

Q—Not the whole book?

A—I don't know. We read most of the book.

Q—Do you know what was read?

A—That was read, that page was read.

Q—Take the table?

A—Well, I don't know about that.

Mr. Darrow—Do you claim anything on the table? If you don't I will not incumber the record.

(Thereupon counsel conferred out of the hearing of the jury and the shorthand reporters.)

Mr. Darrow—That is followed by table. "Over 500,000 species of animals are known to exist today, as the following table shows:

Protozoa	8,000
Sponges	2,500
Coelenterates	4,500

I would rather you read this, I don't know whether you can read it?

Gen. Stewart—I don't care whether you read it at all, or not.

Mr. Darrow:

Echinoderme	4,000
Flatworms	5,000
Roundworms	1,500
Annelids	4,000
Insects	360,000
Myriapods	2,000
Arachnids	16,000
Crustaceans	16,000
Mollusks	61,000
Fishes	13,000
Amphibians	1,400
Reptiles	3,500
Birds	13,000
Mammals	3,500
Total	518,900

Q—This part of that he also read?

A—Yes, sir.

Q—And on page 195. "Man's Place in Nature"—Although we know that man is separated mentally by a wide gap from all other animals, in our study of physiology we must ask where we are to place man. If we attempt to classify men, we see at once he must be placed with the vertebrate animals because of his possession of a vertebral column. Vertebral column, you understand is, backbone?

A—Yes, sir.

Mr. Darrow Reading)—"Evidently, too, he is a mammal, because the young are nourished by milk secreted by the mother and because his body has at least a partial covering of hair. Anatomically we find that we must place man with the apelike mammals, because of these numerous points of structural likeness. The group of mammals which includes the monkeys, apes, and man we call the primates. I see another line marked here. I am ashamed to read that, too.

"Mammals are considered the highest vertebrate animals, not only because of their complicated structure, but because their instincts are so well developed. Monkeys certainly seem to have many of the mental attributes of man."

Gen. Stewart—Just go right on.

Mr. Darrow—I am going to. You have underscored part of it. I want to read it too.

Mind of the Monkey

"Prof. Thorndike, of Columbia university, sums up their habits of learning as follows:

"In their method of learning, although monkeys do not reach the human stage of a right life of ideas, yet they carry the animal method of learning, by the selection of impulses and association of them with different sense-impressions to a point beyond that reached by any other of the lower animals. In this, too they resemble man; for he differs from the lower animals not only in the possession of a new sort of intelligence but also in the tremendous extension of that sort which he has in common with them. A fish learns slowly a few simple habits. Man learns quickly an infinitude of habits that may be highly complex. Dogs and cats learn more than the fish, while the monkeys learn more than they.

In the number of things he learns, the complex habits he can form, the variety of lines along which he can learn them and in their permanence when one formed, the monkey justifies his inclusion with man in a separate mental genus."

Q—That is what was read?

A—Yes, sir.

Q—Anything else in that book you can say was read?

A—No.

Q—How many of those did you have for sale?

A—Oh, I have been selling that book for six or seven years.

Q—Have you noticed any mental or moral deterioration growing out of the thing?

Gen. Stewart—How is that?

Mr. Darrow, Q—Have you noticed any mental or moral deterioration growing out of that thing?

Gen. Stewart—Exception.

The Court—I sustain the exception.

Mr. Darrow—Exception.

Q—How do you get them, Mr. Robinson?

A—From the depository at Chattanooga for this county.

Q—What is the depository?

A—The place that the state designates to handle the state books.

Q—You got them from the state authorities and are the only one who handles them in Dayton?

A—In Dayton, yes, yes.

Q—Were they adopted, as you understand it?

A—By the state board of education.

Q—State board of education?

A—Yes, sir.

Q—And state whether or not they got it from you?

A—Yes, sir.

Gen. Stewart—Perhaps as Mr. Darrow has seen fit to read from that part of the biology in question, your honor, I want at this point to read the first two chapters of Genesis in order to get it into the record.

The Court—You may proceed.

Mr. Darrow—No objection to that.

Gen. Stewart—(Reading from the first two chapters in Genesis.)

Gen. Stewart—Are you through with the cross-examination? Come down.

(Witness excused.)

Mr. Darrow—Gentlemen, I don't know how many more you want to put on, but according to my suggestion—you have some other boys, who will testify the same, just give the names?

Gen. Stewart—One little boy in the science class with the little Morgan boy, whose testimony would be the same.

Mr. Darrow—Give us the name.

Gen. Stewart — Morris Stout. Charles Hagley will testify substantially the same as the Shelton boy.

This book was reviewed about the 20th of April.

Mr. Darrow—Very well.

Gen. Stewart—The state rests.

Thereupon the prosecution rested in chief.

Mr. Darrow—Yes, we would like to continue, two or three minutes would be enough.

The Court—Go ahead.

See that these reporters get in and out; they cannot get in and out. Let the witnesses for the defense come forward.

Thereupon witnesses for the defense came forward.

The Court—Let the witnesses be sworn.

Mr. Hays—Your honor, before the witnesses are sworn, it is necessary for us as a matter of procedure to move to dismiss the prosecution's case.

The Court—Let it be overruled.

Mr. Hays—Exception.

Thereupon the witnesses were duly sworn.

The Court—These executive officers have charge of the courtroom, who shall go in and out, keeping the opening of the aisles, who shall occupy a chair or not. It is foolish to expect the judge of the court to provide chairs for everybody, to leave the bench and have somebody move back, appeal to the sheriff and executive officers, because I have as many responsibilities as I can get along with, without having to attend to these details.

* * *

Testimony of Dr. Metcalf, with the jury excluded, was taken as court adjourned.

Metcalf's Testimony Wednesday Afternoon

Maynard M. Metcalf, the first witness for the defense, being sworn and examined, testified:

Direct Examination:

Questions by Clarence Darrow, Esq.:

Q—Give us your name?

A—Maynard M. Metcalf.

Q—Where do you live?

A—My legal residence is Baltimore. I am living—I have been living the last year in Washington. I do not know how to answer your question.

Mr. Darrow—Living here in Dayton now?

Gen. Stewart—Just a moment, I do not mean to interrupt, but I want to impart a little information to you as a matter of procedure. Of course, you know we are going to except to this scientific testimony. But, we have a rule in this state that precludes the defendant from taking the stand if he does not take the stand first.

Mr. Darrow—Well, you have already caught me on it.

The Court—That is a technicality, we have not gone into the merits. I will allow you to withdraw the witness.

Mr. Darrow—Your honor, every single word that was said against this defendant, everything was true.

The Court—So he does not care to go on the stand?

Mr. Darrow—No, what is the use?

The Court—Well, that is all.

Q—(Mr. Darrow) What is your profession or business?

A—I am a zoologist.

Q—And just what is included in that?

A—It is the study of animals.

Q—How long have you been a zoologist?

A—Why, I began special study, with special interests, when I was about 14 years old. I do not know when I became a zoologist. I am now 58, I think—no 57, I think that is right.

Q—You have not learned it all yet, have you?

A—I am afraid not.

Q—Where do you say you began studying?

A—Why, when I was a youngster starting in at Oberlin college, at the age of 14.

Q—That is Oberlin, O.?

A—Yes, sir.

Q—What is the name of that college—Oberlin college?

A—Oberlin college.

Q—How long did you study at Oberlin?

A—Well, I was there—you mean after I was 14, after I began the study of zoology?

Q—Yes?

A—Four years.

Q—Then, what did you do?

A—I went to the Johns Hopkins university for graduate study in zoology.

Q—How long were you there?

A—Four years, the usual time.

Q—Well, that would make you out there at 22 years of age, then what did you do?

A—Well then, I accepted a position as associate professor of biology at the Woman's college, of Baltimore, it was then called, it is now Goucher college.

Q—How long were you there?

A—I was there until the spring of 1906, if you will excuse me from the mental arithmetic, I will state it that way.

Q—All right, that is just as good. And you are teaching zoology there?

A—What is that?

Q—You were teaching zoology there?

A—Yes, sir, I was teaching zoology with a little botany associated with it.

Q—And from there where did you go?

A—I went abroad.

Q—Where?

Worked in Germany

A—Working at the Naples zoological station, spending a year and a half at the zoological institution and then spending about a half year at the Institute Fur in connection with the Virschow hospital in Berlin.

Q—And from that time?

A—I had, already, before I went abroad, accepted a professorship in Oberlin, and I returned then to my Alma Mater in 1908, after this work in Berlin.

Q—What was the professorship you accepted?

A—Zoology, and the head of the department of zoology.

Q—And you have been there up to the present time?

A—No, I resigned in 1914 to give all of my time to research, in order to be free from teaching duties, I resigned.

Q—Where did you work after that?

A—I worked in my own laboratory, which I called the Orchard laboratory, used that name in publication, which was my private laboratory, which I and a few advanced students in Oberlin college were working in. They were working with me, but by sub-rosa arrangement.

Gen. Stewart—What kind of an arrangement?

A—Sub-rosa.

Q—Is that a zoological term?

A—No, straight Latin.

Mr. Darrow—You are thinking of Rosa.

Mr. Stewart—I though maybe it was a cigar of some kind.

Mr. Darrow—It does sound like one.

Q—(Mr. Darrow) And then what did you do next?

A—Well, I continued that work until—it is hard work for me to remember just what year it was, I think it was three years ago, I went to the Johns Hopkins for a year's work in connection with their laboratory, as the guests of the university, then I returned again to my home, and then last year I have been residing in Washington.

Q—As the guest of which university, Oberlin or Johns Hopkins?

A—Beg pardon?

Q—As the guest of which university?

At Johns Hopkins.

A—As the guest of Johns Hopkins university for that one year.

And then you returned to Oberlin—went to Washington?

A—Yes.

Q—What were you connected with in Washington?

A—I had charge of—I was chief of the division of biology and agriculture of the national research committee.

Q—That was carried on by the government?

A—That was instituted by an executive order of President Wilson, immediately after the war. It was really instituted during the war, for the study of scientific problems associated with the war, and after the war was over, by executive order of President Wilson, it was contined, for the study of scientific problems of use to the country in peace time.

Q—Give us your connection there?

A—In the division of biology and agriculture, the appointments of the chairmanship are regularly one-year appointments, and my appointment expired the 30th of June of this year.

Q—Are you out of a job?

A—No, I am afraid not. I wish I were.

Q—What are you doing now?

A—A year and a half ago I accepted a position on the faculty of the Johns Hopkins university, with the plan to go there at such time as my other duties made convenient, and I am to go there next spring after I return from a zoological trip to South America.

Q—And who is connected with the zoological trip to South America?

A—No one, except I have had some financial assistance from the National Academy of Science, that is the only connection.

Q—Are there others going with you?

A—No. No, I go alone.

Q—And what position are you to hold at Johns Hopkins?

A—Associate in research, associate in zoology, a purely research position.

Q—You have received degrees at colleges?

A—Yes, a few.

Q—I do not know whether you know any more on account of degrees, but I will let you mention them?

A—I beg pardon.

Q—I say I do not know whether you know any more on account of degrees, but I will let you mention them?

A—Well, I do not think they mean much. I took A. B. from Oberlin, and took Ph. D. from Johns Hopkins and have been given doctor of science, honorary, by Oberlin, since.

Q—Have you memberships in various organizations in the line of zoology?

A—Yes, I am a member of a number of the research organizations in this country, and I also have some memberships outside of zoology in economic organizations in this country and abroad, and am a member of one or two organizations abroad—two or three.

Q—In that line?

A—In zoology, or one is in economics.

Q—Have you held any offices in scientific organizations.

A—Oh, yes, from time to time.

Q—You might mention any of them that you have held?

Scientific Connections

A—Well, I have been secretary-treasurer of the Zoological society of the American Society of Naturalists. I have been president of the American Society of Zoologists and been president of Section F, zoological section of the American Association for the Advancement of Science, and I have been on the executive committee of both of these organizations. Oh, I do not know, a number of years. I am on the executive committee now of the division of biology and agriculture of the National Research council. There are a lot of those offices which pile up on a man; he cannot avoid them.

Q—Have you written articles or books?

A—Oh, yes. When a man is engaged in research he has got to pass on the results of his work.

Q—What you have done?

A—Surely.

Q—How many pamphlets and articles?

A—Oh, I suppose sixty or seventy. I do not know, I have not any idea.

Q—In the main I suppose these were scientific magazines?

A—Yes, sir.

Q—Or journals?

A—Yes, sir. I was not counting any outside of the scientific journals.

Q—Are you a member of any church organization?

A—Yes.

Q—What one?

Member of Congregationalist Church.

A—The Congregationalist church. Do you want to know the particular church?

Q—Yes?

A—I am now a member of the United church, in Oberlin, which is a Congregationalist church. I have been a member of two other congregationalist churches—no, one Presbyterian and one Congregationalist.

Q—You have been a Presbyterian, too, have you?

A—Well, I joined the Presbyterian church when I was 11 years old, I think—I am not sure.

A—And have you been connected with church activities aside from being a member?

A—Yes.

Q—In what way?

A—Well, in Baltimore I had charge of a Bible class in the church for about three years. I had charge of a Bible class of college students, well, not exclusively college students, mostly college students, in Oberlin. That is all, I think—of course I have had some church offices, but those do not mean much.

Q—Not unless it is treasurer or something like that.

A—No, nothing worse than deacon.

Q—Doctor, do you understand, or at least ever studied and read evolution?

A—Surely.

Q—For how long ?

A—I cannot answer that question. I think I heard the word and the thought was long ago. I could not remember when, and an old brother with whom I used to sleep, used to discuss with me evolutionary subjects until we went to sleep at night, night after night, before I was eight years old. I guess I had been brought up on it.

Q—Did your evolutionary studies include the development and evolution of man, in a general way?

A—I have never been a student of human morphology or human physiology distinctly, but I have been somewhat of a student of evolution, and especially interested in man, and I have given some lectures here and there on prehistoric man, early man.

Q—And you have studied as to the origin of man, have you not?

A—Well, I have not studied firsthand very much as to the origin of man, I have not been an archeologist or anthropologist, but I have read on it, and such lectures as I have given have been compendia from work done by other men, not my own work.

Q—But, you are familiar with that work?

A—Yes, sir, fairly broadly.

Q—And your studies in zoology, they have naturally been connected with the study of evolution?

A—Yes, I have always been particularly interested in the evolution of the individual organism from the egg, and also of the evolution of organisms as a whole from the beginning of life, that has been a sort of peculiar interest of mine, always.

Q—Are you an evolutionist?

A—Surely, under certain circumstances that question would be an insult, under these circumstances I do not regard it as such.

Q—Do you know any scientific man in the world that is not an evolutionist?

Stewart Objects.

Gen. Stewart—We except to that, of course.

The Court—Sustain the exception.

Mr. Stewart—Of course if you want to take a vote—

Mr. Darrow—No, no, we are talking about scientific things.

Q—(Mr. Darrow) or, is it or not accepted by scientific men?

Mr. Stewart—We object.

The Court—Sustain the objection.

Mr. Hays—We want to take an exception.

Mr. Stewart—You are entitled to it. Your honor is ruling on it?

Is Evolution a Guess?

Mr. Hays—Now, your honor, one of our constitutional points was the question of whether this law was within the police power of the state, depended upon material.

The Court—No, I do not think that is whether or not it was a reasonable exercise of the police power. That would depend largely upon whether evolution is a mere guess by a few men, or generally accepted by all scientists. Certainly it is material from that point of view. I do not think you can bring one witness to prove what others believe.

Mr. Hays—If he knows that, if he knows how far the theory is substantiated—

The Court—That would be hearsay testimony.

Mr. Hays—Hearsay testimony is allowed in cases where it is a question of how a scientific theory is substantiated. The question here depends very largely on our ability to determine whether any such exercise of the police power is reasonable, and here we call a witness to say, among other things, that in the profession that is largely accepted, that is the question, your honor.

Mr. Darrow—While that is not material, may I ask, your honor, at least—

Gen. Stewart—Your honor has passed on that.

The Court—Let the gentleman ask and then—

Mr. Hays—Our whole case depends upon proving that evolution is a reasonable scientific theory.

The Court—I do not know how you can prove it reasonable by proving what some other person believes.

Mr. Hayes—We expect to prove what all science says.

The Court—Then bring them here and offer them. I will hear you.

Witness—I am very glad to be examined on my own judgment.

The Court—Sustain the exception.

Mr. Hayes—Exception.

Mr. Darrow—I do not know whether the practice is to state what we expect the answer to be. Of course, I will not state it before the jury, but I want to give it to the reporter.

Mr. Stewart—We want that in the record later.

Mr. Darrow—I will do that later.

Gen. Stewart—Let the reporter go right to them.

(At this point the reporters and attorneys went to the winess chair and the following occurred:)

Q—(Mr. Darrow.) What would you say, practically all scientific men were or were not evolutionists?

Many Scientists Evolutionists

A—I am acquainted with practically all of the zoologists, botanists and geologists of this country who have done any work; that is, any material contribution to knowledge in those fields, and I am absolutely convinced from personal knowledge that any one of these men feel and believe, as a matter of course, that evolution is a fact, but I doubt very much if any two of them agree as to the exact method by which evolution has been brought about, but I think there is—I know there is not a single one among them who has the least doubt of the fact of evolution.

(The following then occurred in the hearing of the jury:)

Gen. Stewart—Of course, we do not want this part of the record to be in the papers. Of course we will have to keep that away from the jury. They will read that.

Gen. McKenzie—Your honor has a right to keep it from the jury.

Metcalf's Testimony Kept from Jury.

The Court—I will instruct the stenographers (reporters) to not give that part of the transcript to the newspapers. Do you object to that? Get this whole issue and then I can excuse the jury and hear from you.

Mr. Stewart—There are different kinds of evolution. We, of course, maintain it is limited to that particular kind described in this law suit.

Mr. Darrow—How is that?

Mr. Stewart—The point I am making is, there may be different kinds of evolution, perhaps there are, but this question we are insisting on in this case is just that one described by the act itself.

Mr. Darrow—By the act itself?

Mr. Stewart—By the act itself—the law.

Mr. Darrow—Well, you are going to object to that, too?

Mr. Stewart—I am objecting to a general question as to what evolution is. I suggest, your honor, that we discuss some points about this. We might ask your honor to retire the jury and thresh it out here.

Mr. Darrow—Suppose I ask one more question.

The Court—Let us get all the issues now that are going to be in dispute.

Mr. Stewart—We can put them in the record after the jury goes out, I take it, we will not object to it in that way.

Mr. Malone—I take it, we will not have the argument in the presence of the jury. If the attorney-general objects, I see no reason why we should not get the point up in the presence of the jury.

The Court—Of course, the question cannot prejudice the case, since there is no answer.

Mr. Darrow—Will you state what evolution is, in regard to the origin of man?

Mr. Stewart—We except to that.

Mr. Darrow—Now we are ready—

Mr. Stewart (Continuing)—On the further ground that we are excepting, your honor, to everything here that pertains to evolution or to anything that tends to show that there might or might not be a conflict between the story of the divine creation and evolution, and on the same theory we will except to this scientific testimony on the ground it is incompetent, because it is, so far as this case is concerned, it invades the province of the court and jury, and ask your honor to exclude the jury while we argue this matter.

Says Jurors Not Informed

Mr. Darrow—Of the jury, only one of whom every read about evolution, is forced to say what evolution is, without his hearing evidence.

Gen. Stewart—We want your honor to exclude the jury.

The Court—I suggest this, now, gentlemen. You have in mind what the issues are going to be on this question. I wish you would ask now such questions as would correctly and fairly make the issues, so that then I will excuse the jury and hear your argument on those questions, and not have to do it over and over again.

Mr. Stewart—And when we exclude the jury now, we do not want any more questions along this line. I think we have a right to insist that the jury not hear any of the rest of those. They have a certain duty to perform.

The Court—If you have—well, gentlemen of the jury, now, I think the radio, perhaps, is in operation, and when I excuse you gentlemen, I excuse you for the purpose that you do not hear the procedings up here. Please do not linger in the courthouse yard, because you might still perhaps stay up here as down there. I excuse you until 9 o'clock in the morning.

Juror Thompson—I just wanted the benefit of the jury there is not a single juryman that has heard a single word pass over the horns out there.

The Court—Thank you, gentlemen.

Juror Thompson—I just wanted you to know, that is all.

(The jury thereupon retired.)

The Court—Now, gentlemen, I anticipate that this is the most difficult thing the court is going to have to pass on. Do you think you have time to argue this question this afternoon? It is 4:35.

Gen. Stewart—No, we would not have time to complete it. We can get on it.

Mr. Darrow—All of us are tired.

The Court—Of course, this is a question I want all the light I can on, because I anticipate that it is extremely important and perhaps difficult.

Mr. Darrow—I will just ask one more question, so as to make the issues plainer.

(The officer rapped for order, saying: "Let us have order in the courtroom; respect the court.)

The Court—I will hear you, gentlemen.

Mr. Darrow—I will put two or three short ones.

Gen. Stewart—The last question you asked him—what was the last question?

Questions by Mr. Darrow.

Q—Now I want to ask the question, is there anything in the theory of evolution in conflict with the account of the creation of man in Genesis or in the Bible?

Gen. Stewart—We except to that.

Mr. Darrow—I just want to add one or two more and then we will let it all go together. It won't take me but a minute.

Q—Is evolution taught in all the leading colleges of the world?

Gen. Stewart—We except to that, of course.

Q—Or the western world—I will exclude the east; I don't know about that.

A—It is in China and Japan and in India.

Gen. Stewart—You want his answers in the record, don't you?

Mr. Darrow—They are all in there, aren't they?

Gen. Stewart—Do you want the witness to answer them now?

Mr. Darrow—Counsel suggests what is probably the right way, we should let him answer these questions.

Gen. Stewart—I thought they wanted the answers in the record, and he hasn't given them, and I thought you wanted them in the record.

The Court—If you want them in the record you may let him answer, and then they can move to exclude the answer.

Mr. Darrow—Well, counsel suggests that you might answer them altogether.

The Witness—I had rather not do that; I had rather answer them seriatum.

Makes Fine Distinction

A—Evolution and the theories of evolution are fundamentally different things. The fact of evolution is a thing that is perfectly and absolutely clear. There are dozens of theories of evolution, some of which are almost wholy absurd, some of which are surely largely mistaken, some of which are perhaps almost wholly true, but there are many points—theoretical points as to the methods by which evolution has been brought about—that we are not yet in possession of scientific knowledge to answer. We are in possession of scientific knowledge to answer directly and fully the question: "Has evolution occurred."

Q—Now, will you tell what it means, the fact of evolution?

A—A definition is perhaps the most difficult thing that a man can ever be asked to engage in, for any definition in order to be accurate and adequate would have to be fearfully prolix. I beg then to be allowed to answer in a way that certainly will not be adequate, but that may be accurate as far as it goes.

Mr. Darrow—Do it that way, then.

A—Evolution, I think, means the change; in the final analysis I think it means the change of an organism from one character into a different character, and by character I mean its structure, or its behavior, or its function, or its method of development from the egg or anything else—the change of an organism from one set characteristic which characterizes it into a different condition, characterized by a different set of characteristics either structural or functional could be properly called, I think, evolution—to be the evolution of that organism; but the term in general means the whole series of such changes which have taken place during hundreds of millions of years which have producd from lowly beginnings the nature of which is not by any means fully understood to organism of much more complex character, whose structure and functions we are still studying, because we haven't begun to learn what we need to know about them.

Q—Could you briefly sketch what that change is, from inorganic matter on, as far as we know?

A—Well, there must—I can try to do it briefly; you say, including the inorganic?

Q—Yes, starting with the inorganic world.

A—We have all sorts of changes, but leaving that all out of account there has been a tremendous series of changes with the inorganic world by which the universe has been brought into existence and has been molded into its present characteristics. The sun is comparatively young and the earth has gone through a long course of development and change. That is a matter—those two matters are for astronomers and geologists to talk about. I am not an expert in the field of inorganic evolution, although there is a tremendous field of phenomena there, but we are inclined to evolution of living things, of organic evolution, as it is called, we have to conceive of the earliest living things as being able to live upon inorganic food. We have only plants today with that ability. No animal is able to sustain life on the basis of inorganic feed. They have to have other plants and animals to live upon.

Q—Would it bother you for me to interrupt you for one question for the purpose of the record?

A—No, indeed.

Q—Tell us what you mean by organic and inorganic.

A—Organic evolution is connected with living things, organic things are the subject of living bodies, or things that are made by the living activities of those living bodies. There are certain chemicals found in the bodies of living things that are distinguished as against inorganic things which means like rocks and stones and earth.

Q—Minerals?

A—Yes, minerals, and so on.

Q—How do you classify botany, plants?

A—Organic, of course, because they are a part or bodies of living things. Now from the first living things which could live on inorganic substances, there developed a whole series of forms in the plant group gradually becoming more and more complex. Thy make really a remarkably beautiful series as you study them, and this series of increasing complexity in the plants as we find them shown in the rock, the actual plants themselves whose bodies we study in the fossil condition in the rock.

Q—Can you estimate the age of those?

A—Why, no; it takes a chemist to estimate the age of some of these things, for it is a determination of the processes of disintegration in the rock which have been caused largely by chemical forces, aided by the activities of certain bacteria, and I am not an expert in that field, and I would rather not answer.

Q—Could you make any estimate how long from the beginning of organic matter?

A—No, for this reason: I am inclined to believe that there may have been whole series of animals and plants living at certain times upon the earth, which have been completely wiped out, to be succeeded—not completely, but been almost completely wiped out by changes in the earth—to be succeeded by other faunas and floras—other groups of animals and plants reaching development and then in a large measure disappearing. We do not know how many times the different processes in connection with the change upon the earth's surface may have wiped out practically all or wholly certain faunas and floras, and on that account I don't think we are in a position to say when the earliest organisms appeared upon the earth. We do know that there was a very abundant fauna and flora as early as the Cambrin period.

Q—How long ago was that?

Does Not Talk in Years

A—Oh, that is an awfully hard question to answer in years. No geologist talks years—it is ages—and they are beginning now in such matters as the changes in the metals especially, the relations between uranium and lead—to get some idea of the numbers of millions of years that have passed since certain strata which contained fossils were formed, but I am not familiar with that field, I am not a chemist and I do not like to answer scientific questions outside of the field where I know a little of what I am talking about. I would have to be answering what I have heard from others, and I don't like to testify to that kind of stuff.

Q—More than 6,000 years ago, wasn't it?

600,000,000 Years Modest Guess.

A—Well, 600,000,000 years ago is a very modest guess.

Q—Well, just go on where I interrupted you.

A—Well, at the same time that this tremendous series of plants was developing from a lowly condition into the more or less elaborate condition which we now find, there was also developing alongside them a series of animal forms, or differences between the animals and plants, which caused their divergence in their evolution, being largely due to their different habits in connection with food. The plants standing still and letting the food come to them for the most part, while animals hustled and got their food, and that rather fundamental difference between animals and plants has led to the animals developing locomotor organs and grasping organs and other things which have led to still other things which the plants have not developed. The necessities of life have been different under the two food habits and they have been met by a different series of adaptation. Does that sufficiently answer for a sort of general outline of evolution as a fact, and not of the causes of evolution at all?

A—No, I don't.

Life Began on the Borderline

Q—No, I don't quite understand the causes—I might ask something about it. Could you tell us something about the order of plant life and animal life?

A—Well, it isn't quite so easy to tell about the order in which plants evolved with certainty, possibly, as it is to tell about some of the higher animals. There is a rather interesting index difficult to explain that tells us something about the different periods in the earth's history when different kinds of animals emerged from the sea and came into the land. I don't know that we have any similar record, any similar index for the plant, so the only thing we can say about the plants is that there is this series of complexities and that that corresponds to the record in the rock.

Q—Can you say where—I mean within a reasonable certainty—where animal life began, whether in the sea or on the land?

A—I think probably that animal life and plant life both began at the border line between the water and land where were conditions a little more complex—a little more likely to be productive of such a remarkable substance as a living substance but for long periods or over long periods in the earth's history there probably was no such thing as land life, either plant or animal, but all living things were marine.

Q—And what about the development of life in the sea—sea animals becoming land animals and land animals coming out of the sea?

A—The conditions of life in the seas are very simple and very easy for an organism which has this green coloring matter in it which we call chlorophyll and which is able on this account to absorb energy from the sun. You see green plants microscopic unicellular plants living in the ocean are both in a solution containing all of the mineral constituents which they need for their food and they also are exposed to sunlight whose energy they absorb by means of chlorophyll. It is, therefore, somewhat advantageous for them to remain small and unicellular and not to divide into cells and then keep those cells in groups because they cannot then do as well—be surrounded on all sides by their nutriment media and are exposed on all sides to its sources of energy the sunlight, but when terrestrial life began there were conditions of difficulty and in order to meet those conditions of difficulty it would be necessary in order to be successful to develop means adequate to meet those difficulties and the needs of such life have been the occasion—not the cause—have been the occasion for the development of the structures needed to meet conditions of existence there.

Q—Some animal life have gone from the earth to the sea, have they not?

A—Yes, some complex animals have gone back into the ocean, whales and the seal, and a great many of the water birds that spend a considerable portion of their life on the sea have gone back from the land. Of that we are entirely confident on abundant evidence.

Q—The whale (and I am diverting just a little because of some other matter that came up), the whale suckles its young, does it not?

A—Yes.

Q—And how is the whale classified?

A—The whale is a mammal.

Q—Will you give us the definition of mammal?

A—There again I hate to give definitions, but I can tell you some characteristics of mammals.

Q—All right.

Mammals Described

A—Mammals, all of them, have hair—either developed or rudimentary—on some part of their body. The possession of hair is a mammalian characteristic, hair not being known outside the group of mammals. The little hair-like feathers of birds are true feathers and not hair. They differ fundamentally in their structure from hair, and mammals also suckle their young. The mammals all have a vertebra column—a backbone; they all have two pairs of limbs unless they have secondarily lost those through adaptation to conditions of life. The fore limbs and hind limbs—those limbs always have a shoulder or hip girdle. The bone in the trunk attached to a linear series of bones, running out in the arm or leg, finally coming to a group of transversally arranged bones in the wrist or ankle, succeeded again by almost uniformly—I think—in the mammals except through degeneration—five digits and that is a rather—there are other series of characteristics, but the mammalian eyes have certain characteristics and different glands in connection with the body and I might, if I stopped to think up my lesson, tell you fifty points that are characteristic of the order of the mammals in distinction from other organisms.

Q—Now in the classification of the scientist-zoologist, where does man come?

A—He is classed among the primates. Man is not a very highly evolved animal in his body. He isn't as highly specialized as a great many organisms. His hand, for example, is a very generalized structure, nowhere near as much specialized as the hand of a bird, but he clearly belongs among the mammals. A group well up, I think, toward what we could call the well elaborated members of that group physically.

Q—You might tell us just what you mean by primate, for the benefit of us lawyers?

A—Well, I think because the group has been regarded as including man, the group has been given the primacy, I suppose that some of the insects, if they were sufficiently intelligent, might question that, but we do not question it.

The primates mean that order of organisms which include the lemurs, the tailed monkeys of this hemisphere, the tailless monkeys, the ape and baboon and so on of the eastern hemisphere and man and also quite a large number of forms of whose—of whom we have a satisfactory fossil record which we may class as apes or may class as men. It is a little hard to say, it is a little hard work to say over half a dozen or so forms about which there can be legitimate differences of opinion as to where they should be classified, whether as man or as ape.

The Court—Col. Darrow, will this extend very much further? It has been a pretty hard day for me.

Mr. Darrow—(After conferring with Gen. Stewart). I might ask three or four more questions for the benefit of counsel.

Q—Will you give us some of the evidences of the evolution of man from a lower organism?

Evidences of Evolution.

A—The great fundamental series, and I use that word in the plural—of evidences, and there are far more than one series—are found not in man himself, but in the whole organic world. The whole plan of evolution indicated so clearly throughout the whole realm of organic life paralleling as it does the whole plan of evolution seen so clearly in the universe as a whole makes a tremendous probability in favor of the evolution of man. When then we find just such differences among species and different varieties of men as we find among animals and when we find what we may fairly call the more lowly genera, species and varieties of human kind appearing earlier in the geological series just as do the simpler animals, among the lower forms appearing in the lower rocks, that inherent compulsion toward belief in evolution which is found in all of the universe is tremendously reinforced for man. The series is so convincing that I think it would be entirely impossible for any normal human being who was conversant with the phenomena to have even for a moment the least doubt even for the fact of evolution, but he might have tremendous doubt as to the truth of any hypothesis—as to the methods of the evolution which this or that or the other man—even great men of science—might bring up.

Q—And you say that evolution as you speak of it means including man?

A—Surely.

The Policeman—Now, folks, tomorrow we will continue this trial and there is not going to be anybody let in here only to be seated, not going to have any standing room at all, they can go on the outside where they can hear what is going on here right on the lawn.

The Court—We will adjourn until 9 o'clock tomorrow morning.

CHAPTER V.

FIFTH DAY'S PROCEEDINGS—THURSDAY, JULY 16, 1925.

Court met pursuant to adjournment.

Present as before.

Whereupon:

Court—(Raps for order.) Everybody stand up. Dr. Allen, whose name has been—who has been named by the pastors' association to open the court this morning.

Dr. Allen—(Dr. J. A. Allen, pastor, Glensley Avenue Church of Christ, Nashville, Tenn.)—"Our Father who art in Heaven, hallowed be Thy name. We thank Thee for thy blessings upon us all, and for Thy watch, care and protection over us; we pray Thy blessings upon the deliberations of this court, to the end that Thy Word may be vindicated, and that Thy truth may be spread in the earth. We pray Thee to bless and to guide all to Thy Name's honor and glory, to the accomplishment of good in the name of Jesus. Amen.

Court—Open court, Mr. Sheriff.

Bailiff—Oyez, oyez, this honorable circuit court is now open, pursuant to adjournment. Sit down.

Court—Are there any preliminary matters this morning?

Mr. Hays—If your honor please, we are prepared to make our motion on the admissibility of the evidence.

Mr. Darrow—Well, I wanted to ask one or two more questions.

Court—That's a big question. I thought, perhaps, there might be some preliminary matters to get out of the way.

Mr. Darrow—I want to ask just two or three more questions of Dr. Metcalf.

Court—Is any of the jury in the courtroom? If so, let them retire.

(Dr. Metcalf takes the witness stand.)

Questions by Mr. Darrow

Q—Doctor, will you please give us, rather briefly, any other evidence of evolution. The evolution of man.

Gen. Stewart—We want to confine this, so far as the record is concerned. This is done for the purpose of making a record for the supreme court if the defendant should appeal —in order that the defendant may have the benefit of this evidence. It is the insistence of the state that no theory of evolution is competent for the record, before the jury or anybody else, except that theory that teaches that man descended from a lower order of animals. This gentleman (Dr. Metcalf) said yesterday, in a very fair statement, that there were different theories, some true, some perhaps not true, and so forth, but to that particular theory, about which the act itself speaks we want this inquiry confined.

Court—Well, of course, this evidence is going in the record so that in the event the case goes up to the appellate court, they may see what the character and nature of the evidence was that was excluded, if it is excluded, from the jury, so I am inclined to let them get the full testimony of this witness in the record. Of course, I may put some limitations on the number of witnesses that go on the stand if I conclude this evidence is not admissible then I will let you proceed.

Gen. Stewart—Now, your honor, we prefer to proceed in the regular order.

Court—Yes.

Gen. Stewart—The jury was dismissed yesterday for the purpose of asking these questions.

Court—Yes.

Gen. Stewart—And in order that the court might ascertain if this testimony, in the mind of the court, was admissible. Now, your honor, must we spend the morning here—

Court—No, not the morning, I think.

Gen. Stewart—In determining this—whether or not the evidence is admissible. All this is supposed to do is to get before the court just what the evidence is and then, your honor will pass upon it. And if your honor holds it competent, the jury will be brought back, and this man will proceed to testify, and if it isn't competent—

Court—Let me see what the question was.

Gen. Stewart—I say if it isn't competent, now is that we ought to get at once to the issues and let the court pass on the proposition of whether or not it is admissible.

Court—Your plan is, if I was to exclude the evidence, you would want this witness back and have him re-examined.

Gen. Stewart—Yes, sir, that is our procedure, your honor, always as I understand it.

Court—There was a mix-up here by some kind of an agreement or suggestion yesterday.

Gen. Stewart—That was with the suggestion and understanding as I had it that Mr. Darrow would put before the court sufficient of this evidence to let the court and attorneys on the other side intelligibly understand just what he insisted upon.

Mr. Darrow—I don't think we need lose any time—if counsel says the understanding is if the court sustains this objection, we may call them back to prove what they would say. I was proceeding upon a different idea.

Court—I thought if I excluded the evidence, you would put this evidence in; then if I excluded the evidence it would all be in the record, and that would be final so far as this proof is concerned.

Gen. Stewart—No, I didn't so understand it.

Metcalf Called from Stand.

Court—Then, all right. You may stand aside, Dr. Metcalf.

Mr. Darrow—I am inclined to think that is the best way.

Court—We just didn't understand each other. That's all. I didn't know the witness was to be called back in the event the evidence is excluded.

Gen. Stewart—Will you gentlemen just state what you expect to show and let us make our exception?

Mr. Darrow—I don't think we need to do that because we have asked him questions and they are objected to, and after the court passes upon it and the court excludes it, then we will say what we expect to show.

Gen. Stewart—Well, knowing just what is before the court—(confers with Darrow in undertone).

Mr. Darrow—We expect to show by men of science and learning—both scientists and real scholars of the Bible—men who know what they are talking about—who have made some investigation—expect to show first what evolution is, and, secondly, that any interpretation of the Bible that intelligent men could possibly make is not in conflict with any story of creation, while the Bible, in many ways, is in conflict with every known science, and there isn't a human being on earth believes it literally. We expect to show that it isn't in conflict with the theory of evolution. We expect to show what evolution is, and the interpretation of the Bible that prevails with men of intelligence who have studied it. This is an evolutionist who has shown amply that he knows his subject and is competent to speak, and we insist that a jury cannot decide this important question which means the final battle ground between science and religion—according to our friend here—without knowing both what evolution is and the interpretation of the story of creation. And Mr. Hays is prepared with authorities on that subject.

Court—Now I have a great regard for the opinion of great lawyers, gentlemen, but I have—if I had an opinion of the courts of last resort, I have greater regard for them than I do the words of any lawyer on either side. That is my remarks for the record.

Mr. Hays—But I intend to support my argument with authorities, your honor.

Gen. Stewart—Of course, in this matter, the rules of procedure are the same as you made the other day, and the state has the opening and closing.

Mr. Hays—I am not sure of that. Not because your statement of procedure may be wrong, but you will perhaps remember that we had an agreement that we might make a motion to receive this scientific testimony, and I didn't understand—

Gen. Stewart—That agreement was withdrawn.

Mr. Hays—Let me finish, will you? We did not for a moment, suppose that you had any idea in your minds by changing the procedure you would have the opening and closing, thereby taking advantage of us in that way, and, therefore, we insist, as a matter of good faith, we should be permitted to argue this matter.

Lawyers Argue Over Agreement

Gen. Stewart—That, of course—the agreement was mutually withdrawn because we found—

Mr. Hays—Pardon me, we never withdrew the agreement.

Mr. Malone—I was a party to it, and it was not mutually withdrawn.

The Court—I won't stand for any discussion between you gentlemen addressing yourselves to each other. You must address yourselves to the court. Let me hear the attorney-general's statement, and then I will hear you.

Gen. Stewart—This was to be brought up in the regular and usual way by objection made when the witness went on the stand. Now out of perhaps being overzealous to accommodate these gentlemen, I said to them on last Friday, I would take this up out of order—that is, on Monday, we would discuss this proposition as to whether the evidence of these witnesses would be competent, and upon reflection I found that that could not be done, and Mr. Malone and Mr. Neal and myself agreed that that was right, and that the matter would simply come up in its regular order. Later in the day—an hour later—Mr. Neal came to me and said that other counsel did not agree to that and I told him I felt it was an agreement that should stand, but, regardless of that—it doesn't make any difference about an agreement—it is a matter of procedure, and the record can not properly be made up, except in this way—we cannot make up a moot record—we cannot require the judge to give us an advisory opinion in advance—

Mr. Hays—Before the attorney-general starts to make his argument I wish to be heard on the question of the stipulation.

Court—Are you through with your statement, general?

Gen. Stewart—I was just fixing to make a motion to exclude this evidence.

Court—Then, I will hear your motion.

Gen. Stewart—By the way, I want to reduce this to writing.

Court—Do you want to do it now?

Gen. Stewart—Well, we can file this at noon.

Mr. Malone—May I suggest, before you pass upon this motion, that you hear—

Court—Oh, I will hear you, Colonel, but I cannot hear more than one at a time.

Mr. Malone—I don't want you to hear more than one at a time, your honor.

Court—Well, I think—go ahead, judge.

State Moves to Exclude Evidence

Gen. Stewart—The state moves to exclude the testimony of the scientists by which the counsel for the defendant claim that they may be able to show that there is no conflict between science and religion, or in question, and the story of divine creation of man, on the grounds that under the wording of the act and interpretation of the act, which we insist interprets itself, this evidence would be entirely incompetent.

The act states that should be unlawful, that this theory that denies the divine story of creation, and to teach instead thereof that man descended from a lower order of animals, with that expression, and they have admitted that Mr. Scopes taught that man descended from a lower order of animals, the act under what we insist is a proper construction thereof, would preclude any evidence from any scientist, any expert, or any person, that there is no conflict between the story of divine creation, as taught in the Bible, and proof that a teacher tells his scholars that man descended from a lower order of animals.

The act says that they shall not teach that man descended from a lower order of animals according to our construction, and for these reasons this testimony would be incompetent.

In other words, the act does say that it shall be a violation of the law to teach such a theory, and, therefore, they cannot come in here and try to prove that what is the law is not the law. That would be the effect of it.

The Court—That is your motion, general?

Gen. Stewart—That is part of it, your honor.

The Court—Be careful not to get any argument into it.

Gen. Stewart—No, sir.

Another thing, your honor, is that this testimony undertakes to present to the jury the opinion of certain men who claim to be expert on this question of evolution, to give to the jury their opinion, when we insist that is the only issue now left to the jury to determine. There is no defense presented here or undertaken to be presented except by these scientific witnesses.

We have proved and have admitted yesterday—

The Court—Wait a minute, General, you are getting into argument.

Gen. Stewart—No, sir; I am not.

The Court—You say if you prove and they admit it would not be any part of your motion?

Gen. Stewart—Yes, sir; it is part of the motion, your honor, to show that there is no issue left except the issue as to whether or not this conflicts with the Bible.

The Court—I think you are making an argument.

Mr. Malone—I am sure he is, your honor.

Gen. Stewart—Now, then, we insist, if the court please, this is incompetent, because it invades the province of the court and the jury. It is not material to the issue here. It cannot be material to the issues. It is for the jury to say whether or not this conflicts, and that is an invasion of their rights, and of the right of the court. I think those are the true principal questions that I want to raise by this motion, that the act does prohibit it. And that under the rules of evidence it is an invasion of the province of the jury and the court.

Mr. Malone—Your honor, I would like to be heard very briefly, about the stipulation. There is no agreement between attorney-general, Dr. Neal and myself. It is question of fact.

Malone Reminds Court of Promise

Your honor will remember that for the convenience of our witnesses, and for the convenience of witnesses your honor agreed that this matter would be taken out of the usual order, and it was to have been heard on Monday last, and we worked over the week-end and were prepared to be heard on Monday last. But the vicissitudes of the trial interfered with it. On Sunday Gen. Stewart came to our house.

Gen. Stewart—On Saturday.

Mr. Darrow—He doesn't go on Sunday.

Mr. Malone—And saw—

Mr. Darrow—Wouldn't expect to find you there on Sunday, Mr. Malone?

Mr. Malone—No, I probably would not be there on Sunday, but I was at this time. Came out to see Judge Neal and myself, and for a personal reason stated that it would be better as a matter of public policy to revert to the original order. I didn't think it necessary and the General will probably not consider it necessary to state the reason in addition, but Judge Neal and I were sympathetic to his point of view, and then we went into the house—of course, we have other counsel, and the father of our house is Mr. Darrow.

Darrow Grandfather

Mr. Darrow—Grandfather.

Mr. Malone—And then we sat down and conferred on this matter. And it was determined that we should not go back to the regular order. But as we had considered the questions and given our time in the law library in the preparation of cases, briefs and citations, we should stand by the stipulation that was made.

Now, your honor, we were laboring under a delusion that when a stipulation was entered into in open court, in the presence of the court and the court thereafter set a time for hearing upon it, it was a binding stipulation. We afterwards found out it was the custom in Tennessee that a stipulation should be in writing. We had no idea that the stipulation should be in writing. When we found we had no legal rights, when the the prosecution decided to change its mind again, we did not insist upon the stipulation.

We wish to be fair and we wish to act as lawyers, when we are in Tennessee to act like the people of Tennessee, and when in Tennessee we are bound to know the theory of law, though not a question of fact, it is the theory of your state. We have the right, if we had argued on Monday, according to our ethical stipulation, to open and close this argument.

Gen. Stewart—What right? To open and close it?

Mr. Malone—It was our motion.

Gen. Stewart—For what?

Mr. Malone—Our motion that this evidence be heard. You objected to it.

Gen. Stewart—No, there was no objection. This was all just friendly conjecture.

Mr. Malone—After all, it is for the court to decide. We believe there is an ethical situation here. We have not insisted upon it, because we have been technically barred.

Gen. Stewart—I don't want them to feel that they have been technically barred. I feel this way about a matter of that sort. They don't need a stipulation of the court to hold me in line. A stipulation is a stipulation, with me wherever it is made. I think they should take the same position about it. Dr. Neal and Mr. Malone agreed—

Mr. Malone—We agreed there was merit in your contention.

Gen. Stewart—You agreed, and I came to your house to see you, and saw Dr. Neal.

Mr. Malone—What happened immediately after, when we went into conference? Dr. Neal went into—

Gen. Stewart—It seems to resolve itself into the question of who is in authority.

Mr. Malone—We know who is in authority. Mr. Darrow is in authority.

Gen. Stewart—You should have called him in conference when we went three miles out there to see you.

Mr. Malone—We came back three miles to tell you the truth.

(Laughter in the court room.)

Mr. Hays—General, we are visitors; why not let us go ahead?

The Court—As the court sees it, there is not much at issue.

Mr. Malone—Excepting the opening and closing, your honor.

The Court—It is immaterial, when you address this court, whether you open or close, no jury being present, the court seeking light and truth, and whether you speak in the beginning, or speak in the middle or at the back end, does not make any difference to me. I will hear you just as patiently and give what you say the same consideration.

I would not have counsel from foreign states feel that they have been taken advantage of. As I understood the stipulation a few days ago, for the convenience of counsel for the defendant, there was some negotiations that this question be raised without them bringing their witnesses to Tennessee, or some reason that was consummated. The court could only have acquiesced in it, no objection to it. Since it has been called off the court has no further concern as to that. This motion having been made by the state's counsel, to exclude this testimony, the court feels, under the rule of procedure in Tennessee, that the state is entitled to open and close. I do not see that that gives any advantage to either party myself, because I shall hear both sides alike.

To which ruling defendant duly excepts.

Gen. Stewart—Mr. Bryan, Jr., will present the opening.

The Court—I will hear you, Col. Bryan?

Bryan's Son Pleads Against Expert Testimony

Mr. Bryan—If the court please.

The attorney-general has requested me on this discussion to divide the time on the expert testimony. It is, I think, apparent to all that we have now reached the heart of this case, upon your honor's ruling, as to whether this expert testimony will be admitted largely determines the question of whether this trial from now on, will be an orderly effort to try the case upon the issues, raised by the indictment and by the plea or whether it will degenerate into a joint debate upon the merits or demerits of someone's views upon evolution.

Mr. Neal—We are very anxious to hear every word. Can you speak a little louder?

Mr. Bryan—This expert evidence is being offered for the avowed purpose of showing that the theory of evolution as understood by the witness, offering the testimony does not contradict the Biblical account of creation, as understood by the witness. All of which, the state contends, is wholly immaterial, incompetent and inadmissible for many reasons since the beginning of time, at least since the beginning of time, since we have had courts and juries and experts to testify, this particular class of testimony has been regarded of all testimony the weakest and most capable of abuse and the most dangerous.

No Way to Get Expert for Perjury

If a man testifies as to a fact his testimony may be met, or contradicted by other facts. If he testifies falsely, he can be punished for perjury. But if a man gives a false opinion there is no way that you can contradict him. There is no way he can be punished. There has scarcely been a trial in recent years where the material issues have been testified to by experts, but that the public has again been convinced of the utter futility of that testimony.

The Court—Mr. Bryan, I am sure everyone is anxious to hear every word you say. Will you speak a little louder?

Mr. Bryan—I will try to speak a little louder, yes. I have heard a good many harsh things, said about experts. I believe it was my good friend, Mr. Darrow, who, in the Loeb trial characterized one of the experts there used, as a purveyor of perjury. He was probably justified in so characterizing him. But it is a fact, I have not been able in the examination of the books to find any statement as strong as that—but it is a fact, that the courts have unfavorably regarded this sort of evidence, and received it with extreme caution, and investigated it with every care. Our courts have held that the testimony of expert witnesses should be received with caution and investigated with every care.

This rule is stated in Jones on Evidence, and in every work of authority upon evdence. In Volume II, page 374, it is well-stated as follows:

(Reading beginning with the words, "It is the general disposition of the courts to restrict the admission of expert testimony within the strict bonds" to "is desired.")

And the same authority goes on to quote from remarks of Justice Early in the case of Ferguson vs. Hubbell, 97 N. Y., 507, which refers to the famous Tarduie case and early English cases upon this particular subject. Early said as follows:

"The rules admitting the opinions of experts should not be unnecessarily extended. Experience has shown that it is much safer to confine the testimony of witnesses to the facts in all cases where that is practicable, and to leave the jury to exercise their judgment and their experience upon the facts, proved. Where witnesses testify to facts they may be specially contradicted. If they testify falsely they are liable to punishment for perjury, but they may give false opinions without fear of punishment. It is generally safer to take the judgment of unskilled jurors than the opinions of hired and generally biased experts."

Now, this rule has been repeatedly recognized and followed by the courts in this state. In the case of Wilcox vs. State, 94, Tenn., at 112, your own supreme court speaking in regard to this subject, held that it was no error to charge the jury as follows:

Expert Testimony Field of Speculation

"While expert testimony is sometimes the only means of, or the best way to reach the truth, yet it is largely a field of speculation besought with pitfalls and uncertainties, and requires patient and intelligent investigation to reach the truth."

The same rule is stated and followed in Persons vs. State, 6 Pickle, 291, and Adkins vs. State, 119 Tenn., at 458. The following quotations, if the court please, on this point, are taken from corpus juris, Volume 22, page 498, and following, and are merely the expressions of opinions that have received such widespread recognition and have been followed and cited until they have become axioms of the law.

Experts Endanger Case.

"The danger involved in receiving the opinion of the witness is that the jury may substitute such an opinion for their own. But courts will not require the parties to encounter this danger unless necessity therefore appears. The jury should not be influenced by the opinion of anyone who is not any more competent to form one than they themselves are. The verdict should express the jurors' own independent conclusions from the facts and circumstances in evidence, and not be the echo of witnesses, perhaps not unbiased."

Of course, if the court please, I do not mean to argue that there are not cases where it is absolutely necessary to have opinions of experts, where the matters in issue are of such a technical or involved nature that expert opinion is the only way by which the truth as to facts may be arrived at. These exceptions will, no doubt, be fully argued to you by the counsel for the defense. However, that may be, the courts are unanimous in adhering to the rule that expert testimony can be introduced only under the stress of necessity. In other words, the court will seek the aid of opinion evidence only where the issues involved or facts are of such a complex nature that the man of ordinary understanding is not competent or qualified to form an opinion, but, if the court please, even this exception is limited by the rule of law to which I shall refer later in my argument, that prohibits in any event the introduction of expert testimony upon the very facts that the jury are to pass upon.

The first test that the court should apply to determine whether expert testimony is admissible in any event, is, whether the facts relevant to the issues are such that they can be introduced into evidence, and whether the jury are competent to draw a reasonable inference therefrom—not necessarily the inference that the court would draw, or that I would draw, or that the expert would draw; and are they competent to draw a reasonable inference of their own. It is the rule supported by the weight of authority, I think, in almost every state of this Union that where all relevant facts can be introduced in evidence and the jury are competent to draw their reasonable inferences, therefrom, that opinion evidence may not be received. This is the law in the state of Tennessee.

In the case of Cumberland Telephone and Telegraph company vs. Dooley, 110 Tenn., page 109, it was sought to introduce the opinion of a witness as to whether or not a fire could have been stopped and controlled with the apparatus then and there at hand; and, it was held that such evidence was not properly a subject of expert opinion, inasmuch as every fact constituting an element of the opinion of such witnesses was capable of being presented to the jury.

Again, in the case of Nashville & Chattanooga Railway vs. J. N. Carroll, 43 Tenn., 368, it was urged that the court erred in refusing to allow an expert to testify what was meant by an obstruction. It was a railroad accident case, and one of the allegations was that the railroad had permitted an obstruction to remain upon the tracks, thus causing the wreck. The obstruction being a hand-car, I believe. The court held that there was no error in excluding the evidence of an expert as to what constituted an obstruction saying: "What is or is not an obstruction, is a simple question of fact which could be determined by the jury as well as the expert."

Now, what are the issues in this case, if the court please? The indictment simply charges that John Scopes taught, in violation of law, that man has descended from a lower order of animals, and the state has offered evidence tending to prove that he did so teach. As a matter of fact, this evidence has not been controverted by the defendant. There is no issue of fact raised by evidence, the facts are agreed upon both sides. Under this state of evidence, if the court please, if this were a civil case instead of a criminal case, your honor would be compelled to take the case from the jury and find for the plaintiff. What issue of fact is there left for the experts to express an opinion upon? There is no issue of fact upon which expert testimony is either proper or necessary. The only question in this case is, whether or not the jury believes that the admitted facts show a violation of the law, and this, I submit, is one of those mixed questions of law and fact to be determined by the jury under the proper instruction of the court, and can never be a proper subject of expert testimony.

And now, if the court please, I come to the limitation I adverted to a moment ago; and that is, that opinion evidence may not, under any circumstances be received to determine the fact in issue; in other words, to invade the province of the jury. The rule is stated in 22 Corpus Juris, 502, and the hundreds of citations supporting it as follows:

"As the opinion evidence rule against admissability is to provide against the mischief of the invasion of the province of the jury, a court should exclude the inference, conclusion or judgment of a witness as to the ultimate fact in issue, and this is true, even though the circumstances presented are such as might warrant a relaxation of excluding the opinion, but for this one circumstance."

In other words, it matters not how technical the subject, how involved the issue may be, there is one place where expert testimony may never, in any event, be received; and that is where it is upon the very issue that the jury is to determine, and that is the situation in this case, if the court please. This has always been the law in Tennessee, as well as other states.

The Court—What case do you read from?

Mr. Bryan—I will read from the case of Bruce vs. Beall, 99 Tennessee, 313. This was, if I remember rightly, a case for personal injuries received in the fall of an elevator, and one of the questions at issue was whether the defendant had been negligent in permitting the cables to be used for a certain period of time, and the court excluded certain questions asked the expert as to whether or not the use for that length of time was safe or not. The court used this language:

"While the general rule is that witnesses must speak the facts, yet, upon questions of skill and science, experts are competent to give their opinions in evidence, but they will not be permitted to state their opinion upon any point the jury has to decide. Deductions from facts belong to the jury, and when the examination extends so far as to substitute the opinion of the witness upon the very issue in controversy, for that of the jury, the province of that tribunal is unwarrantedly invaded. We think it is clear that in no case can the witness be allowed to give an opinion upon the very issue involved, a danger from this would be to substitute the opinion of the experts for that of the jury themselves, whose duty it is to find the facts and whose verdict in only an expression of their deductions from the facts."

This case also cites the case of Gibson vs. Gibson in 9, Yeager, 329, which is one of the early cases, and which is to the same effect.

And again, in the case of Cumberland Telephone and Telegraph Company vs. Mill Company, 109 Tennessee, 381, the court said it is an accepted rule that while experts may testify as to what, in their opinion, may or may not have been the cause of a given result or condition, it is not permissible for them to give their opinion as to the only fact that the jury was organized to determine, the question now under consideration required the witness to enter the domain of the jury and to pass upon one of the ultimate propositions inhering in the verdict.

Precedents Are Cited.

Now, this same position, if the court please, has been followed in the case of Cumberland Telephone and Telegraph Company vs. Mill Company—the one I have just cited, in Railroad Company vs. Brangee, which is a strong case, 114 Tennessee, 35, and in Kirkpatrick vs. Kirkpatrick, 1 Tennessee Cases, at 257; Owen vs. Jackson, 1 Appealed Cases, 413, where the court stated:

"Upon the facts to be determined by the jury no witness, expert or nonexpert, should be asked his conclusion upon any material fact that is to be passed upon by the jury."

In the case of Memphis Street Railway vs. Hicks, 1 Cates, File 13, it is said: "It is not permissable to ask a witness, expert or otherwise, his opinion upon issues which are to be determined by the jury. It is proper to propound to a witness a question that calls for an expression of opinion as to any point that the jury will, of necessity, have to determine."

Now, if the court please, as the state sees this case, the only issue this jury has to pass upon is whether or not what John Scopes taught is a violation of the law. That is the issue, and it is the only issue that the jury is to pass upon, and we maintain that this cannot be the subject of expert testimony. To permit an expert to testify upon this issue would be to substantiate trial by experts for trial by jury, and to announce to the world your honor's belief that this jury is too stupid to determine a simple question of fact. Admission of this testimony would be followed and, in our opinion, it would be reversible error. I, therefore, respectfully urge your honor to sustain the objection of the state to the introduction of this testimony.

The Court—Be at ease for two or three minutes.

(After a recess of fifteen minutes the hearing of this case was resumed.)

The Court—We have some lawyers in the case who, at times, indulge in a lot of wit. I do not know who is going to argue the case, and I do not know whether they are going to display their wit or not, but if they do, I don't want any manifestations in the courtroom, for two reasons: The first reason is that it is improper; the second reason is that this floor of the courthouse building is heavily burdened with weight. I do not want to alarm you; I do not know myself, for I am not a mechanic, but I do know that the floor is heavily weighted and the least vibration might cause something to happen, and applause might start trouble.

Mr. Hays—If your honor please, I am rather embarrassed by your allusion that there will be such thunderous applause that the building might come down.

The Court—I believe the other vibrations won't cause it. I will say to you lawyers, gentlemen, that this is, of course a big question. I don't want any lawyer to feel that he has to be in a hurry. Take your time. Of course, I do not want you to occupy unreasonable time, but I want the information.

Hays is Astounded.

Mr. Hays—If your honor please, I am learning every day more about the procedure in the state of Tennessee. First, our opponents object to the jury hearing the law; now, they are objecting to the jury hearing the facts. The jury is to pass on questions that are agitated not only in this country, but, I dare say, in the whole world. There is one proposition made by the opposition, which I believe is unusual; that is, the insistence by the prosecution of trying the case for the defense; for they are continually telling your honor their theory in this case. And, when we have tried to present our theory of this case, they have objected. The learned attorney-general started his argument this morning by saying, we admit Mr. Scopes taught something contrary to the law, while we admit that Mr. Scopes taught what the witnesses said that he did, but as to whether that is contrary to the theory of the Bible should be a matter of evidence. Possibly the prosecution are without evidence. There are other rather unusual propositions of law I have heard this morning and I think they are based on possible differences in fact. One thing appeals to me in this case; that is, that my mind is so constituted that while I concede all the law the other side presents, I cannot see how it is in point. I concede anything Mr. Bryan said on that subject, yet it does not bear on the questions before us. Certainly no court has ever held it to be dangerous to admit the opinions of scientific men in testimony. Jurors cannot pass upon debatable scientific questions without hearing the facts from men who know. Is there anything in Anglo-Saxon law that insists that the determination of either court or jury must be made in ignorance? Somebody once said that God has bountifully provided expert witnesses on both sides of every case. But, in this case, I believe all our expert witnesses, all the scientists in the country are only on one side of the question; and they are not here, your honor, to give opinions; they are here to state facts. For instance, in Mr. Bryan's Tennessee case, where it was concluded that an expert could not give an opinion as to whether the fall of an elevator was caused by negligence. Of course, he could not. Even I, coming from New York, would know that. But an expert could state the facts with reference to the control of a hydraulic elevator. On that point, the expert did not give only opinion evidence. Experts state facts, but, of course, so far as the weight of their authority is concerned, we want to point to your honor that not a single expert in this case is a paid expert, and every scientist who comes here comes in the interest of science, with no promise of compensation. Which leads me to be sure we can warrant theirs being impartial testimony.

With respect to the remark made by Gen. McKenzie the other day, when he said that any Tennessee school boy of 16 should understand this law, I wish to say, that if that is so, they forget it by the time they get to the age of Atty.-Gen. Stewart, and do not again acquire it by the time they reach the charming age of Gen. McKenzie.

Now, as to evolution, does your honor know what evolution is? Does anybody know? The title of the act refers to evolution in the schools, but when that is done, you do not know what evolution is. I suppose ultimately, the jury, because under your constitution they are the judges, ultimately, of the law as well as the facts, and they will have to pass on the evidence, and that is a question that has been observed by scientific men for at least two centuries.

I have in my hand a part of a proof of the book by Dr. Newman, whom your honor, I hope, will have an opportunity to hear. I hope your honor will not give up the opportunity to hear him.

Two Darwinisms.

Dr. Newman says:

"The secret of the difficulty lies in the fact that there are two Darwinisms, the popular one and the technic one. The layman uses the term Darwinism as a synonym of evolution in the broadest sense; the evolutionist never uses the word in this sense, but always uses it as a synonym for natural selection, one of Darwin's chief theories. The general principle of evolution has nothing to do with natural selection. The latter might be totally discredited without in the least shaking the validity of the principle. But this situation is not at all understood by the antievolutionists, who believe that Darwinism (the principle of evolution) is inextricably bound up with Darwinism (the theory of natural selection).

Well, there is a short statement, but of course, it is a comprehensive statement and your honor would want the facts to show how experts—how the scientists came to their opinion, and if your honor says that opinion evidence cannot be introduced, at least evidence of the facts may be introduced, so you gentlemen can determine the facts, and then draw your opinion as to what this statute means. Any boy of sixteen can understand this law, you say, why any boy of sixteen, without special study doesn't even understand the term "lower order of animals" and neither does the prosecution. Their theory seemed to be at the beginning that Prof. Scopes taught and that evolution teaches that man has descended from a monkey. If Prof. Scopes taught that, he would not be violating this law. Now, you will need evidence to prove that that is a fact, because the orders of animals were classified by Linnaeus about 200 years ago, which was an artificial classification. In the first order—the primate order, was man, monkeys, apes and lemurs. That is the first order. To prove that man was descended from a monkey would not prove that man was descended from a lower order of animals, because they are all in the same order of animals—the first order—and that is the use of the term "order of animals" by zoologists and I suppose we have got to interpret this term according to its usual use and so even if Prof. Scopes taught what the prosecution thinks, even then according to our theory, they would not prove that Scopes taught that man descended from a lower order of animals. They might say that man came from a different genus but not a lower order of animals. Perhaps that is new to you, gentlemen, and I confess it was new to me and yet these men had the audacity to come into court and ask the court to pass upon these questions without offering any evidence. What are the questions of fact in this case? Before I get to that I should like to read to your honor this quotation from 22 Corpus Juris, page 165. I don't think I need cite the authorities, because it is almost hornbook law.

The Court—Will you furnish me the memorandum?

Mr. Hays—Yes, sir.

Mr. Hays (Reading)—"It is no objection to the admissibility to a party's testimony that is competent only on his theory of the case; he has a right to have the case submitted to the jury on his theory if there is any testimony to support it."

Hays Says Prosecution Wants One Side Only

When these gentlemen tell your honor what their theory of the case is, and then say, "the defense should put in no evidence because this is our theory" they immediately suggest to your honor that you should hear one side of the case only. Your honor may know of the occasion some time ago when a man argued a question for the plaintiff before a judge who had a very Irish wit and after he had finished the judge turned to the defendant and said, "I don't care to hear anything from the defendant, to hear both sides has a tendency to confuse the court" (Laughter in the courtroom). These people cannot bind us by their theory of what our case is. Now then we start at the beginning with a very simple proposition of evidence.

The Court—Have you a paperweight there?

Mr. Hays—I have lots of them, your honor. Where did they get the idea that in a court of law evidence is not admissible to elucidate and explain what it is about? Is the court and jury to pass on a question without knowing what these questions involve, particularly when they are scientific questions? Apparently the gentlemen of Tennessee believe that testimony in a law court has only to do with direct evidence—that nothing is relevant that is indirect and introduced for the purpose of explanation or elucidation. Of course, your honor knows that isn't the law—that under the law anything is relevant that tends to throw light on the subject and particularly in a case like this, where such great elucidation is necessary. What are the questions of fact? A man is quilty of a violation of the law if he teaches any theory different from the theory taught in the Bible. Has the judge a right to know what the Bible is? Does that law say that anything is contrary to the Bible that does not interpret the Bible literally—every word interpreted literally? Oh, no, the law says that he must teach a theory that denies the story as stated in the Bible. Are we able to say what is stated in the Bible? Or is it a matter of words interpreted literally? Is your honor going to put into that statute any theory contrary to creation as stated in the Bible with the words "literally interpreted word by word" because if you are the statute doesn't say so. Are we entitled to show what the Bible is? Are we entitled to show its meaning? Are we entitled to show what evolution is?

Entitled to Show What Evolution Is

We are entitled to show that, if for no other reason than to determine whether the title is germane to the act. Are we entitled to show that the development of man from a cell does not make him a lower order of animals? I know that every human being develops from a cell in the very beginning of life. I know that in the womb of the mother the very first thing is a cell and that cell grows and it subdivides and it grows into a human being and a human being is born. Does that statement, as the boy stated on the stand, that he was taught that man comes from a cell—is that a theory that man descended from a lower order of animals? I don't know and I dare say your honor has some doubt about it. Are we entitled to find out whether it is or not in presenting this case to the jury? Further than that, how well substantiated is the doctrine of evolution? I presented your honor in opening this case, with what I conceive to be a parallel statute and a great many people smiled. You remember my supposed statute concerning the Copernican theory and my friend, the attorney-general proposed another statute concerning the rights of teachers. I would like to say the only difference between the attorney-general and myself is that I believe such statutes are unconstitutional—I believe his was unconstitutional, as well as my own and this. The only difference between the parallel I proposed and the law we are discussing, humorous as my parallel may have been—is that the Copernican theory is accepted by everybody today—we know the earth and the planets revolve about the sun. Now, I claim, and it is the contention of the defense these things we are showing are just as legitimate facts, just as well substantiated as the Copernican theory and if that is so, your honor, then we say at the very beginning that this law is an unreasonable restraint on the liberty of the citizens and is not within the police power of the state. Apparently, my opponents have the idea that just as long as the question is one of law for the court, then no evidence is required. There was never anything further from the truth. They had apparently the idea that the court takes judicial knowledge of a subject, such as matters of science, and that then no evidence need be introduced. If your honor is interested in my personal opinion I should like to say if on no other ground even though your honor thinks these are questions of law and even if the court believes that the court takes judicial knowledge—if on no other ground, this testimony would be admissible, in order to inform the court, because the court must be informed as to what the issues are and what these things mean. In Jones' Commentary on Evidence, Vol. 1, page 26—and your honor will realize that this is no reflection on the court—that the author said:

"Courts should observe the utmost caution to avoid assuming knowledge of natural facts and laws that are beyond the scope of common, positive knowledge."

And, in Dumphrey vs. St. Joseph Stock Yards company, 118 Mo., App., 506, the court said:

"The mysteries of nature are so manifold, deep and subtle, that the finite man cannot indulge in dogmatic conclusions affecting them without falling into error. Human nature being microcosmic, is not certainly known save in its prominent outlines."

Jones says further:

"It goes without saying that every judge upon the bench would disclaim such an encyclopedic knowledge added to a phenomenal memory, as would serve him on every application that the court should take judicial cognizance of a given fact. However wide his reading the suggestions frequently make a demand upon him, to which, without some means of reference or refreshing his knowledge, he might not be able to respond."

Points Out Duty of Judge

And further:

"The judge has no right to act upon his personal or special knowledge of facts as distinct from that general knowledge which might properly be important to other persons of importance."

Your honor well knows that there are occasions on which a judge takes what is called perhaps unfortunately judicial knowledge, because they are presumed not to be ignorant of what everybody else knows. I take that statement from Commonwealth vs. Peckham, 2 Gray, Mass., 514.

When we come to the proposition of judicial notice the taking of judicial notice has always favored a party litigant. A court is never bound to take judicial notice except possibly the laws of the statutes. If a matter is not of such common knowledge as to be known by everybody, a court may take judicial cognizance, in which case evidence must be introduced to inform the court, but doesn't take judicial notice in the sense that no evidence is required. Do I make myself clear?

The Court—You might just review your statement.

Mr. Hays—I like the term judicial cognizance better than the term judicial notice, because the court even takes judicial cognizance of facts of which it has no actual judicial knowledge at all. If the court takes judicial cognizance of matters, since the court is merely human and the bounds of knowledge are limited somewhat, the court must take testimony and evidence on facts which are not matters of common knowledge in order to inform itself, because there is nothing more important than that the court should not fall into error on questions of fact as well as of law. Perhaps this statement makes it clearer and this is supported by any number of federal cases and I think it is such sound law that my opponents won't require any further elucidation.

"The court is not bound to receive evidence as to a matter of which it takes judicial notice, but it is, of course, bound to notice facts merely as the facts as to those matters of law upon which an issue of fact cannot be made."

The Court—Such as matters of common knowledge?

Court Bound to Notice Facts

Mr. Hays—Yes, sir. But it is, of course, bound to notice facts correctly. In other words if your honor does take judicial cognizance you are bound to notice the facts correctly. It is not prejudicial error to receive evidence in such cases and even as to these matters the court may seek information—that is as to common ordinary matters it has been held will require the production of evidence. If your honor says, "I will take judicial notice of all science, I will take judicial notice of evolution in the field of geology, zoology, embryology and everything else, "you would be doing us a great favor, but we assume that the court won't take that position, but even if it is a question of law and involves only a question of judicial knowledge, your honor must receive the evidence and I take it if your honor does receive the evidence—this being a criminal case—the evidence must be given in the presence of the jury. This author says "Proof may be required of facts of which the court entertains doubt, even though they are subject for judicial notice. Especially may this be so when to the court's doubt is added denial of such facts." And in connection with that cites Marshall vs. Middleborough and Commonwealth vs. King, 150 Mass., 221. I am stressing this point not because I have any doubt that there are questions of fact, but because if your honor should confine us to the narrow ground in your judgment as to whether the evidence should be required, yet we are entitled to put in this evidence and it would be error to refuse to receive it. May I read that again, your honor? (Reading.)

"Proof may be required of fact of which the court entertains doubt, even though they are proper subjects for judicial notice. Especially may this be so when to the court's doubt is added denial of such facts."

Now there is another very interesting phase of this situation, which shows the necessity for evidence. The state here prosecutes Scopes—it is a crime as I understand it not to use school books prescribed by the state and to use a school book as Prof. Scopes used it, is also a crime. I assume that the state of Tennessee did not intend to make it a crime if the teacher used it and likewise make it a crime if the teacher didn't use it. I cannot imagine two laws, one of which compels a man to do a thing and another which makes it a misdemeanor for him to do it.

The Court—Let's see if I understand the proof a while ago on that, Mr. Hays. I understood Prof. White to say that the contract whereby it was provided that this Hunter's Biology was to be used, expired in August, 1924.

Mr. Hays—I understand that, but I understood until a new textbook was prescribed the state used the same book, but there will be further evidence on that subject. Of course he did not undertake to testify what the law was. I was merely using this for the purpose of illustration.

Crime Either Way.

If your new law intended to amend the old one it would have said so. I say as I look at it there are two laws in this state, one of which compels a teacher to use the book and the other of which makes it a crime for him to use the book. I don't think the Tennessee legislature meant by their statute to say something quite different from what was taught in the book, because in the meaning of the term, what is evolution, what is stated in the Bible, is a matter that requires evidence. If your state of Tennessee intended to make it a crime to teach things in that book at the same time compelling the teacher to use that book, well, it has done something I believe no other state in the Union has ever done since the Union was founded and I don't think the state has done it, and I think the reason why those two statutes can be reconciled will come out in the evidence. When you gentlemen find out what evolution is we think you are compelled to take our theory because of those two laws which are diametrically opposed, unless you say which is evidence and find out what these facts are.

Now, your honor, one thing has rather surprised me about this motion on the evidence. I believe that when we all were lawyers and none of us were advocates that we all agreed upon this proposition. I refer, of course, to our opponents as well as, I may say, to your honor, we all agree upon the proposition that evidence was admissible. Mr. Bryan—I should not, perhaps, mention the name—the distinguished leader of the prosecution—

Court and Hays Argue Over Bryan's Name

The Court—There is no reason why you should not mention counsel's name.

Mr. Hays—Mr. Bryan, the distinguished leader of the prosecution.

The Court—Do you mean young Bryan?

Mr. Hays—Mr. W. J. Bryan.

The Court—He has not appeared as counsel, yet?

Mr. Hays—What?

The Court—When I say that, he is counsel, but I mean that he has not made any argument.

Mr. Hays—May I put it this way: That the prosecution gave us to understand before we came down here—

Mr. Darrow—Is his appearance entered?

Mr. Malone—Is Mr. W. J. Bryan's name entered in this court as counsel on that side?

The Court—I just stated that he appears as counsel, but he has made no argument and I thought the lawyer was referring to something he said. Of course something he said on the outside, you should not refer to. But any reference you make to young Mr. Bryan, who has made an argument is an entirely different thing.

Mr. William Jennings Bryan—Your honor, may we not, as well in the beginning, recognize that however much interested the attorneys for the defense are in making me this case, they ought to recognize the attorney-general is in charge of this case, and they ought to recognize this, about which they speak so honestly and knowingly, when it comes to this fact, that the attorney-general is in charge of the case, and I am associate counsel.

Gen. Stewart—As a matter of personal privilege, your honor, I will state that in law, the attorney-general has charge, but in the presence of such a distinguished person as Mr. Bryan, that lawyers bear him respect.

Mr. Hays—May I say this?

The Court—You may proceed.

Mr. Hays—On this point, on the admission of evidence, I should be justified in stating the opinion of anybody and your honor would accept it according to its legal worth.

I assume that if I state the opinion of one of the counsel for the prosecution I am stating the opinion of a lawyer which your honor will recognize for what it is worth.

Should Obey Golden Rule.

I am stating the opinion of a lawyer made when he was merely a lawyer and not an advocate. Of course, we men in New York, when we read the opinion of this distinguished lawyer to the effect that this was a duel to the death, to the effect that this case was a duel to the death without evidence, was evidence to be given? We relied then upon the opinion of that distinguished lawyer and we have spent thousands of dollars bringing witnesses here. And I have heard that men, even though charged with more religion than I am, ordinarily obey the golden rule and there is a proposition of ethics in that.

But, wholly aside from that, I assume that was his opinion as a lawyer when he was not an advocate.

Now, your honor, you have heard the opinion of the defense as lawyers. And finally I shall refer to the opinion as a lawyer of one who plays a far more important part in this case.

Your honor said, before this matter came up, that the only difference—this statement was made, and if the statement is incorrect, your honor will correct me. I am reading:

"The only difference between the attitude of Judge Raulston and those of either side is that he calls the case an investigation."

"A judge should begin all investigations with an open mind and should never hastily and rashly rush to conclusions.

"So long as there is any question of either law or fact in doubt he should diligently inquire for the truth."

I am quoting that and I think it is sound.

Certainly, if your honor determined this case is an investigation it was because your honor had in mind, it could mean nothing else, when speaking as a lawyer, that you would require evidence, on these facts.

You said on one occasion, that the case would warrant one of three decisions: First, one of not guilty; second, that the defendant taught evolution and, third, that the law was unconstitutional. Either that the law was unconstitutional, but that there was nothing in the subject of evolution when the subject was properly understood, to break down religious faith. Can we take that position, your honor, without showing what evolution is, without showing what the subject is?

Doesn't that require evidence?

And, finally, with your honor so ably stating the duties of a judge, that a judge should begin all investigations with an open mind, and never hastily or rashly rush to a conclusion, so long as there is any question of either law or fact in doubt he should diligently inquire for the truth.

When your honor said that, had you any doubt, as a lawyer, that in this investigation you wanted to hear the facts and the law to the fullest extent?

Who is afraid of the statement of facts? Or do our friends on the prosecution feel that our scientists merely state opinions, and give no evidence of facts? But if this is to be an investigation facts are necessary. If this is to be an investigation, your honor, as a lawyer, knows it is necessary to properly introduce that evidence.

It may be your view was made up from the fact that the court has a right to inform himself. It may be your view is narrower than mine? Or your honor's duties, as the court, to inform the court, but if you, as a lawyer, had a mind that this evidence was admissible, there is no doubt whatever, and shall take it not only as a lawyer, but also as a judge, because yesterday your honor stated that the caption of the act was germane to the body. "In my conception of the terms employed in the caption of the body."

Judge Should Have Open Mind

That was your conception before you heard the evidence. Now, the evidence is to be produced, and I assume that, when later we make a motion to dismiss, or a motion in arrest of judgment, and argue again, your honor will take it up and hear us with an open mind. Am I right about that?

The Court—Oh, yes.

Mr. Hays—That your honor's position would be the same unless you permitted the introduction of evidence.

Now, then, I assume when all of us were lawyers and not advocates, we agreed that the evidence was admissible.

Your honor, this is a serious thing. It is an important case. The eyes of the country, in fact of the world, are upon you here. This is not a case where the sole fact at issue is whether or not Mr. Scopes taught Howard Morgan that life was evolved from a single cell.

The Court—We will take a few minutes recess.

Whereupon a few minutes were taken. After which the following proceedings were had:

(Following recess.)

The Court—I will hear you, Mr. Malone.

Mr. Bryan—No, Mr. Malone is entitled to speak after Mr. Hicks and Gen. McKenzie.

The Court—Oh, I see.

Mr. Bryan—They are only to have two arguments, we want to use two more.

Mr. Hicks—If your honor, please, in this case, as we understand, they will only have one more argument for the defense, I think it would be proper that the general go ahead and present his arguments at this time, and leave me out.

The Court—No, I will hear you all.

Mr. Hicks—If your honor please—

The Court—Come around.

Mr. Darrow—We want to hear you.

Mr. Malone—You are the best looking man on that side.

Mr. Hicks—If your honor pleases, it is now insisted by the defense that they have the right to inject into this lawsuit a large number of theologians and scientists from different parts of the United States, who will come in here and testify that science and the Bible are not in conflict, that the subject that was taught by J. T. Scopes does not conflict with the Bible.

Now, in regard to the gentlemen for the defense; they have put me in the position which I have experienced as a gun pointer in the navy trying to fire upon a submarine. You will see the periscope at one place, and it will go down and in another moment it will be here, and in another moment it will be there. Mr. Hays has said that these experts are paying their own expenses to come here to testify in this case.

The Court—I am not interested in that, Mr. Hicks, at all. I do not care whether they are or not.

Mr. Hicks—If your honor please, they admit that those experts who are coming here are greatly interested in this trial, in the outcome of this trial, and I just want to call your honor's attention to the fact that this is the position that they are in, and to the regard which the higher courts of the state of Tennessee take in regard to the admission of expert testimony in any case. Our higher courts have said that it is largely a field of speculation, and that it is full of pitfalls, that it is full of danger, and must be received with great caution.

Now, in every other case which has been called to the minds of the courts of Tennessee, how much more so must it be in the case at bar, because the theory of evolution itself is unproven and such an eminent scientist as Bateson accepts evolution because he cannot find any better theory to advocate as to the creation of animal life upon earth.

Mr. Darrow—When did he state that?

Mr. Hicks—In his speech at Toronto.

Mr. Darrow—Oh, no, we have that speech.

Mr. Hicks—It was something to that effect.

Mr. Malone—Oh, well, something to the effect.

The Court—Address any objection you have to the court, gentlemen.

Mr. Hicks—That is all right, I don't care. If your honor please, the words of the statute itself preclude the introduction of such testimony as they are trying to bring into the case. I call your honor's attention to the last clause of this act, they are very careful to admit that—they are very careful to leave out even any mention of Section 1, and this law reads: "Be it enacted by the general assembly of the state of Tennessee that it shall be unlawful for any teacher in any of the universities, normals or other public schools of the state, which are supported in whole or in part by the public school funds of the state, to teach any theory that denies the story of the divine creation of man as taught in the Bible, and to teach instead"—instead of what?—"instead of the story of divine creation as taught in the Bible that man has descended from a lower order of animals."

Now, this proof is amply shown, that Mr. Scopes taught that man descended from a lower order of animals—

The Court—Do you think that that meets the requirements of the statute?

Mr. Hicks—Absolutely. There is no question as to that, your honor. In other words, instead of the Bible theory of creation, he taught that man descended from the lower order of animals. Now, on the construction of any statute, our courts hold this, that if one clause of that statute, one part of it is vague, not definitely understood, that you must construe the whole statute together, that you must look at the other part of that statute and see what is the character, what is the intention which our legislature intended to put into that act. Now, that the last part defines that first part. It says what this evolution, or law is, to teach instead—instead of what?—instead of the Bible story of creation, that man has descended from a lower order of animals.

Know What We Want.

Now, in regard to that very feature of it, your honor, I would like to review just a little Tennessee law—down here, in Tennessee, we believe in Tennessee law, and when our leading courts, our courts of last resort, pass upon a question, we do not think you need to go outside of Tennessee to find law, when it is upon the very issues involved in the case, in regard to the construction of statutes. I would like to read from 142 Tennessee, ex rel Thomason vs. Temple, it says:

"A few elemental rules in the construction of statutes support our conclusions.

"A statute is to be construed so as to give effect and meaning to every part of the statute"—

They can not take the first part of the statute and leave off the last, which Mr. Darrow endeavored to do here the other day in his great speech—

"—And words may be modified, altered, or supplied so as to obviate any repugnancy or inconsistencies."

Now, if our legislature had the intent to prohibit teaching in our schools that man descended from the lower order of animals, they would not have to put that last clause on there, that explains the whole thing, and from that the court can, and could, define the section, as to what the intent of the legislature is. Reading further from Thomason vs. Temple:

"In 36 Cyc., 1111, it is said; 'For the purpose of determining the meaning, although not the validity of a statute, recourse may be had to considerations of public policy, and to the established policy of the legislature as disclosed by a general course of legislation.'

"And in Grannis vs. Superior court, 146 Cal., 247, 79 Fac., 893, 106 Am. St. Rep. 26, it is said: 'The provision of the code must be construed with a view to effect its objects, and when the language used is not entirely clear, the court may, to determine the meaning, and in aid of the interpretation, consider the spirit, intention and purpose of a law, and to ascertain such object and purpose.'"

What is the purpose of this law? It is to prevent the teaching in our schools that man descended from a lower order of animals, and when he taught that, as has been proven by our proof in chief, he violated the law, and cannot get around it.

"'Consider the spirit, intention and purpose of a law, and to ascertain such object and purpose may look into contemporaneous and prior legislation on the same subject, and the external and hisorical facts and conditions which led to its enactment.'"

Now, in the case of Norris vs. People, Fourth Colorado Appeals, 136, a statute was construed which penalized any person who should, by false representations, "obtain a credit, thereby defraud any person." It was held that the word "and" should be supplied before the word "thereby," the court saying:

Construing a Statute

"An insignificant alteration in the phraseology, or the omission of a word of this description in the adoption of a statute of another state, or in the revision of a statute, does not necessarily imply any intention to alter the construction of the act. It is equally settled that wherever there is an apparent mistake on the face of a statute the character of the error may often be determined by reference to other parts of the enactment, which may always be legitimately referred to in order to determine its legitimate construction."

In other words, in that last clause of this act, the legislature set forth their intention what they intended to do; that is just as plain as can be.

The Court—Now, if I understand you correctly, Mr. Hicks, you say when the state proved that he taught —that you insist that the state proved that he taught that man descended from a lower order of animals, and that by implication this proof meets the requirement of the first clause of the act?

Mr. Hicks—Absolutely. In other words, in construing that first clause, "to teach," where it prohibits any teacher in any public school, or schools supported in whole or in part by the state, to teach any theory which denies the story of the divine creation as taught in the Bible and then our legislature goes on and explains what that is—"and to teach instead"—instead of what?—that is my point.

The Court—What does the proof show, Mr. Hicks? Does the proof show Mr. Scopes taught that this little cell of life first evolved into a lower order of animals; is that your insistence?

Mr. Hicks—It says that it began in the sea.

The Court—That it began in the sea?

Mr. Hicks—As a little one-celled animal, and it continued to evolve on up through different stages of life until it culminated in man himself.

The Court—Before it culminated in man, if it went directly from that one cell and never crystallized into a lower animal—

Mr. Hicks—That is not the proof.

The Court—What is the proof?

Mr. Hicks—The proof shows it started as a one-celled animal, and then developed along for a while in the sea.

The Court—Does he call it a one-celled animal, or a one-celled life, or what?

Life Began as One Cell in the Sea.

Mr. Hicks—As I remember, he stated that life, animal life, began as one cell in the sea, and that it lived in the sea for a time, and it developed up and crawled out on the bank.

The Court—And developed into what?

Mr. Hicks—Into a higher form of life.

Gen. Stewart—That all animal life developed from one cell, from the same egg, the man, the monkey, the horse, the cow, everything.

Dr. Darrow—That is what it is, all animal life began in that one cell.

The Court—Is that the state's insistence, that this witness swore—

Mr. Hicks—Yes, sir.

The Court—That it never did develop into the different animals, but came direct to man?

Mr. Hicks—No, sir.

The Court—I am trying to get your theory.

Mr. Hicks—Our theory is, he taught it developed into the different animals, and came from one animal to another, and passed on up until it culminated into man himself.

The Court—It might be of one common origin, and from that one common origin fowl, beast, fish and man came. Now, do you understand them to say that from this one cell it developed directly into man without first having become a different kind of animal?

Mr. Hicks—No, that is not the proof.

The Court—But that it developed into different animal life, and from that animal life into man?

Gen. Stewart—Through all different kinds of animal life.

The Court—Well, all right.

Mr. Hicks—Now, if your honor please, the only issue here in this case—

The Court—A little louder.

What Did Scopes Teach?

Mr. Hicks—The issue of fact for the jury to determine is whether or not Prof. Scopes taught man descended from the lower order of animals. Now, if your honor is going to permit them to make a special issue of these experts, if you are going to permit them to come in here as a secondary jury, which they are endeavoring to do, that is an unheard of procedure in the courts of Tennessee. We are not endeavoring to run here a teachers' institute; we do not want to make out of this a high school or college; we do not object for these foreign gentlemen, as they please to call themselves—

The Court—Do not call them that.

Mr. Hicks—They call themselves that.

Mr. Malone—That is all right.

The Court—That is all right.

Mr. Hicks—We do not object to them coming into Tennessee and putting up a college, we will give them the ground to put the college on. If they want to educate the people of Tennessee as they say they do, but this a court of law, it is not a court of instruction for the mass of humanity at large. They, themselves, admit that it is their purpose, your honor, to enlighten the people of Tennessee. Now, your honor, how can these experts qualify as jurors?

I would like to be given the right to challenge these men, to pass upon them before they come into this court and give their opinions upon the facts which are in issue; the very province of the jury is invaded by the gentlemen we do not have the right to pass upon. I would like to be given the right to challenge three without cause, because they are without the state of Tennessee, and they come in to interpret our law, of our legislature. What do they know about the Bible? They have to qualify in both the Bible and science before they can.

Mr. Malone—May it please your honor, I do not know whether he is talking about the attorneys or the expert witnesses.

Mr. Hicks—I am talking about the expert witnesses. I will talk about you gentlemen later.

Mr. Hays—We want you to hear them first, before you decide.

Mr. Darrow—After they testified, the motion would be to strike their testimony, if you do not know.

Mr. Neal—I might say, we have a very distinguished Tennessean, the state gologist, Wilbur Nelson.

Gen. Stewart—I expect we would get along better if there were less heckling.

The Court—Proceed.

Mr. Hicks—Go to it. Any question you would like to ask.

Mr. Darrow—There is one question I would like to call your attention to.

Mr. Hicks—All right, Mr. Darrow.

Mr. Darrow—A question of law. I would like to have your view on it, and anybody else that speaks afterward. The caption of this act, as has been so often said, is entitled, "An act to prevent the teaching of evolution in public schools." The body of the act says: "Whoever teaches any doctrine as to the origin of man, contrary to that contained in the divine account in the Bible, and that he descended from some lower organism, is guilty," and so on. Now then, in order to make your act constitutional, the court must hold that the body of the act describes evolution. Does the court get me?

The Court—Yes.

Mr. Darrow—Do you?

Mr. Hicks—Yes.

Mr. Darrow—Unless the act itself is an act against evolution, then it is not constitutional, and, therefore, you must assume that this act forbidding the teaching of evolution, the body of the act not mentioning evolution, and the caption of the act does not present anything else, so, to say it is constitutional, you must say the body of the act means evolution.

Mr. Hicks—If your honor please, I do not care to take up that. Your honor has held that the act is constitutional.

The Court—Proceed with your argument, Mr. Hicks.

Experts Must Qualify Both as Scientists and Bible Authorities.

Mr. Hicks—Now, if your honor please, I insist this, when the experts come in they have to qualify upon two subjects, as experts upon the Bible and experts upon the particular branch of science, which they are supposed to know about. Now, why should these experts know anything more about the Bible than some of the jurors? There is one on there I will match against any of the theologians they will bring down, on the jury; he knows more of the Bible than all of them do.

Mr. Malone—How do you know?

Mr. Hicks—What is the interpretation of the Bible? Some of the experts whom they have brought here do not believe in God; the great majority, the leading ones, do not believe in God; they have different ideas—

Mr. Malone—If your honor please, how does he know until he gets them on the stand, what they believe? We object.

The Court—Sustain the objection; you cannot assume what they believe.

Mr. Malone—We would prefer for the sake of speed to have discussed only the witnesses whom we have called, and not the ones we may have called, but have not.

The Court—Sustain the objection. You cannot anticipate what they will say.

Mr. Hicks—I say this, this witness, when asked the hypothetical question as to whether or not what Prof. Scopes taught denies the story of the divine creation as taught in the Bible, is absolutely usurping the place of the jury. He is taking the place of the jury. He is invading it. Now, all these Tennessee decisions hold it is a kind of evidence that should be received with great caution—it is a matter of speculation—these scientists differ over it—Mr. Darrow said in his speech not long ago, that evolution is a mystery. Therefore, if expert testimony is full of pitfalls or dangers, or uncertainties in any issue, how much more so must it be in this issue; how much more so must it be in this issue in regard to evolution when Mr. Darrow himself says that evolution is a mystery. So, why admit these experts? Why admit them? It is not necessary. Why admit them? They invade the province of the jury. Why admit them, because the ones that they have introduced so far have not qualified as experts; he has only qualified in one line, and that is in the line of biology. If they want to make a school down here in Tennessee to educate our poor ignorant people, let them establish a school out here; let them bring down their great experts. The people of Tennessee do not object to that, but we do object to them making a school house or a teachers' institute out of this court. Such procedure in Tennessee is unknown. I do not know how about where these foreign gentlemen come from, but I say this in defense of the state, although I think it is unnecessary, the most ignorant man of Tennessee is a highly educated, polished gentleman compared to the most ignorant man in some of our northern states, because of the fact that the ignorant man of Tennessee is a man without an opportunity, but the men in our northern states, the northern man in some of our larger northern cities have the opportunity without the brain. (Laughter.)

The Court—Let me understand the arrangement; Mr. Malone and Col. Darrow are both to speak, are they?

Mr. Darrow—No, your honor, we have arranged with the attorneys that Mr. Bryan and Gen. McKenzie will speak, then Mr. Malone and Mr. Stewart, I am not going to speak—I am saving up.

The Court—I will hear you, Gen. McKenzie, and will adjourn for the noon hour.

Mr. Darrow—Your honor, cannot we get through, because we have some witnesses here from a great distance, some have to get away, it is a very great hardship?

The Court—I think it highly probable the court will not pass on this question today—I don't know.

Mr. Darrow—I think you ought to pass on it immediately, even if you pass on it wrong. It is a very great hardship for these men to wait here, some of them have to go.

The Court—I will hear you general.

General McKenzie Confesses Love-at-First-Sight for Darrow.

Gen. McKenzie—May it please your honor, I do not want to be heard but a very few moments. I want to say this, since the beginning of this lawsuit, and since I began to meet these distinguished gentlemen, I have begun to love them—everyone—and it is a very easy task, in fact, it was a case, when I met Col. Darrow—a case of love at first sight. These other gentlemen come right on, but you know they wiggled around so rapidly that I could not get my lover turned loose on them until I got a chance, but I love the great men. The newspapers have some of them said, that McKenzie is waving the bloody shirt. I just want to make this explanation, I have referred to the great metropolitan city, and of these distinguished gentlemen being from New York, for this reason, we have some of our own boys up there.

Mr. Malone—You bet you have.

Talks of Littleton and Carlisle.

Gen. McKenzie—From the South, we have Martin W. Littleton, I guess these gentlemen admire him. We do. We feel proud of him. We think he is so smart that he scintillates—stands at the very head of his profession, and I thought that I was paying the gentlemen a compliment, I never meant anything about it. This is our country from one ocean to the other, and from New York to that section away down where we can bathe our feet in the Gulf of Mexico and all our possessions, and you know this, the thing of bathing your feet ought to be a good thing, it would save the use of selling so much of this antifoot sweat.

Then we had another great man up there from the South, considered a pretty fair lawyer, John S. Carlisle. He had a great big sign up there, it said: "Counsellor of Counsellors"—a powerful, good man to resort to if you happened to get into a pinch, in a tight place for knowledge. We love him. We love these distinguished gentlemen, and love our local counsel, they are one of us, among us.

But, to the question in controversy in this case, if the honorable court please, as earnestly as I have believed any proposition of law to be established in this state, I believe that this act construes itself; that there is not a thing on the face of the earth that is ambiguous about it.

We Have Done Crossed the Rubicon.

We have done crossed the Rubicon. Your honor has held that the act was reasonable, within the powers of the legislature; that it was not vague, indefinite and void as it was insisted as one of their grounds for motion in this case. That has been passed over, that it was a valid exercise of the police power of the state of Tennessee and that Tennessee had the right to regulate its common schools and prescribe any common school curriculum it desired. That never left anything on the face of the earth to determine, except as to the guilt or the innocence of the defendant at bar in violating that act. The theory of evolution, as to whether it contradicts the Bible, your honor has allowed and correctly so, to introduce that Bible on the stand and it has been read to the jury. It is the duty of your honor to construe all writings if it gives any constructtion, that is the oldest principle of law in every state in this Union, it is a primary principle of law. What is there to construe? Another thing, is there any ambiguity about it, that these distinguished gentlemen through their experts can explain, that is competent in evidence in this case? No, a thousand times no, if it has a single bit of ambiguity bearing on the face of the instrument, there is no remedy for it. It can not be, as the old language of the law is, helped by expert proof, that is the language, it has been held a thousand times in regard to wills and deeds, and other instruments. I have an authority right here, it is an old one, your honor knows all about it, if it is obsolete on its face, too void for enforcement, you can not make a new contract by shooting in your proof, and it must fall only if there is a case of latent ambiguity; that is, if it says, "I bequeath to my good friend Col. Darrow, of New York, my shotgun," and there happens to be two Col. Darrows up there, they say you can introduce proof to show which Col. Darrow I have reference to.

Not Opposing Bible.

They do not undertake to destroy the Bible, or set up a story in contradiction of it, but attempt to reconcile, that is the point I want your honor to catch, and I know your honor does.

Says God Made Man Complete.

The Court—General, let me ask you a question. Is this your position, that the story of the divine creation is so clearly set forth in the Bible, in Genesis, that no reasonable minds could differ as to the method of creation, that is, that man was created, complete by God?

Gen. McKenzie—Yes.

The Court—And in one act, and not by a method of growth or development; is that your position?

Gen. McKenzie—From lower animals—yes, that is exactly right.

The Court—That God created Adam first as a complete man, did not create a single cell of life.

Gen. McKenzie—That is right.

The Court—The cell of life did not develop in time.

Gen. McKenzie—That is right, and man did not descend from a lower order of animals that originated in the sea and then turned from one animal to another and finally man's head shot up.

The Court—Here is what I want to get, the act says it shall be unlawful to teach any theory that denies the divine story of the creation of man; that is one issue. Or teach or instead thereof—

Mr. Malone—"And" is the word.

The Court—And teach instead thereof that man descended from a lower order of animals. Now, in order to make a case, does the state have to prove that the defendant Scopes taught a theory denying the divine creation, and then go further and prove that he taught that man descended from a lower order of animals; or do you claim that if you meet the second clause, by implication of law you have met the requirement of the first?

Gen. McKenzie—Yes, that is exactly it. I want to read this, you may look to the caption as well as the body of the act to resolve any ambiguity. Let us read the act.

It being an act of the state of Tennessee that it shall be unlawful for any teacher in the universities, normals or other public schools of the state, which are supported in whole or in part by the public funds to teach a theory that denies the story of the divine creation of man as told in the Bible—

The Court—Now, General, just suppose he stopped there, and the other clause were stricken out, would this proof be competent for the purpose offered, or not?

Mr. B. G. McKenzie—I think not. No, sir, I do not.

The Court—You think the divine story is so clearly told, it is not ambiguous and should be accepted by any one of reasonable fairness?

Mr. B. G. McKenzie—I do. But it goes further, and leaves it out of the proposition, and says, and teach instead thereof that man is descended from a lower order of animals, and, therefore, defines the other proposition. It tells exactly what it means, in both the caption and the body of the act. And our supreme court, in case after case, in Tennessee, has sustained our contention as to the interpretation of statutes. Now, if your Honor please, as said a minute ago, they don't want to destroy that account.

The Court—They want to reconcile—

Evolutionists Would Have Man Descended from Soft Dish Rag.

Mr. B. G. McKenzie—They are seeking to reconcile it, if your honor please, and come right along and prove by the mouth of their scientist that when he said God created man in His own image, in His own image created He him out of the dust of the ground and blew into him the breath of life, and he became as a living creature they want to put words into God's mouth, and have Him to say that He issued some sort of protoplasm, or soft dish rag, and put it in the ocean and said, "Old boy, if you wait around about 6,000 years, I will make something out of you." (Laughter.) And they tell me there is no ambiguity about that.

Mr. Darrow—Let me ask a question. When it said, "in His own image," did you think that meant the physical man?

B. G. McKenzie—I am taking the Divine account—"He is like unto me."

Mr. Darrow—Do you think it is so?

B. G. McKenzie—I say that, although I know it is awfully hard on our Maker to look like a lot of fellows who are profusely ugly, to say he favored the Master.

Mr. Darrow—You think then that you do?

Mr. McKenzie—You are all right. I don't mind your favoring Him, but when one commits acts against the law, there ought to be some remedy for it.

Mr. Darrow—Wait a minute, colonel. You do think the physical man is like God?

B. G. McKenzie—Why, yes, I do and I will give you my reason.

Mr. Darrow—I think God knows better. You think men must believe that to believe the Bible, that the physical man as we see him looks like God.

B. G. McKenzie—Yes, sir, and I will give you my reasons as soon as you want them.

Mr. Darrow—And when you see man, you see a picture of God.

Believes Bible Story

B. M. McKenzie—Like unto Him and made in His image; and the reason why I believe that firmly is because the Bible teaches it. When Christ came to earth—and I believe in the virgin birth of Christ.

Mr. Neal—Mr. McKenzie?

B. G. McKenzie—What is it, Mr. Neal? Do you want to ask a question?

Mr. Neal—Do you think if a teacher in the Tennessee schools if he failed to teach that man is physically like God, would be violating the statute?

B. G. McKenzie—Well, we will try that law suit when we get to it. Let us talk about the matters involved in this case.

Mr. Darrow—Let me ask another question?

B. G. McKenzie—All right.

Mr. Darrow—I don't think we will have any trouble as long as he gives me the title of colonel. He is calling everybody else colonel. You spoke about it taking a good many thousand years to get man under our theory. You said there was the first day, the second day, the third day, the fourth day, the fifth day, the sixth day, and so on. Do you think they were literal days?

B. G. McKenzie—Colonel, we didn't have any sun until the fourth day. I believe the Biblical account. Now, in regard to Christ being just a man, walking around looking like us. I believe He was the same, a man of sorrow and grief, crucified for us. And I believe that still. And when He was here, He was like other men, but he was in the image of God. And that is why I believe He was in the image of man.

Mr. Malone—Your honor, I am objecting, on this ground. I don't know whether the general is arguing now, or testifying as an expert witness on the other side.

B. G. McKenzie—He is objecting to me, yet, Mr. Malone said a speech of an hour yesterday, presenting their theories of the case; it was on evolution, and it was not competent.

Mr. Malone—The court admitted it.

Hints Few Darrow Disciples in Rhea County.

B. G. McKenzie—Yes, and he is the best judge in the world. Now, if the court please, I say they are seeking to put words into the mouth of God, and substitute another story, entirely different to God's word. They bring in a distinguished gentleman, and I believe he is absolutely a disciple of Col. Darrow. He says evolution is an established fact, and that there are a lot of them in this country. But I tell you one thing, no great number of them grow on the mountain sides and in the valleys of Rhea. Then, after they get all their testimoney in, and the issues were drawn, they didn't throw light on the proposition. They introduced sixty witnesses, and have a lot of hypotheses, but they don't know anything about the things that are to be testified about. They can't read scientific works for us and put them in evidence.

Mr. Darrow—I think you misunderstand our position. What we claim is that there is no question among intelligent men about the fact of evolution. As to how it came about, there is a great deal of difference.

B. G. McKenzie—That is it. Yes, you are now coming back to the point in the defense in which you say you want us to recognize your theory, and yet you just absolutely jangle along, going in one door and out the same door. I wonder if that man has ever read the Bible.

Mr. Darrow—He had one with him.

McKenzie Cracks Jokes.

Mr. McKenzie—That may be. But it is not competent for anything after they get all the witnesses in court, and then want to charge the jury after you submit it to him. It reminds me of the shape that the old Dutch judge was in, when there were a lot of witnesses swearing different tales. They say they know that man is both of the animal and vegetable kingdoms, coming from the same source. If that is so, this great array has been eating up their relations — they are depopulating their relatives very rapidly.

But that is another proposition. That judge, when he went to charge the jury, he said, "Now, gentlemen of the jury"—He was a new judge—"If the plaintiff and his witnesses have sworn the truth about this matter, you will find, of course, for the plaintiff; but, if, on the other hand, the defendant and his witnesses have sworn the truth, you will, of course, find for the defendant. But if you are like me and believe that they are all swearing lies, I don't know what the debble you will do."

I don't know where they got their evidence, but they are putting it up against the Word of God. I reckon the next thing will be to—

Mr. Hays—May I interrupt you for a moment?

B. G. McKenzie—Yes, sir.

Mr. Hays—You seem so sure as to what our witnesses are going to testify. We have not brought our witnesses out; how is it that you are in a position to know what they are going to say?

Mr. McKenzie—You know no expert testimony is competent in this case, but I think this is competent.

The Court—He asked you how you knew what they were going to testify.

Mr. McKenzie—I think his witness swore the truth when he said none of them knew. He said they didn't know, and I think they will tell the truth. Do you believe the story of divine creation?

"None of Your Business."

Mr. Hays—That is none of your business.

Mr. McKenzie—Then don't ask me any more impertinent questions.

Mr. Malone—General, will you give me the law?

The Court—I do not think that Col. Hays' answer to Gen. McKenzie was as courteous an answer as he should give in this court.

Apologies

Mr. Hays—That is so. Instead of those words, I will say I think it doesn't concern Gen. McKenzie.

Mr. McKenzie—I will say to you that I have as little concern as to where you emanated from, or as to where you are going, as any man I ever met.

Mr. Hays—Now, may I ask for an apology, your honor?

The Court—Yes, sir.

Mr. McKenzie—I didn't mean to give offense; I beg your pardon.

Mr. Hays—It is like old sweethearts made up.

The Court—Col. Bryan, it is only fifteen minutes to noon. Can you complete your argument in that time?

Mr. Bryan—What time is it now?

The Court—A quarter of twelve.

Mr. Darrow—Although it is a short while—

Mr. Malone—Can't we continue a little longer?

The Court—That is what I am getting at.

Mr. Malone—I am not referring to Col. Bryan's time; I am asking for Court to continue longer.

Mr. Gordon McKenzie—We have some ceiling fans coming. I want to ask your honor to adjourn a little early and let them put the fans in.

The Court—I have information that the sheriff wants to put ceiling fans in during the noon hour. I think you all will like to be cooled off. Will they be put in during the noon hour?

Mr. McKenzie—Yes, sir; they will be.

The Court—We will adjourn until 1:30.

AFTERNOON SESSION

The Court—Open court, Mr. Sheriff. Everybody stand up.

The Bailiff—Oyez oyez, this honorable circuit court is now open pursuant to adjournment. Sit down.

The Court—Now as I announced this morning, the floor on which we are now assembled is burdened with a great weight. I do not know how well it is supported, but sometimes buildings and floors give away when they are unduly burdened, so I suggest to you to be as quiet in the courtroom as you can; have no more emotion than you can avoid; especially have no applause, because it isn't proper in the courtroom. Now I regret very much that there are many people here who cannot get inside and hear the speaking, but, of course, it isn't within my power, physical power, to enlarge the courtroom. Mr. Counsel for the defendant—Mr. Counsel for the defendant, have you—has Mr. Darrow decided to speak or not?

Mr. Darrow—No, Mr. Malone is the only other.

The Court—The only other counsel to speak for that side?

Mr. Darrow—Yes.

The Court—Well, I believe Mr. Bryan then will speak next for the state.

William Jennings Bryan's Speech.

If the court please we are now approaching the end of the first week of this trial and I haven't thought it proper until this time to take part in the discussions that have been dealing with phases of this question, or case, where the state laws and the state rules of practice were under discussion and I feel that those who are versed in the law of the state and who are used to the customs of the court might better take the burden of the case, but today we come to the discussion of a very important part of this case, a question so important that upon its decision will determine the length of this trial. If the court holds, as we believe the court should hold, that the testimony that the defense is now offering is not competent and not proper testimony, then I assume we are near the end of this trial and because the question involved is not confined to local questions, but is the broadest that will possibly arise, I have felt justified in submitting my views on the case for the consideration of the court. I have been tempted to speak at former times, but I have been able to withstand the temptation. I have been drawn into the case by, I think nearly all the lawyers on the other side. The principal attorney has often suggested that I am the arch-conspirator and that I am responsible for the presence of this case and I have almost been credited with leadership of the ignorance and bigotry which he thinks could alone inspire a law like this. Then Mr. Malone has seen fit to honor me by quoting my opinion on religious liberty. I assume he means that that is the most important opinion on religious liberty that he has been able to find in this country and I feel complimented that I should be picked out from all the men living and dead as the one whose expressions are most vital to the welfare of our country. And this morning I was credited with being the cause of the presence of these so-called experts.

Duel to the Death?

Mr. Hays says that before he got here he read that I said this was to be a duel to the death, between science—was it? and revealed religion. I don't know who the other duelist was, but I was representing one of them and because of that they went to the trouble and the expense of several thousand dollars to bring down their witnesses. Well, my friend, if you said that this was important enough to be regarded as a duel between two great ideas or groups I certainly will be given credit for foreseeing what I could not then know and that is that this question is so important between religion and irreligion that even the involking of the divine blessing upon it might seem partisan and partial. I think when we come to consider the importance of this question, that all of us who are interested as lawyers on either side, could claim what we—what your honor so graciously grants—a hearing. I have got down here for fear I might forget them, certain points that I desire to present for your honor's consideration. In the first place, the statute—our position is that the statute is sufficient. The statute defines exactly what the people of Tennessee desired and intended and did declare unlawful and it needs no interpretation. The caption speaks of the evolutionary theory and the statute specifically states that teachers are forbidden to teach in the schools supported by taxation in this state, any theory of creation of man that denies the divine record of man's creation as found in the Bible, and that there might be no difference of opinion—there might be no ambiguity—that there might be no such confusion of thought as our learned friends attempt to inject into it, the legislature was careful to define what it meant by the first part of the statute. It says to teach that man is a descendant of any lower form of life—if that had not been there—if the first sentence had been the only sentence in the statute, then these gentlemen might come and ask to define what that meant or to explain whether the thing that was taught was contrary to the language of the statute in the first sentence, but the second sentence removes all doubt, as has been stated by my colleague. The second sentence points out specifically what is meant, and that is the teaching that man is the descendant of any lower form of life, and if the defendant taught that as we have proven by the textbook that he used and as we have proven by the students that went to hear him—if he taught that man is a descendant of any lower form of life, he violated the statute, and more than that we have his own confession that he knew he was violating the statute. We have the testimony here of Mr. White, the superintendent of schools, who says that Mr. Scopes told him he could not teach that book without violating the law. We have the testimony of Mr. Robertson—Robinson—the head of the Board of Education, who talked with Mr. Scopes just at the time the schools closed, or a day or two afterward, and Mr. Scopes told him that he had reviewed that book just before the school closed, and that he could not teach it without teaching evolution and without violating the law, and we have Mr. Robinson's statement that Mr. Scopes told him that he and one of the teachers, Mr. Ferguson, had talked it over after the law was passed and had decided that they could not teach it without the violation of the law, and yet while Mr. Scopes knew what the law was and knew what evolution was, and knew that it violated the law, he proceeded to violate the law. That is the evidence before this court, and we do not need any expert to tell us what that law means. An expert cannot be permitted to come in here and try to defeat the enforcement of a law by testifying that it isn't a bad law and it isn't—I mean a bad doctrine—no matter how these people phrase the doctrine—no matter how they eulogize it. This is not the place to try to prove that the law ought never to have been passed. The place to prove that, or teach that, was to the legislature. If these people were so anxious to keep the state of Tennessee from disgracing itself, if they were so afraid that by this action taken by the legislature, the state would put itself before the people of the nation as ignorant people and bigoted people—if they had half the affection for Tennessee that you would think they had as they come here to testify, they would have come at a time when their testimony would have been valuable and not at this time to ask you to refuse to enforce a law because they did not think the law ought to have been passed. And, my friends, if the people of Tennessee were to go into a state like New York—the one from which this impulse comes to resist this law, or go into any state—if they went into any state and tried to convince the people that a law they had passed ought not to be enforced, just because the people who went there didn't think it ought to have been passed, don't you think it would be resented as an impertinence? They passed a law up in New York repealing the enforcement of prohibition. Suppose the people of Tennessee had sent attorneys up there to fight that law, or to oppose it after it was passed, and experts to testify how good a thing prohibition is to New York and to the nation, I wonder if there would have been any lack of determination in the papers in speaking out against the offensiveness of such testimony. The people of this state passed this law, the people of this state knew what they were doing when they passed the law, and they knew the dangers of the doctrine—that they did not want it taught to their children, and my friends, it isn't—your honor, it isn't proper to bring experts in here to try to defeat the purpose of the people of this state by trying to show that this thing that they denounce and outlaw is a beautiful thing that everybody ought to believe in. If, for instance—I think this is a fair illustration—if a man had made a contract with somebody to bring rain in a dry season down here, and if he was to have $500 for an inch of rain, and if the rain did not come and he sued to enforce his contract and collect the money, could he bring experts in to prove that a drought was better than a rain? (Laughter in the courtroom.) And get pay for bringing a drought when he contracted to bring rain. These people want to come here with experts to make your honor believe that the law should never have been passed and because in their opinion it ought not to have been passed, it ought not to be enforced. It isn't a place for expert testimony. We have sufficient proof in the book —doesn't the book state the very thing that is objected to, and outlawed in this state? Who has a copy of that book?

The Court—Do you mean the Bible?

Mr. Bryan—No, sir; the biology. (Laughter in the courtroom.)

A Voice—Here it is; Hunter's Biology.

Cannot Teach Bible in State.

Mr. Bryan—No, not the Bible, you see in this state they cannot teach the Bible. They can only teach things that declare it to be a lie, according to the learned counsel. These people in the state—Christian people—have tied their hands by their constitution. They say we all believe in the Bible for it is the overwhelming belief in the state, but we will not teach that Bible, which we believe even to our children through teachers that we pay with our money. No, no, it isn't the teaching of the Bible, and we are not asking it. The question is can a minority in this state come in and compel a teacher to teach that the Bible is not true and make the parents of these children pay the expenses of the teacher to tell their children what these people believe is false and dangerous? Has it come to a time when the minority can take charge of a state like Tennessee and compel the majority to pay their teachers while they take religion out of the heart of the children of the parents who pay the teachers? This is the book that is outlawed if we can judge from the questions asked by the counsel for the defense. They think that because the board of education selected this book, four or five years ago, that, therefore, he had to teach it, that he would be guilty if he didn't teach it and punished if he does. Certainly not one of these gentlemen is unlearned in the law and if I, your honor, who have not practiced law for twenty-eight years, know enough to know it, I think those who have been as conspicuous in the practice as these gentlemen have been, certainly ought to know it and that is no matter when that law was passed; no matter what the board of education has done; no matter whether they put their stamp of approval upon this book or not, the moment that law became a law anything in these books contrary to that law was prohibited and nobody knew it better than Mr. Scopes himself. It doesn't matter anything about who ordered these books—the law supercedes all boards of education for the legislature is the supreme court on this subject from which there is no appeal. What does this law teach, my friends? We have little—what is the Morgan boy's first name?

Howard Morgan Understands Subject Better Than Darrow.

A Voice—Howard.

Mr. Bryan—Little Howard Morgan—and, your honor, that boy is going to make a great lawyer some day. I didn't realize it until I saw how a 14-year-old boy understood the subject so much better than a distinguished lawyer who attempted to quiz him. The little boy understood what he was talking about and to my surprise the attorney's didn't seem to catch the significance of the theory of evolution and the thought—and I'm sure he wouldn't have said it if he hadn't had thought it—he thought that little boy was talking about the individuals coming up from one cell. That wouldn't be evolution—that is growth, and one trouble about evolution is that it has been used in so many different ways that people are confused about it, but I am not surprised that the gentleman from New York—Mr. Hays, was confused, the National Education association even is confused, for if you noticed the other day they had a meeting in Indianapolis and it was said that they were going to tell Tennessee where to head in. We had several flaming advance notices of how the ignorance and bigotry of Tennessee was to be scored by the educational association—the teachers of the United States. Well, during the early days we would have flaming announcements of what was going to be done and then we had a very mild report. The chairman of the committee on resolutions reported that there would be no resolution passed—no, they were not going to say a word. Why? Well, there were so many different kinds of evolution or so many definitions of evolution that if they made a general statement it would be useless and if they went into detail it would excite controversy. (Laughter in the courtroom.) No wonder the gentleman from New York was not able to distinguish by just hearing it once, between the evolution of life that began in the ocean away down in the bottom and evolved up through animals bigger and bigger, until finally they got a land animal some way and then when it got on the land where it had a firmer footing it kept on evolving more and more and then finally man was the climax. That little boy could understand that and I wonder if the lawyers cannot understand it by this time. (Laughter in the courtroom.) That is evolution and that is what he taught. Not the growth of an individual from one cell, but the growth of all life from one cell and while I am on this point I might call attention to another thing that the distinguished lawyer who spoke this morning—Mr. Hays, said. He quotes, I think, from Linnaeus, if I am not mistaken. I may not be as familiar with these scientific experts as he is, but I know some of them even besides those already brought here and Linnaeus I think was the one he referred to who gave us the classification and put man among the primates. Am I correct? Was it Linnaeus? And the monkeys were also among the primates, and he says if he taught that man came from a monkey he didn't violate the law in this state, because the monkey is in the same class of primates with man.

Mr. Hays—No, I didn't say that. I beg your pardon.

Mr. Bryan—What did you say?

Mr. Hays—I said the term order of animals was a scientific term and that they were in the same order and that the words should have been the words you used. They are of a different class, but they are of the same order.

Mr. Bryan—Then are there ranks in an order or all one rank?

Mr. Hays—No, there are various ranks in the order. They should have used your words, should have used the words "class" or "families" —that is what I said.

Mr. Bryan—No matter what you said it wouldn't make much difference because the answer would be just the same. (Laughter in the courtroom.) I want to remind your honor that if men and monkeys are in the same class, called primates, that doesn't settle the question, for it is possible that some of those primates are the descendants of other primates, but if it were true that every primate was in a class by itself and was not descended from any other primate, therefore, according to evolution all the primates in that class descended from other animals, evolved from that class, and you go back to the primates, to the one evolved until you get to the one-cell animal in the bottom of the sea.

Christian Believes Man from Above— Evolutionist from Below.

So, my friends, if that were true, if man and monkey were in the same class, called primates, it would mean they did not come up from the same order. It might mean that instead of one being the ancestor of the other they were all cousins. But it does not mean that they did not come up from the lower animals, if this is the only place they could come from, and the Christian believes man came from above, but the evolutionist believes he must have come from below.

(Laughter in the courtroom.)

And that is from a lower order of animals.

Your honor, I want to show you that we have evidence enough here, we do not need any experts to come in here and tell us about this thing. Here we have Mr. Hunter. Mr. Hunter is the author of this biology and this is the man who wrote the book Mr. Scopes was teaching. And here we have the diagram. Has the court seen this diagram?

The Court—No, sir, I have not.

Bryan Shows "Tree of Life" to Court

Mr. Bryan—Well, you must see it (handing book to the court.)

(Laughter in the courtroom.)

On page 194—I take it for granted that counsel for the defense have examined it carefully?

Mr. Darrow—We have examined it.

Mr. Bryan—On page 194, we have a diagram, and this diagram purports to give some one's family tree. Not only his ancestors but his collateral relatives. We are told just how many animal species there are, 518,900. And in this diagram, beginning with protozoa we have the animals classified. We have circles differing in size according to the number of species in them and we have the guess that they give.

Of course, it is only a guess, and I don't suppose it is carried to a one or even to ten. I see they are round numbers, and I don't think all of these animals breed in round numbers, and so I think it must be a generalization of them.

(Laughter in the courtroom.)

The Court—Let us have order.

Mr. Bryan—8,000 protozoa, 3,500 sponges.

Must Be More Than 35,000 Sponges.

I am satisfied from some I have seen there must be more than 35,000 sponges.

(Laughter in the courtroom.)

Mr. Bryan—And then we run down to the insects, 360,000 insects. Two-thirds of all the species of all the animal world are insects. And sometimes, in the summer time we feel that we become intimately acquainted with them—a large percentage of the species are mollusks and fishes. Now, we are getting up near our kinfolks, 13,000 fishes. Then there are the amphibia. I don't know whether they have not yet decided to come out, or have almost decided to go back.

(Laughter in the courtroom.)

But they seem to be somewhat at home in both elements. And then we have the reptiles, 3,500; and then we have 13,000 birds. Strange that this should be exactly the same as the number of fishes, round numbers. And then we have mammals, 3,500, and there is a little circle and man is in the circle, find him, find man.

There is that book! There is the book they were teaching your children that man was a mammal and so indistinguishable among the mammals that they leave him there with thirty-four hundred and ninety-nine other mammals.

(Laughter and applause.)

Including elephants?

Has Daniel Story Beaten.

Talk about putting Daniel in the lion's den? How dared those scientists put man in a little ring like that with lions and tigers and everything that is bad! Not only the evolution is possible, but the scientists possibly think of shutting man up in a little circle like that with all these animals, that have an odor, that extends beyond the circumference of this circle, my friends.

(Extended laughter.)

He tells the children to copy this, copy this diagram. In the notebook, children are to copy this diagram and take it home in their notebooks. To show their parents that you cannot find man. That is the great game to put in the public schools to find man among animals, if you can.

Tell me that the parents of this day have not any right to declare that children are not to be taught this doctrine? Shall not be taken down from the high plane upon which God put man? Shall be detached from the throne of God and be compelled to link their ancestors with the jungle, tell that to these children? Why, my friend, if they believe it, they go back to scoff at the religion of their parents! And the parents have a right to say that no teacher paid by their money shall rob their children of faith in God and send them back to their homes, skeptical, infidels, or agnostics, or atheists.

This doctrine that they want taught, this doctrine that they would force upon the schools, where they will not let the Bible be read!

Why, up in the state of New York they are now trying to keep the schools from adjourning for one hour in the afternoon, not that any teacher shall teach them the Bible, but that the children may go to the churches to which they belong and there have instruction in the work. And they are refusing to let the school do that. These lawyers who are trying to force Darwinism and evolution on your children, do not go back to protect the children of New York in their right to even have religion taught to them outside of the schoolroom, and they want to bring their experts in here.

As we have one family tree this morning given to us, I think you are entitled to have a more authentic one. My friend, my esteemed friend from New York, gave you the family tree according to Linnaeus.

Mr. Malone—Beg pardon, Mr. Bryan?

Hits at Darwinism.

Mr. Bryan—I will give you the family tree according to Darwin. If we are going to have family trees here, let us have something that is reliable. I will give you the only family tree that any believer in evolution has ever dared to outline—no other family tree that any evolutionist has ever proposed, has as many believers as Darwin has in his family tree. Some of them have discarded his explanations. Natural selections! People confuse evolution with Darwinism. They did not use to complain. It was not until Darwin was brought out into the open, it was not until the absurdities of Darwin had made his explanations the laughing stock, that they began to try to distinguish between Darwinism and evolution. They explained that evolutionists had discarded Darwin's idea of sexual selection—I should think they would discard it, and they are discarding the doctrine of natural selection.

But, my friends, when they discard his explanations, they still teach his doctrines. Not one of these evolutionists have discarded Darwin's doctrine that makes life begin with one cell in the sea and continue in one unbroken line to man. Not one of them has discarded that.

Let me read you what Darwin says, if you will pardon me. If I have to use some of these long words —I have been trying all my life to use short words, and it is kind of hard to turn scientist for a moment.

(Laughter in the courtroom.)

And try to express myself in their language.

Here is the family tree of Darwin and remember that is the Darwin that is spoken of in Hunter's biology, that is Darwin he has praised. That is the Darwin who has series—

Mr. Malone—What is the book, Mr. Bryan?

Mr. Bryan—"The Descent of Man," by Charles Darwin.

Mr. Malone—That has not been offered as evidence?

Mr. Bryan—I should be glad to offer it.

Mr. Malone—No, no, no. No, no.

Mr. Bryan—Let me know if you want it, and it will go in.

Mr. Malone—I would be glad to have it go in.

(Laughter in the courtroom.)

Mr. Bryan—Let us have it put in now so that there will be no doubt about it.

Mr. Malone—If you will let us put our witnesses on to show what the works are—

Mr. Hays—If you will let us put evidence in about it, perhaps we can settle the questions of what it is. I would be satisfied.

Mr. Bryan—If you attach that condition to it, I may not be willing.

Mr. Hays—No.

Mr. Bryan—You seemed to be so anxious about Darwin, I thought you would be content.

Mr. Malone—I merely wanted to know whether it was a book offered by the prosecution; that was the purpose of my question.

Mr. Bryan—No. It was just referred to and Mr. Hays quoted from Linnaeus on the family tree. I will read this.

Reads from "Descent of Man."

"The most ancient projenitors in the kingdom of the Vertebrata, at which we are able to obtain an obscure glance, apparently consisted of a group of marine animals, resembling the larvae of existing Ascidians. These animals probably gave rise to a group of fishes, as lowely organized as the lancelet, and from these the Ganoids, and other fishes like the Lepidosiren must have been developed. From such fish a very small advance would carry us on to the amphibians. We have seen that birds and reptiles were once intimately connected together; and the Monotremata now connect mammals with reptiles in a slight degree. But no one can at present say by what line of descent the three higher and related classes, namely, mammals, birds and reptiles were derived from the two lower vertebrate classes, namely, amphibians and fishes. In the class of mammals the steps are not difficult to conceive which led from the ancient Monotremata to the ancient Marsupials, and from these to the early progenitors of the placental mammals. We may thus ascend to the Lemuridae, and the interval is not very wide from these to the Simiadae. The Simiadae then branched off into two great stems, the new world and the old world monkeys, and from the latter, at a remote period, man, the wonder and glory of the universe, proceeded."

"Not Even from American Monkeys."

Not even from American monkeys, but from old world monkeys. (Laughter.) Now, here we have our glorious pedigree, and each child is expected to copy the family tree and take it home to his family to be submitted for the Bible family tree —that is what Darwin says. Now, my friends—I beg pardon, if the court please, I have been so in the habit of talknig to an audience instead of a court, that I will sometimes say "my friends," although I happen to know not all of them are my friends. (Laughter.)

The Court—Let me ask you a question: Do you understand the evolution theory to involve the divine birth of divinity, or Christ's virgin birth, in any way or not?

Mr. Bryan—I am perfectly willing to answer the question. My contention is that the evolutionary hypothesis is not a theory, your honor.

The Court—Well, hypothesis.

Mr. Bryan—The legislature paid evolution a higher honor than it deserves. Evolution is not a theory, but a hypothesis. Huxley said it could not raise to the dignity of a theory until they found some species that had developed according to the hypothesis, and at that time, Huxley's time, there had never been found a single species, the origin of which could be traced to another species. Darwin himself said he thought it was strange that with two or three million species they had not been able to find one that they could trace to another. About three years ago, Bateson, of London, who came all the way to Toronto at the invitation of the American Academy for the Advancement of Sciences—which, if the gentlemen will brace themselves for a moment, I will say I am a member of the American Academy for the Advancement of Science—they invited Mr. Bateson to come over and speak to them on evolution, and he came, and his speech on evolution was printed in Science magazine, and Science is the organ of the society and I suppose is the outstanding organ of science in this country, and I bought a copy so that if any of the learned counsel for the plaintiff had not had the pleasure of reading Bateson's speech that they could regale themselves during the odd hours. And Bateson told those people after having taken up every effort that had been made to show the origin of species and find it, he declared that every one had failed—every one—every one. And it is true today; never have they traced one single species to any other, and that is why it was that this so-called expert stated that while the fact of evolution, they think, is established, that the various theories of how it come about, that every theory has failed, and today there is not a scientist in all the world who can trace one single species to any other, and yet they call us ignoramouses and bigots because we do not throw away our Bible and accept it as proved that out of two or three millian species not a one is traceable to another. And they say that evolution is a fact when they cannot prove that one species came from another, and if there is such a thing, all species must have come, commencing as they say, commencing in that one lonely cell down there in the bottom of the ocean that just evolved and evolved until it got to be a man. And they cannot find a single species that came from another, and yet they demand that we allow them to teach this stuff to our children, that they may come home with their imaginary family tree and scoff at their mother's and father's Bible.

Bryan Refers to Own Degrees.

Now, my friends, I want you to know that they not only have no proof, but they cannot find the beginning. I suppose this distinguished scholar who came here shamed them all by his number of degrees—he did not shame me, for I have more than he has, but I can understand how my friends felt when he unrolled degree after degree. Did he tell you where life began? Did he tell you that back of all these that there was a God? Not a word about it. Did he tell you how life began? Not a word; and not one of them can tell you how life began. The atheists say it came some way without a God; the agnostics say it came in some way, they know not whether with a God or not. And the Christian evolutionists say we come away back there somewhere, but they do not know how far back—they do not give you the beginning—not that gentleman that tried to qualify as an expert; he did not tell you how life began. He did not tell you whether it began with God or how. No, they take up life as a mystery that nobody can explain, and they want you to let them commence there and ask no questions. They want to come in with their little padded up evolution that commences with nothing and ends nowhere. They do not dare to tell you that it ended with God. They come here with this bunch of stuff that they call evolution, that they tell you that everybody believes in, but do not know that everybody knows as a fact, and nobody can tell how it came, and they do not explain the great riddle of the universe—they do not deal with the problems of life—they do not teach the great science of how to live—and yet they would undermine the faith of these little children in that God who stands back of everything and whose promise we have that we shall live with Him forever bye and bye. They shut God out of the world. They do not talk about God. Darwin says the beginning of all things is a mystery unsolvable by us. He does not pretend to say how these things started.

The Court—Well, if the theory is, Col. Bryan, that God did not create the cell, then it could not be reconcilable with the Bible?

Mr. Bryan—Of course, it could not be reconcilable with the Bible.

The Court—Before it could be reconcilable with the Bible it would have to be admitted that God created the cell?

Evolution Not Reconcilable with Bible.

Mr. Bryan—There would be no contention about that, but our contention is, even if they put God back there, it does not make it harmonious with the Bible. The court is right that unless they put God back there, it must dispute the Bible, and this witness who has been questioned, whether he has qualified or not, and they could ask him every question they wanted to, but they did not ask him how life began, they did not ask whether back of it all, whether if in the beginning there was God. They did not tell us where immortality began. They did not tell us where in this long period of time, between the cell at the bottom of the sea and man, where man became endowed with the hope of immortality. They did not, if you please, and most of them do not go to the place to hunt for it, because more than half of the scientists of this country—Prof. James H. Labell, one of them, and he bases it on thousands of letters they sent to him, says more than half do not believe there is a God or personal immortality, and they want to teach that to these children, and take that from them, to take from them their belief in a God who stands ready to welcome his children.

Discusses Virgin Birth, Resurrection, and Atonement.

And your honor asked me whether it has anything to do with the principle of the virgin birth. Yes, because this principle of evolution disputes the miracle; there is no place for the miracle in this train of evolution, and the Old Testament and the New are filled with miracles, and if this doctrine is true, this logic eliminates every mystery in the Old Testament and the New, and eliminates everything supernatural, and that means they eliminate the virgin birth—that means that they eliminate the resurrection of the body—that means that they eliminate the doctrine of atonement and they believe man has been rising all the time, that man never fell, that when the Savior came there was not any reason for His coming, there was no reason why He should not go as soon as He could, that He was born of Joseph or some other co-respondent, and that He lies in his grave, and when the Christians of this state have tied their hands and said we will not take advantage of our power to teach religion to our children, by teachers paid by us, these people come in from the outside of the state and force upon the people of this state and upon the children of the taxpayers of this state a doctrine that refutes not only their belief in God, but their belief in a Savior and belief in heaven, and takes from them every moral standard that the Bible gives us. It is this doctrine that gives us Nietzsche, the only great author who tried to carry this to its logical conclusion, and we have the testimony of my distinguished friend from Chicago in his speech in the Loeb and Leopold case that 50,000 volumes had been written about Nietzsche, and he is the greatest philosopher in the last hundred years, and have him pleading that because Leopold read Nietzsche and adopted Nietzsche's philosophy of the superman, that he is not responsible for the taking of human life. We have the doctrine—I should not characterize it as I should like to characterize it—the doctrine that the universities that had it taught, and the professors who taught it, are much more responsible for the crime that Leopold committed than Leopold himself. That is the doctrine, my friends, that they have tried to bring into existence, they commence in the high schools with their foundation in the evolutionary theory, and we have the word of the distinguished lawyer that this is more read than any other in a hundred years, and the statement of that distinguished man that the teachings of Nietzsche made Leopold a murderer.

Mr. Darrow—Your honor, I want to object; there is not a word of truth in it. Nietzsche never taught that. Anyhow, there was not a word of criticism of the professors, nor of the colleges in reference to that, nor was there a word of criticism of the theological colleges when that clergyman in southern Illinois killed his wife in order to marry someone else. But, again, I say, the statement is not correct, and I object.

Mr. Bryan—We do not ask to have taught in the schools any doctrine that teaches a clergyman killed his wife—

The Court—Of course, I can not pass on the question of fact.

Mr. Darrow—I want to take an exception.

Mr. Bryan—I will read you what you said in that speech here.

Mr. Darrow—If you will read it all.

Mr. Bryan—I will read that part I want; you read the rest. (Laughter.) This book is for sale.

Mr. Darrow—First, of all I want to say, of course this argument is presumed to be made to the court, but it is not, I want to object to injecting any other case into this proceeding, no matter what the case is. I want to take exception to it, if the court permits it.

The Court—Well, Col. Bryan, I doubt you are making reference to what Col. Darrow has said in any other case, since, since he has not argued this case, except to verify what you have said, it can not be an issue here, perhaps you have the right—

Mr. Bryan—Yes, I would like very much to give you this.

Mr. Darrow—If your honor permits, I want to take an exception.

The Court—You may do so.

Mr. Bryan—If I do not find what I say, I want to tender an apology, because I have never in my life misquoted a man intentionally.

Mr. Darrow—I am intimating you did. Mr. Bryan, but you will find a thorough explanation in it. I am willing for him to refer to what he wants, to look it up, and I will refer the court to what I want to later.

The Court—All right.

Mr. Darrow—It will only take up time.

Mr. Bryan—I want to find what he said, where he says the professors and universities were more responsible than Leopold was.

Mr. Darrow—All right, I will show you what I said, that the professors and the universities were not responsible at all.

Mr. Bryan—You added after that you did not believe in excluding the reading of it, that you thought that was one of the things—

Mr. Darrow—The fellow that invented the printing press did some mischief as well as some good.

Mr. Bryan—Here it is, Page 84, and this is on sale here in town. I got four copies the other day; cost me $2; anybody can get it for 50 cents apiece, but he cannot buy mine. They are valuable.

Mr. Malone—I will pay $1.50 for yours. (Laughter.)

Bryan Quotes Darrow in Loeb-Leopold Case.

Mr. Bryan (Reading)—"I will guarantee that you can go down to the University of Chicago today—into its big library and find over 1,000 volumes of Nietzsche, and I am sure I speak moderately. If this boy is to blame for this, where did he get it? Is there any blame attached because somebody took

Nietzsche's philosophy seriously and fashioned his life on it? And there is no question in this case but what it is true. Then who is to blame? The university would be more to blame than he is. The scholars of the world would be more to blame than he is. The publishers of the world—and Nietzsche's books are published by one of the biggest publishers in the world—are more to blame than he. Your honor, it is hardly fair to hang a 19-year-old boy for the philosophy that was taught him at the university." Now, there is the university and there is the scholar.

Mr. Darrow—Will you let me see it?

Mr. Bryan—Oh, yes, but let me have it back.

Mr. Darrow—I'll give you a new one autographed for you. (Laughter.)

Mr. Bryan—Now, my friends, Mr. Darrow asked Howard Morgan, "Did it hurt you? Did it do you any harm? Did it do you any harm?" Why did he not ask the boy's mother?

Mr. Darrow—She did not testify.

Mr. Bryan—No, but why did you not bring her here to testify?

Mr. Darrow—I fancy that his mother might have hurt him.

Mr. Bryan—Your honor, it is the mothers who find out what is being done, and it is the fathers who find out what is being done. It is not necessary that a boy, whose mind is poisoned by this stuff, poisoned by the stuff administered without ever having the precaution to write poison on the outside, it is the parents that are doing that, and here we have the testimony of the greatest criminal lawyer in the United States, stating that the universities—

Mr. Darrow—I object, your honor, to an injection of that case into this one.

The Court—It is argument before the court period. I do not see how—

Mr. Darrow—If it does not prejudice you, it does not do any good.

The Court—No, sir; it does not prejudice me.

Mr. Darrow—Then, it does not do any good.

The Court—Well. (Loud laughter and great applause.)

Mr. Bryan—If your honor, please, let me submit, we have a different idea of the purpose of argument, my idea is that it is to inform the court, not merely to prejudice the court.

The Court—Yes.

Mr. Darrow—I am speaking of this particular matter.

The Court—Suppose you get through with Col. Darrow as soon as you can, Mr. Bryan.

Mr. Bryan—Yes, I will. I think I am through with the colonel now. The gentleman was called as an expert, I say, did not tell us where life began, or how. He did not tell us anything about the end of this series, he did not tell us about the logical consequences of it, and the implications based upon it. He did not qualify even as an expert in science, and not at all as an expert in the Bible. If a man is going to come as an expert to reconcile this definition of evolution with the Bible, he must be an expert on the Bible also, as well as on evolution, and he did not qualify as an expert on the Bible, except to say he taught a Sunday School class.

Mr. Malone—We were not offering him for that purpose. We expect to be able to call experts on the Bible.

Mr. Bryan—Oh, you did not count him as an expert?

Mr. Malone—We count him as a Christian, possibly not as good as Mr. Bryan.

Mr. Bryan—Oh, you have three kinds to be called.

Mr. Malone—No, just Americans. It is not a question of citizenship and not a distinction.

Mr. Bryan—We are to have three kinds of people called. We are to have the expert scientist, the expert Bible men and then just Christians.

Mr. Malone—We will give you all the information you want, Mr. Bryan.

Mr. Bryan—Thank you, sir. I think we have all we want now. (Applause.) Now, your honor, when it comes to Bible experts, do they think that they can bring them in here to instruct the members of the jury, eleven of whom are members of the church? I submit that of the eleven members of the jury, more of the jurors are experts on what the Bible is than any Bible expert who does not subscribe to the true spiritual influences or spiritual discernments of what our Bible says.

Voice in audience, "Amen!"

Must Be a Christian to Understand the Bible.

Mr. Bryan—(Continuing) and the man may discuss the Bible all he wants to, but he does not find out anything about the Bible until he accepts God and the Christ of whom He tells.

Mr. Darrow—I hope the reporters got the amens in the record. I want somewhere, at some point, to find some court where a picture of this will be painted. (Laughter.)

Mr. Bryan—Your honor, we first pointed out that we do not need any experts in science. Here is one plain fact, and the statute defines itself, and it tells the kind of evolution it does not want taught, and the evidence says that this is the kind of evolution that was taught, and no number of scientists could come in here, my friends, and override that statute or take from the jury its right to decide this question, so that all the experts that they could bring would mean nothing. And, when it comes to Bible experts, every member of the jury is as good an expert on the Bible as any man that they could bring, or that we could bring. The one beauty about the Word of God is, it does not take an expert to understand it. They have translated that Bible into five hundred languages, they have carried it into nations where but few can read a word, or write, to people who never saw a book, who never read, and yet can understand that Bible, and they can accept the salvation that that Bible offers, and they can know more about that book by accepting Jesus and feeling in their hearts the sense of their sins forgiven than all of the skeptical outside Bible experts that could come in here to talk to the people of Tennessee about the construction that they place upon the Bible, that is foreign to the construction that the people here place upon it. Therefore, your honor, we believe that this evidence is not competent, it is not a mock trial, this is not a convocation brought here to allow men to come and stand for a time in the limelight, and speak to the world from the platform at Dayton. If we must have a mock trial to give these people a chance to get before the public with their views, then let us convene it after this case is over, and let people stay as long as they want to listen, but let this court, which is here supported by the law, and by the taxpayers, pass upon this law, and when the legislature passes a law and makes it so plain that even though a fool need not err therein, let us sustain it in our interpretations. We have a book here that shows everything that is needed to make one understand evolution, and to show that the man violated the law. Then why should we prolong this case. We can bring our experts here for the Christians; for every one they can bring who does not believe in Chrstianity, we can bring more than one who believes in the Bible and rejects evolution, and our witnesses will be just as good experts as theirs on a question of that kind. We could have a thousand or a million witnesses, but this case as to whether evolution is true or not, is not going to be tried here, within this city; if it is carried to the state's courts, it will not be tried there, and if it is taken to the great court at Washington, it will not be tried there. No, my friends, no court or the law, and no jury, great or small, is going to destroy the issue between the believer and the unbeliever. The Bible is the Word of God; the Bible is the only expression of man's hope of salvation. The Bible, the record of the Son of God, the Savior of the world, born of the virgin Mary, crucified and risen again. That Bible is not going to be driven out of this court by experts who come hundreds of miles to testify that they can reconcile evolution, with its ancestor in the jungle, with man made by God in His image, and put here for purposes as a part of the divine plan. No, we are not going to settle that question here, and I think we ought to confine ourselves to the law and to the evidence that can be admitted in accordance with the law. Your court is an office of this state, and we who represent the state as counsel are officers of the state, and we cannot humiliate the great state of Tennessee by admitting for a moment that people can come from anywhere and protest against the enforcement of this state's laws on the ground that it does not conform with their ideas, or because it banishes from our schools a thing that they believe in and think ought to be taught in spite of the protest of those who employ the teacher and pay him his salary.

The facts are simple, the case is plain, and if those gentlemen want to enter upon a larger field of educational work on the subject of evolution, let us get through with this case and then convene a mock court for it will deserve the title of mock court if its purpose is to banish from the hearts of the people the Word of God as revealed. (Great applause.)

The Court—We will take a short recess.

Darrow's Statement.

The Court—Col. Darrow, did you say you had a statement you wanted to make.

Mr. Darrow—I want to read what I said. I shan't include an argument.

The Court—There is no objection, colonel.

Mr. Darrow—I shan't include argument; I don't think I have the right. Following what Mr. Bryan said—(Commotion in courtroom near judge's stand.)

Court Officer—Just a picture machine fallen over.

Mr. Darrow—Following what he used in a paragraph explanatory of it that I want to quote:

"Now, I do not want to be misunderstood about this. Even for the sake of saving the lives of my clients, I do not want to be dishonest, and tell the court something I do not honestly think in this case. I do not believe that the universities are to blame. I do not think they should be held responsible. I do think, however, that they are too large, and that they should keep a closer watch, if possible, upon the individual. But you cannot destroy thought because, forsooth, some brain may be deranged by thought. It is the duty of the university, as I conceive it, to be the great storehouse of the wisdom of the ages, and to let students go there, and learn, and choose. I have no doubt but that it has meant the death of many; that we cannot help. Every changed idea in the world has had its consequences. Every new religious doctrine has created its victims. Every new philosophy has caused suffering and death. Every new machine has carved up men while it served the world. No railroad can be built without the destruction of human life. No great building can be erected but that unfortunate workmen fall to the earth and die. No great movement that does not bear is toll of life and death; no great ideal but does good and harm, and we cannot stop because it may do harm.

In connection with Nietzsche, he was not connected with a university at all; he was a disciple of the doctrine of the superman.

W. J. Bryan—I want to show that Nietzsche did praise Darwin. He put him as one of the three great men of his century. He put Napoleon first, because Napoleon had made war respectable. And he put Darwin among the three great men, his supermen were merely the logical outgrowth of the survival of the fittest with will and power, the only natural, logical outcome of evolution. And Nietzsche, himself, became an atheist following that doctrine, and became insane, and his father and mother and an uncle were among the people he tried to kill.

Darrow—He didn't make half as many insane people as Jonathan Edwards, your great theologian. And he did not preach the doctrine of evolution. He said that Darwin had a great mind. I suppose Col. Bryan would say that. And Napoleon, though neither Mr. Bryan nor I adore Napoleon—I know I don't, and I don't think he does. He did not teach the doctrine of evolution.

Court—All right, colonel, be certain to return the book.

Malone Replies to Bryan.

Dudley Field Malone—If the court please, it does seem to me that we have gone far afield in this discussion. However, probably this is the time to discuss everything that bears on the issues that have been raised in this case, because after all, whether Mr. Bryan knows it or not, he is a mammal, he is an animal and he is a man. But, your honor, I would like to advert to the law, and to remind the court that the heart of the matter is the question of whether there is liability under this law.

I have been puzzled and interested at one and the same time at the psychology of the prosecution and I find it difficult to distinguish between Mr. Bryan, the lawyer in this case; Mr. Bryan, the propagandist outside of this case, and the Mr. Bryan who made a speech against science and for religion just now—Mr. Bryan my old chief and friend. I know Mr. Bryan. I don't know Mr. Bryan as well as Mr. Bryan knows Mr. Bryan, but I know this, that he does believe—and Mr. Bryan, your honor, is not the only one who believes in the Bible. As a matter of fact there has been much criticism, by indirection and implication, of this text, or synopsis, if you please, that does not agree with their ideas. If we depended on the agreement of theologians, we would all be infidels. I think it is in poor taste for the leader of the prosecution to cast reflection or aspersions upon the men and women of the teaching profession in this country. God knows, the poorest paid profession in America is the teaching profession, who devote themselves to science, forego the gifts of God, consecrate their brains to study, and eke out their lives as pioneers in the fields of duty, finally hoping that mankind will profit by his efforts, and to open the doors of truth.

Mr. Bryan quoted Mr. Darwin. That theory was evolved and explained by Mr. Darwin seventy-five years ago. Have we learned nothing in seventy-five years? Here we have learned the truth of biology, we have learned the truth of anthropology, and we have learned more of archeology? Not very long since the archeological museum in London established that a city existed, showing a high degree of civilization in Egypt 14,000 years old, showing that on the banks of the Nile River there was a civilization much older than ours. Are we to hold mankind to a literal understanding of the claim that the world is 6,000 years old, because of the limited vision of men who believed the world was flat, and that the earth was the center of the universe, and that man is the center of the earth. It is a dignified position for man to be the center of the universe, that the earth is the center of the universe, and that the heavens revolve about us. And the theory of ignorance and superstition for which they stood are identical, a psychology and ignorance which made it possible for theologians to take old and learned Galileo, who proposed to prove the theory of Copernicus, that the earth was round and did not stand still, and to bring old Galileo to trial—for what purpose? For the purpose of proving a literal construction of the Bible against truth, which is revealed. Haven't we learned anything in seventy-five years? Are we to have our children know nothing about science except what the church says they shall know? I have never seen harm in learning and understanding, in humility and open-mindedness, and I have never seen clearer the need of that learning than when I see the attitude of the prosecution, who attack and refuse to accept the information and intelligence, which expert witnesses will give them. Mr. Bryan may be satisfactory to thousands of people. It is in so many ways that he is satisfactory to me; his enthusiasm, his vigor, his courage, his fighting ability these long years for the things he thought were right. And many a time I have fought with him, and for him; and when I did not think he was right, I fought just as hard against him. This is not a conflict of personages; it is a conflict of ideas, and I think this case has been developed by men of two frames or mind. Your honor, there is a dif-

Theological and Scientific Minds Differ.

ference between theological and scientific men. Theology deals with something that is established and revealed; it seeks to gather material, which they claim should not be changed. It is the Word of God, and that cannot be changed; it is literal, it is not to be interpreted. That is the theological mind. It deals with theology. The scientific is a modern thing, your honor. I am not sure that Galileo was the one who brought relief to the scientific mind; because, theretofore, Aristotle and Plato had reached their conclusions and processes, by metaphysical reasoning, because they had no telescope and no microscope. These were things that were invented by Galileo. The difference between the theological mind and the scientific mind is that the theological mind is closed, because that is what is revealed and is settled. But the scientist says no, the Bible is the book of revealed religion, with rules of conduct, and with aspirations—that is the Bible. The scientist says, take the Bible as guide, as an inspiration, as a set of philosophies and preachments in the world of theology.

And what does this law do? We have been told here that this was not a religious question. I defy anybody, after Mr. Bryan's speech, to believe that this was not a religious question. Mr. Bryan brought all of the foreigners into this case. Mr. Bryan had offered his services from Miami, Fla.; he does not belong in Tennessee. If it be wrong for American citizens from other parts of this country to come to Tennessee to discuss issues which we believe, then Mr. Bryan has no right here, either. But it was only when Mr. Darrow and I had heard that Mr. Bryan had offered his name and his reputation to the prosecution of this young teacher, that we said, Well, we will offer our services to the defense. And, as I said in the beginning, we feel at home in Tennessee; we have been received with hospitality, personally. Our ideas have not taken effect yet; we have corrupted no morals so far as I know, and I would like to ask the court if there was any evidence in the witnesses produced by the prosecution, of moral deterioration due to the course of biology which Prof. Scopes taught these children—the little boy who said he had not been hurt by it, and who slipped out of the chair possibly and went to the swimming pool; and the other who said that the theory he was taught had not taken him out of the church. This theory of evolution, in one form or another, has been up in Tennessee since 1832, and I think it is incumbent on the prosecution to introduce at least one person in the state of Tennessee whose morals have been affected by the teaching of this theory.

After all, we of the defense contend, and it has been my experience, your honor, in my twenty years, as Mr. Bryan said, as a criminal lawyer, that the prosecution had to prove its case; that the defense did not have to prove it for them. We have a defendant here charged with a crime. The prosecution is trying to get your honor to take the theory of the prosecution as the theory of our defense. We maintain our right to present our own defense, and present our own theory of our defense, and to present our own theory of this law, because we maintain, your honor, that if everything that the state has said in its testimony be true—and we admit it is true—that under this law the defendant Scopes has not violated that statute. Haven't we the right to prove it by our witnesses if that is our theory, if that is so. Moreover, let us take the law—Be it enacted by the state of Tennessee that it shall be unlawful for any teacher in any universities, normals or any other schools in the state which are supported in whole or in part by public funds of the state, to teach any theory that denies the story of divine creation of man as taught in the Bible, and to teach him that man is descended from a lower order of animals. If that word had been "or" instead of "and," then the prosecution would only have to prove half of its case. But it must prove, according to our contention, that Scopes not only taught a theory that man had descended from a lower order of animal life, but at the same time, instead of that theory, he must teach the theory which denies the story of divine creation set forth in the Bible. And we maintain that we have a right to introduce evidence by these witnesses that the theory of the defendant is not in conflict with the theory of creation in the Bible. And, moreover, your honor, we maintain we have the right to call witnesses to show that there is more than one theory of the creation in the Bible. Mr. Bryan is not the only one who has spoken for the Bible; Judge McKenzie is not the only defender of the word of God. There are other people in this country who have given their whole lives to God. Mr. Bryan, to my knowledge, with a very passionate spirit and enthusiasm, has given most of his life to politics. We believe—(Applause.)

The Court—Mr.—

Bible Not Book of Science.

Mr. Malone—I would like to say your honor, as personal information, that probably no man in the United States has done more to establish certain standards of conduct in the mechanics and world of politic than Mr. Bryan. But is that any reason that I should fall down when Bryan speaks of theology? Is he the last word on the subject of theology?

Well do I remember in my history the story of the burning of the great library at Alexandria, and just before it was burned to the ground that the heathen, the Mohamedians and the Egyptians, went to the hostile general and said, "Your honor, do not destroy this great library, because it contains all the truth that has been gathered," and the Mohamedian general said, but the Koran contains all the truth. If the library contains the truth that the Koran contains we do not need the library and if the library does not contain the truth that the Koran contains then we must destroy the library anyway."

But these gentlemen say the Bible contains the truth—if the world of science can produce any truth or facts not in the Bible as we understand it, then destroy science, but keep our Bible." And we say "keep your Bible." Keep it as your consolation, keep it as your guide, but keep it where it belongs, in the world of your own conscience, in the world of your individual judgment, in the world of the Protestant conscience that I heard so much about when I was a boy, keep your Bible in the world of theology where it belongs and do not try to tell an intelligent world and the intelligence of this country that these books written by men who knew none of the accepted fundamental facts of science can be put into a course of science, because what are they doing here? This law says what? It says that no theory of creation can be taught in a course of science, except one which conforms with the theory of divine creation as set forth in the Bible. In other words, it says that only the Bible shall be taken as an authority on the subject of evolution in a course on biology.

The Court—Let me ask you a question, colonel? It is not within the province of this court to determine which is true is it?

Mr. Malone—No, but it is within the province of the court to listen to the evidence we wish to submit to make up its own mind, because here is the issue—

The Court—I was going to follow that with another question. Is it your theory—is it your opinion that the theory of evolution is reconcilable with the story of the divine creation as taught in the Bible?

Mr. Malone—Yes.

Scientists Are God-Fearing Men.

The Court—In other words, you believe—when it says—when the Bible says that God created man, you believe that God created the life cells and that then out of that one single life cell the God created man by a process of growth or development—is that your theory?

Mr. Malone—Yes.

The Court—And in that you think that it doesn't mean that he just completed him, complete all at once?

Mr. Malone—Yes, I might think that and I might think he created him serially—I might think he created him anyway. Our opinion is this—we have the right, it seems to us, to submit evidence to the court of men without question who are God-fearing and believe in the Bible and who are students of the Bible and authorities on the Bible and authorities on the scientific world—they have a right to be allowed to testify in support of our view that the Bible is not to be taken literally as an authority in a court of science.

The Court—That is what I was trying to get, your position on. Here was my idea. I wanted to get your theory as to whether you thought it was in the province of the court to determine which was true, or whether it was your theory that there was no conflict and that you had a right to introduce proof to show what the Bible—what the true construction or interpretation of the Bible story was.

Mr. Malone—Yes.

The Court—That is your opinion.

Mr. Malone—Yes. And also from scientists who believe in the Bible and belong to churches and who are God-fearing men—what they think about this subject, of the reconcilement of science and religion—of all science and the Bible—your honor, because yesterday I made a remark, your honor, which might have been interpreted as personal to Mr. Bryan. I said that the defense believed we must keep a clear distinction between the Bible, the church, religion and Mr. Bryan. Mr. Bryan, like all of us, is just an individual, but like himself he is a great leader. The danger from the viewpoint of the defense is this, that when any great leader goes out of his field and speaks as an authority on other subjects his doctrines are quite likely to be far more dangerous than the doctrines of experts in their field who are ready and willing to follow, but what I don't understand is this, your honor, the prosecution inside and outside of the court has been ready to try the case and this is the case. What is the issue that has gained the attention not only of the American people, but people everywhere? Is it a mere technical question as to whether the defendant Scopes taught the paragraph in the book of science? You think, your honor, that the News Association in London, which sent you that very complimentary telegram you were good enough to show me in this case, because the issue is whether John Scopes taught a couple of paragraphs out of his book? Oh, no, the issue is as broad as Mr. Bryan himself has made it. The issue is as broad as Mr. Bryan has published it and why the fear? If the issue is as broad as they make it why the fear of meeting the issue? Why, where issues are drawn by evidence, where the truth and nothing but the truth are scrutinized and where statements can be answered by expert witnesses on the other side—what is this psychology of fear? I don't understand it. My old chief—I never saw him back away from a great issue before. I feel that the prosecution here is filled with a needless fear. I believe that if they withdraw their objection and hear the evidence of our experts their minds would not only be improved but their souls would be purified. I believe and we believe that men who are God-fearing, who are giving their lives to study and observation, to the teaching of the young—are the teachers and scientists of this country in a combination to destroy the morals of the children to whom they have dedicated their lives? Are preachers the only ones in America who care about our youth? Is the church the only source of morality in this country? And I would like to say something for the children of the country. We have no fears about the young people of America. They are a pretty smart generation. Any teacher who teaches the boys or the girls today, an incredible theory

No Need to Worry About Children.

—we need not worry about those children of this generation paying much attention to it. The children of this generation are pretty wise. People, as a matter of fact I feel that the children of this generation are probably much wiser than many of their elders. The least that this generation can do, your honor, is to give the next generation all the facts, all the available data, all the theories, all the information that learning, that study, that observation has produced—give it to the children in the hope of heaven that they will make a better world of this than we have been able to make it. We have just had a war with twenty-million dead. Civilization is not so proud of the work of the adults. Civilization need not be so proud of what the grown ups have done. For God's sake let the children have their minds kept open—close no doors to their knowledge; shut no door from them. Make the distinction between theology and science. Let them have both. Let them both be taught. Let them both live. Let them be reverent, but we come here to say that the defendant is not guilty of violating this law. We have a defendant whom we contend could not violate this law. We have a defendant whom we can prove by witnesses whom we have brought here and are proud to have brought here, to prove, we say, that there is no conflict between the Bible and whatever he taught. Your honor, in a criminal case we think the defendant has a right to put in his own case, on his own theory, in his on way. Why! because your honor, after you hear the evidence, if it is inadmissible if it is not informing to the court and informing to the jury, what can you do? You can exclude it—you can strike it out. What is the jury system that Mr. Bryan talked so correctly about just about a week ago, when he spoke of this jury system, when he said it was a seal of freedom for free men, in a free state? Who has been excluding the jury for fear it would learn something? Have we? Who has been making the motions to take the jury out of the courtroom? Have we? We want everything we have to say on religion and on science told and we are ready to submit our theories to the direct and cross-examination of the prosecution. We have come in here ready for a battle. We have come in here for this duel. I don't know anything about dueling, your honor. It is against the law of God. It is against the church. It is against the law of Tennessee, but does the opposition mean by duel that our defendant shall be strapped to a board and that they alone shall carry the sword, is our only weapon the witnesses who shall testify to the accuracy of our theory—is our weapon to be taken from us, so that the duel will be entirely one-sided? That isn't my idea of a duel. Moreover it isn't going to be a duel.

Truth Is Imperishable and Eternal.

There is never a duel with the truth. The truth always wins and we are not afraid of it. The truth is no coward. The truth does not need the law. The truth does not need the forces of government. The truth does not need Mr. Bryan. The truth is imperishable, eternal and immortal and needs no human agency to support it. We are ready to tell the truth as we understand it and we do not fear all the truth that they can present as facts. We are ready. We are ready. We feel we stand with progress. We feel we stand with science. We feel we stand with intelligence. We feel we stand with fundamental freedom in America. We are not afraid. Where is the fear? We meet it, where is the fear? We defy it, we ask your honor to admit the evidence as a matter of correct law, as a matter of sound procedure and as a matter of justice to the defense in this case. (Profound and continued applause.)

The bailiff raps for order.

Is the Rev. Dr. Jones or the Rev. Dr. Cartwright in the house? An old resident of Dayton, Mr. Blevins, has passed away and his funeral will be this afternoon at 4:30, those wishing to attend may do so. Pass out quietly.

The Court—Col. Darrow, did you say you had something you wished to say?

Mr. Darrow—No, I just wanted about that much, to try a little more to specifically answer the questions you asked Mr. Malone. I wouldn't think of trespassing or making a speech as I have explained to the attorney-general. Your question as I understood it was whether the doctrine of evolution was consistent with the story of Genesis that God created man out of the dust of the earth—whether the doctrine of evolution that he came up from below a long period of time is consistent with it. What I want to say won't be more than that much. We say that God created man out of the dust of the earth is simply a figure of speech. The same language is used in reference to brutes many times in the Scriptures and it doesn't mean necessarily that he created him as a boy would roll up a spitball out of dust—out of hand—but Genesis, or the Bible says nothing whatever about the method of creation.

The Court—The processes?

Mr. Darrow—It might have been by any other process, that is all.

The Court—So your theory—your opinion, Colonel, is that God might have created him by a process of growth?

Mr. Darrow—Yes.

The Court—Or development?

Mr. Darrow—Yes.

The Court—The fact that he created him did not manufacture him like a carpenter would a table?

Mr. Darrow—Yes, that is all. That is what we claim.

The Court—You recognize God behind the first spark of life?

Mr. Darrow—You are asking me whether I do?

The Court—Your theory—no, not you.

Mr. Darrow—We expect most of our witnesses to take that view. As to me I don't pretend to have any opinion on it.

The Court—My only concern is that as to your theory of it.

Mr. Darrow—So far as this question is concerned, we claim there is no conflict because it doesn't mean making man like a carpenter would make him, but that it is perfectly consistent to say that he was made by a process—perfectly consistent with the Bible—not inconsistent with it—that he was made out of the dust of the earth. Animals were made out of the dust of the earth and everything was made out of the dust of the earth and that had nothing to do with the process, but simply gives a general statement and there is nothing in the Bible which shows the process.

The Court—Colonel, let me ask you another question. You have stated your theory—is it your theory that man and beast had a common origin of life? Does your theory teach that men developed directly from that common origin without first developing into the form of any other animal or that he developed in the one form of life or one physical existence and then passed from that to another form of physical existence—or what is your theory?

No Such Thing as Species.

Mr. Darrow—The theory of evolution as I understand it, and which I believe—it will only take a moment because I have no right to make any argument—life commenced probably with very low forms, most likely one-celled animals and probably in the sea or on the border of the land and sea. That out of that one form grew another. That there is no such thing as species—that is all nonsense. Science does not talk about species. There are differences—and that the differences came by various processes which perhaps none is certain of, but are easily traced through all the history of life that is now extinct, that life has joined on to it, one linking with another and that man is the highest product of it, having the first stem of all life in a very low organism and one branch growing out and soon another branch in that direction and another branch in that direction until we reach the apex in man, where he stands alone, but connects his whole history with the primal origins of life. We say that is entirely consistent. It is a process we are interested in and the Bible story is not inconsistent with that.

The Court—Let me see if I get you clearly?

Mr. Darrow—Not necessarily; some people might say it was and some not.

The Court—A common source—you say all life came from the one cell?

Mr. Darrow—Well, I am not quite so clear, but I think it did. It all came from protoplasm, which is a bearer of life and probably all came from one cell, but all human life comes from one cell. You came from one and I came from one—nothing else, a single cell. All animal life came that way.

The Court—What I want to be clear on—do you say that man developed directly from that one cell into man or did he develop from that one cell into a lower animal and so on from one form of animal life to another until the apex man was reached and he was man?

Mr. Darrow—One form of animal life grew out of another, beginning below—variation exists—variations of all kinds. All life varies and we are creating those new variations every day. They are not species, they are variations and as you went on up there would be a variation in animal structures on up to man. That is surely consistent with the story that man was created out of the dust of the earth.

The Court—According to your theory where did man become endowed with reason?

Mr. Darrow—Well, judge, I don't suppose there is any scientist today but what knows that the lower order of animals have reason.

The Court—It is just in a higher development in man?

Mr. Darrow—No, reason begins way below man.

The Court—I say man has a greater development?

Mr. Darrow—Oh, yes, much greater—very much greater—very much greater than any other animal.

The Court—Does your theory of evolution speak at all on the question of immortality?

Mr. Darrow—There are a lot of people who believe in evolution and who believe in the theory of immortality and no doubt many who do not. Evolution, as a theory, is concerned with the organism of man. Chemistry does not speak of immortality and hasn't anything to do with it. Geology doesn't know anything about it. It is a separate branch of science. I know there are a lot of evolutionists who believe in immortality.

The Court—Those who believe in immortality, where do they—do they also believe that other animals are indowed with immortality?

Mr. Darrow—John Wesley used to believe it, he was an evolutionist in a way. He expected to meet his dog and his horse in the future world. Indians believe it. It has been very common all through the ages, but I don't know—I couldn't say exactly how all evolutionists believe. As to where the idea of immortality came from and as for me I am an agnostic on that. I don't claim to know. I have been looking for evidence all my life and never found it.

Mr. Hays—Might I not ask the court, don't your very inquiry show the necessity of evidence in this case? We have witnesses who can testify to all of this and all that we ask is a chance.

The Court—I was just endeavoring to get Col. Darrow's conception of the theory. I will hear you, Gen. Stewart.

Gen. Stewart Disclaims Kin of Monkey or Ass.

Gen. Stewart—This discussion, which is supposed to be a purely legal discussion, has assumed many and varied aspects. Young Haggard, with the prosecution, suggests to me that it would be necessary that I preach a sermon in order to answer what has been said. My views of things—it has been my nature to always be progressive and liberal in the use of the word evolution. The word evolution, as Mr. Bryan stated, has been misunderstood. The word has been misused. I am not an evolutionist. I don't believe that I came from the same cell with the monkey and the ass, and I don't believe they do as much as they appear anxious to be so classified. I believe that civilization was one time at a very low ebb. I believe that it was in an embryonic stage, so to speak. I believe there was a little civilization. I believe that man is more or less a cave-man and I think sometimes when our tempers get ruffled that we have sufficient evidence of that fact, as I am sure Mr. Hayes will agree with me. I do not ascribe to this theory of evolution, however, which undertakes to teach in defiance of the law of the state of Tennessee, that man descended from a lower order of animals. This is an argument being presented to your honor for the purpose of aiding or assisting your honor, if such be possible, from these gentlemen interested on both sides of this case, in determining whether or not scientific testimony shall be introduced here. The primary purpose of which is to show that there is no conflict between science and the Bible, or strictly speaking, that there is nothing in the theory of evolution that man came from a lower order of animals which conflicts with the story of divine creation. I think, your honor, this turns on an entirely legal question. Mr. Bryan, Jr.—William Jennings Bryan, Jr., very ably presented to the court this morning, even though he was sick and hardly able to do it, a splendid brief that he had prepared on the subject of such testimony being an invasion of the right of the court and jury. That having been so ably handled, I only care, your honor, to discuss this feature, and that is the construction of the act. Who has a copy of that act, please?

Mr. Malone—I have a copy of it. (Mr. Malone gives copy of act to Gen. Stewart.)

Gen. Stewart—We are all familiar at this time with the wording of the act, but it is well to have it before us.

(Counsel thereupon read the act in question.)

Your honor is familiar with the citations, and above all of them, that the cardinal rule of construction in all instruments, and this includes legislative acts, is that the court shall always endeavor to construe the instrument in full accord with the intention of the maker thereof.

Must Determine Intention of the Legislature.

The general assembly, the legislators, convened at Nashville, the last session, that is the spring of 1925—passed this act. It was passed on the 21st. It was signed or approved by the governor on the 21st of March. According to the working of the act it takes effect from and after its passage, which means the date of approval, March 21st. The intention being the test which the court always placed upon written instruments.

Then we have a broad latitude of discussion in undertaking to ascertain what was the intention of the legislature in the passage of this act. To determine this intention the whole act is looked to. The caption, the body and all of the act, and as your honor well knows, outside matter, except under very peculiar circumstances, is inadmissable to determine what the intention of the legislature might be.

I think, if your honor please, that this act—that a correct construction of this act, as a matter of law, prohibits the introduction of this scientific testimony. If you place scientists on the witness stand, men who claim to know and who say they are versed and who no doubt are, no doubt you have many splendid and eminent gentlemen here, who say they are versed in matters of science and particularly in that branch of science which devoted itself to this theory of evolution.

If you place them upon the witness stand, they must confine themselves to that branch or theory of evolution which teaches that man descended from a lower order of animals. That is because the act says so; that is because the act states in so many words, that they shall not teach that man descended from a lower order of animals. I think under the construction, what I concede to be a proper construction of the act that any other theory of evolution might be lawfully taught. Perhaps, but the theory of evolution that we deal with is, whether or not man descended from a lower order of animals, and none other. And we have no right to discuss any other, and the scientists, according to my opinion, would have no right, if the court please, to undertake to talk about any other theory, if they were allowed to talk about any, they would have to qualify as to their familiarity with this particular evolution, the particular kind of evolution that teaches that man descended from a lower order of animals. That is true, I think, your honor, on that question.

That being true, then, if the court please, I think we have proved it sufficiently. Our proof shows it beyond any question. I think the book read shows it, and I think the words from the mouths of witnesses shows it beyond a question that the defendant here did teach to the children in Rhea county High school, that man descended from a lower order of animals. And that they taught that theory.

What Could Scientists Testify To?

Now what could these scientists testify to? They could only say as an expert, qualified as an expert upon this subject, I have made a study of these things and from my standpoint as such an expert, I say that this does not deny the story of divine creation. That is what they would testify to, isn't it? That is all they could testify about.

Now, then, I say under the correct construction of the act, that they cannot testify as to that. Why? Because in the wording of this act the legislature itself construed this instrument according to their intention. Now, says, that any theory that teaches that man descended from a lower order of animals, necessarily—necessarily, denies the story of divine creation. They say it denies it, and, therefore, who can come here to say what is the law is not the law? Who can come here to testify from the witness stand that it does not deny the story of divine creation, when the act says it does?

Your honor, I feel as confident that that is the correct construction as I live by faith, I mean emphasis when I make such an expression, I mean—I do not mean, if the court please, to show any disrespect when I say that. But I mean, if the court please, that as much as I can believe anything under the sun, that that is the correct construction of that act.

Mr. Hays—May I ask you a question, general.

Gen. Stewart—I don't want to be disrespectful; but I cannot keep my line of thought.

Mr. Hays—When you get through?

Gen. Stewart—When I have completed this, all right.

It would be unlawful to teach any theory that denies the story of the divine creation of man, as taught by the Bible, in the Bible, and to teach instead thereof, that man descended from a lower order of animals, instead—instead of what? Instead of the story of divine creation. It shall be unlawful to teach instead of the divine story of creation that man has descended from a lower order of animals. That is what the legislature meant? That is a correct construction.

It is a rule of construction in the state of Tennessee, a rule of the court, if your honor please, that in construing an act of the legislature, that it is the duty of the court to never place an absurd construction upon the act. And I submit that the construction, as I understand it, they insist upon would be absurd.

Would it be necessary to say that the school-teacher brings his class in and says to them: Now, children, I proceed to instruct you as to the story of the divine creation of man as told by the Bible. But I am not going to do that, or that is not true, or something to that effect, and instead of that, I will teach you and instruct you that man descended from a lower order of animals.

Now, he don't have to do that. You don't have to do that. Under a correct construction of the act, if the court please, when this teacher teaches to the children of the high schools of Rhea county, that they are descended from a lower order of animals, he has done all that is necessary to violate this act. He has at the same time taught a theory that denies the divine story of the creation of man.

Why? Because the act says so. Instead—instead of what? Instead of the story of divine creation. Instead of the story of divine creation, and I submit, your honor, that with the application of reason, no other construction can be placed upon this.

What will these scientists testify? They will say, no, this was simply the method by which God created man. I don't care. This act says you cannot testify concerning that, because it denies the literal story that the Bible teaches, and that is what we are restricted to. That is what the legislature had in mind.

Why did they pass that act? They passed it because they wanted to prohibit teaching in the public schools of the state of Tennessee a theory that taught that man was descended from a lower order of animals. Why did they want to pass an act which would deny the right of science to teach this in the schools? Because it denies the story of divine creation. That is why they wanted it passed. And that is why they did pass it. And I submit, your honor, that no other construction can be placed upon it, and no other reasonable reply can be made to the construction.

That is why the legislature, if your honor please, passed this statute. That is why, because this act in so many words says that when you teach that man descended from a lower order of animals you have taught a theory that denies God's Bible, that is what they are driving at.

And to bring experts here to testify upon a construction of the Bible, is (pounding with his hand on the shorthand reporter's table) I submit, respectfully, to your honor, that would be a prostitution upon the courts of the state of Tennessee, and I believe it. It is not admissible, if the court please, under any construction they can place upon it. I know your honor's honest desire to do right about this, and your honor knows that I want to make a correct and proper argument, and not to misstate what I conceive to be the law. And my only purpose is to tell your honor what I conceive to be the everlasting truth about the matter.

Your honor knows if I were to undertake to place a captious construction upon this, your honor knows that when I say that I believe that construction, that I think I am right about it.

I have studied the act. I do not undertake to say that I am right and everybody else is wrong. I do not take that position, but I have studied that act and I believe I am right and I have never believed anything any stronger yet. That is how much emphasis I can put on it. The cardinal rule of construction in Tennessee, as I stated, your honor, is that the intention of the legislature shall govern your honor in construing the statute.

The Legislature Knew What It Meant.

Do you suppose, your honor, that the legislature intended to open the doors to an unending and everlasting argument about whether there is a conflict? Did they have such a thought in mind? How could they? How could the legislature of this state, a body of such splendid men, as we had there last year—how could they design such a thought—how could they hope to place upon the people of this commonwealth such a dangerous law?

They did not have that in mind. How do we know? The act says they did not. They had no thought that the doors would be opened to religious argument and that men would be brought upon the witness stand to testify as to their opinion whether there was a conflict or not a conflict. They determined themselves, this question: Whether there was a conflict between the story of divine creation and the theory that man came from a lower order of animal—monkey, rat, or what not. They say so. And therefore you stand here in the face of this act and undertake to put this on.

Some of the authorities I have cited, your honor, I would like to read to show you how strong the courts make this.

First—A statute should never be given an absurd construction, but must always be construed, if possible, so as to make them effective and carry out the purposes for which they are enacted. The legislative intent will prevail over the literal or strict language used. And, in order to carry into effect this intent, general terms will be limited and those that are narrow expanded.

How much stronger could they make it? General terms will be limited and those that are narrow will be expanded. How eager are the courts that the act shall be construed so as to carry out the intention of the legislature into the court? That is 117 Tennessee, 381 and 134 Tennessee 577.

"Uncertainty of sense does not alone spring from uncertainty of expression. It is always presumed, in regard to a statute, that no absurd or unreasonable result was intended by the legislature. Hence, if viewing a statute from the standpoint of the literal sense of its language it is unreasonable or absurd and obscurity of meaning exists, calling for judicial construction, we must, in that event look to the act as a whole, to the subject with which it deals, to the reason and spirit of the enactment, and thereby, if possible, discover its real purposes; if such purposes can reasonably be said to be within the scope of the language used, it must be taken to be a part of the law, the same as if it were expressed by the literal sense of the words used. In that way while courts do not and cannot bend words, properly, out of their reasonable meaning to effect a legislative purpose, they do give to words a strict or liberal interpretation within the bounds of reason, sacrificing literal sense and rejecting interpretation not in harmony with the evident intent of the lawmakers rather than that such intent should fail."—134 Tennessee, 577."

Another excerpt from the same case. "In construing a statute the meaning is to be determined, not from special words in a single sentence or section but from the act taken as a whole, comparing one section with another and viewing the legislation in the light of its general purpose.—134 Tennessee, 612"

What was the general purpose of the legislature here? It was to prevent teaching in the public schools of any county in Tennessee that theory which says that man is descended from a lower order of animals. That is the intent and nobody can dispute it under the shining sun of this day. That was the purpose of it. Because it denies the story of the divine creation of the Bible. That is the intent, and to bring men, mere men here, made of mud and clay, common mud and clay, to say, that God's word is not contravened by this act. Your honor, there would never be an end to such inquiry as that, there would never be an end, because American citizens to the extent of 100,000,000 abroad in the land of the age of discretion, all have their own opinion, about these things.

Therefore, therefore, what good does the opinion do? We get back to the act every time. Under a construction of this law it is not admissible.

"The fundamental rule, says, Judge Cooper, speaking for the court in the case of Brown vs. Hamlet in 8th Lea 735, 'of construction of all instruments is that the intention shall prevail, and for this purpose the whole of the instrument will be looked to. The real intention will always prevail over the literal use of terms. Legislative acts fall within the rule, and it has been well said that a thing which is within the letter of the statute is not within the statute unless it be within the intention of the law makers.'"

Many cases are cited but it is not necessary for me to read all of these, your honor is familiar with that principle, I know.

The Court—General, as I understand your position, there are two set qualities—

Gen. Stewart—Yes, sir.

The Court—You say when you meet the requirements of the second clause and prove that it is violated, that by necessity by implication of law, meets the first section.

Gen. Stewart—Yes, sir.

Mr. Hays—May I ask your honor to ask Gen. Stewart a question?

The Court—Ask him yourself.

Mr. Hays—You construe the statute to be just the same as if the first part were out, that it is only the second part that you have to prove, so the statute must be the same as if the first part were out. Am I right on that?

Gen. Stewart—So far as the evidence is concerned.

Mr. Hays—So far as the evidence is concerned. You also agree with me, do you not, that one rule of construction in Tennessee is that every word or phase in the statute should be given some meaning?

Gen. Stewart—No, sir.

Mr. Hays—You do not.

Gen. Stewart—No, sir. The court has a right, under our rules of construction, to leave out the words that do not express the intention of the legislature.

Mr. Hays—And haven't you a presumption that the legislature intends that those words must mean something, when it puts words into the statute? Why do you leave out the words? Why not leave out the other words as well?

"Intention" of Legislature "Must" Prevail.

Gen. Stewart—The cardinal rules of construction is that the intention of the legislature must prevail, and it must prevail over everything else.

Mr. Hays—But, it must be gathered from the terms of the act.

Gen. Stewart—You cannot—

(A train whistle interrupts for a moment.)

You cannot change the rule of construction with reference to the intention of the legislature by requiring it to give a meaning to every word.

Mr. Hays—No, general, but I should like to know—you cannot ask the court to accept this statute by cutting out one clause.

Gen. Stewart—Which clause?

Mr. Hays—The first part, that any story contrary to the story of creation taught in the Bible, you construe the statute as if it were cut out?

Gen. Stewart—No, sir; only as to the evidence.

Mr. Hays—As to the evidence, yes.

Mr. Darrow—Doesn't the statute show?

Gen. Stewart—That shows the intention of the legislature as clearly as though they had talked for a month.

Mr. Darrow—They don't have to have any intention if it is plain.

Gen. Stewart—There is an intention every time a man does an act. You have to show the intention whether it is plain or ambiguous. You must always show the intention, that is the first thing you come to.

Mr. Darrow—It shall be unlawful to teach that man descended from a lower order of animals and of course, that would follow your argument, but if it would be legal without what precedes it that has to be given a construction.

Gen. Stewart—The meaning is that the legislature conceived in its mind that that theory did deny the story of creation and it wanted to be emphatic. It had in its mind the Bible, and it had in mind no man should teach a theory contrary to the story of the Bible creation.

Mr. Hays—Exactly the same as if the word "and" was "or."

Gen. Stewart—No. It would not. Anything else?

Mr. Hays—Yes, sir. Perhaps we will agree on this. Hasn't the court to determine on a motion to dismiss as to whether this act is a reasonable act under the police power of the state?

Gen. Stewart—The court has passed upon that already.

Mr. Hays—Hasn't the court, with an open mind, met our argument when we move to dismiss and to produce evidence?

Gen. Stewart—A motion to dismiss is unknown in criminal procedure in Tennessee.

Mr. Hays—Or on a motion in arrest of judgment?

Gen. Stewart—It is unknown in criminal procedure, at this state of a trial.

Mr. Hays—At any rate I can bring up the question before the court in some fashion.

The Court—It cannot come until after conviction.

Mr. Hays—Whenever it comes. There will come a time in this case when we can make the argument that this act is unconstitutional, because it is unreasonable. If we may make that argument, we have a right to produce evidence before your honor in the trial of this case to show that the act is unreasonable. If we do not have that right, don't you agree that the court has a right to accept the evidence if it chooses to do so?

Gen. Stewart—No, sir; I absolutely do not.

Mr. Hays—I am sure of both of these propositions. On the same thing that the general is sure.

Stewart Pretty Sure.

Gen. Stewart—I am just as sure you are wrong as I am sure I am right.

(Laughter and applause.)

Mr. Hays—As the man said to his wife, "You are right and I am wrong."

Gen. Stewart—What did she say?

Mr. Darrow—She said something or other.

The Court—Order please.

Gen. Stewart—Is that all?

Mr. Hays—I think those are the two important issues, though.

Gen. Stewart—Now, your honor, the first report, the first volume of the report of the proceedings of the supreme court of Tennessee, is called First Overton, and the second volume is called Second Overton. Digressing a moment from the immediate points at issue, the reports in that day were gotten out after the names of the judges on the supreme bench, and they then had three, perhaps.

Some weeks ago, in searching the books for something to aid a case, like in the construction of this act, I found a case in Second Overton, and, by the way, which is referred to in one of the United States supreme court reports. This was a lawsuit in which the legal question was, whether or not an entry was a special or general entry and the court in order to determine this, had to determine upon the construction of a statute.

I want to read, if your honor please, a part of what the court said:

"The reasoning powers of men differ as much as their faces. Sometimes different premises are assumed. At others different deductions are drawn from the same premises. With some the result of a process of reasoning is believed to be a fair inference from the premises taken, and consonant to the natural order and fitness of things. Whilst others think they see with equal clearness the process distorted, the result absurd and inconvenient. The truth is we are all imperfect beings, imperfection is the lot of humanity; and in this sublonary state of existence, the views of the wisest head are but limited and indistinct. Ours is but the twilight of knowledge, and he who has the strongest mental eye has, by any method of reasoning he may adopt, only a little better chance of seeing matters as they are disposed by the Supreme Arbiter of all things. Laws were made for the better government of societies; particularly for the convenience and happiness of the community on whom they were intended to operate. Where laws are not local in their nature, operate indiscriminately on the individuals of whom such society is composed, and where civil rights are continually growing out of them and men have for a considerable time immediately succeeding the introduction of these laws, thought and acted alike in relation to them, we may safely adopt the general sense of those concerned, as the most exceptionable ground of decision. In this we cannot err. Individuals may make mistakes in selecting their means of happiness in their process of reasoning, but societies rarely found to have settled down in principle unappropriate to their situation.

Judge Depends on Others' Rulings.

The first utensil a lawyer lays hold of in order to ascertain the law arising in any case is the concurrant opinion of judges or sages of the law who have preceded him; he applies it in preference to any reasoning of his own, independent of experience which he has had, in which the experience of the wisest men in all countries and ages shows that he is continually subject to err. In the absence of evidence of this kind as to what shall be considered a ground of interpretation, courts have adopted the general sense of society for a length of time immediately after the enactment of a law, as much more safe and infallible than theoretic reasoning in all cases where the words of a law are not directly and flatly opposed to such consideration. And even this barrier has been broken down by long and inveterate habits. If the individuals of whom society is composed are generally satisfied with an erroneous construction of a statute and have evinced that satisfaction by conforming their actions to it, who has a right to find fault? Surely not the courts. Legal constructions have always in view the happiness of the people. If they are content and happy in the practical construction of any statute, the end is attained."

That is referring, your honor, to a statute that had been for some time the law and a practice had grown up under this particular statute. Now, your honor, a law is passed in Tennessee and it applies to all people who are within the jurisdictional limits of this sovereign state. A law is passed in Kentucky, in Ohio and in New York and it applies to all who are within the boundaries of its jurisdiction.

This Union, composed of different states, necessarily the different states have different laws which are shaped and formed so as to meet the needs, the conveniences and the notions of the people who dwell within each jurisdiction. This law, which is in test at bar, was passed by the Tennessee legislature. It is a Tennessee law, and it applied to all within the boundaries of this commonwealth, the same as it would apply were it the law of any other state to the boundaries of that commonwealth.

This rule of construction says that the court has in mind always the happiness and contentment of the people. What people? All the people upon whom this law is restrictive, upon whom this law may be enforced, which conforms with this nation. The legislature. It was formed and passed by the legislature, because they thought they saw a need for it. And who, forsooth, may interfere?

What is the Thing?

What is that, that is back of this law? What is this thing that comes here to strike within the bounds of this jurisdiction, and to tell the people of this commonwealth that they are doing wrong to prohibit the teaching of this theory in the public schools? From whence does this opposition originate? Who conceived the idea that Tennessee did not know what she was doing? They say it is sponsored by a lot of religious bigots. Mr. Darrow said that, substantially that.

Ignorant—Who said so? A little handful of folks—a mere handful, who bring to you a theory which they, themselves, can never say is anything but a theory. How far back can science go? How can science go? How can science know that man began as a little germ in the bottom of the sea? Science should continue to progress and it should be unhampered in the bounds of reason, and I am proud of the progress that it has made, and I should say, your honor, that when science treads upon holy ground, then science should invade no further. Almighty God, in His conception of things here, did not intend that there should be a clash upon this earth between any of the forces here, except—save and except the forces of good and evil, and I am sorry that there has come a clash between scientific investigation and God's word.

Stewart on the Side of Religion.

If we, if the court please, who live in this sovereign jurisdiction prefer to worship God according to the dictates of our own consciences, and we give everyone that right to do so, and your honor, I would criticize no man for his individual view of things, but, why, if the court please, is this invasion here? Why, if the court please, have we not the right to interpret our Bible as we see fit? Why, have we not the right to bar the door to science when it comes within the four walls of God's church upon this earth? Have we not the right? Who says that we have not? Show me the man who will challenge it. We have the right to pursue knowledge—we have the right to participate in scientific investigation, but, if the court please, when science strikes at that upon which man's eternal hope is founded, then I say the foundation of man's civilization is about to crumble. They say this is a battle between religion and science. If it is, I want to serve notice now, in the name of the great God, that I am on the side of religion. They say it is a battle between religion and science, and in the name of God, I stand with religion because I want to know beyond this world that there may be an eternal happiness for me and for all. Tell me that I would not stand with it. Tell me that I would believe I was a common worm and would writhe in the dust and go no further when my breath had left my body? There should not be any clash between science and religion. I am sorry that there is, but who brought it on? How did it occur? It occured from teaching that infidelity, that agnosticism, that which breeds in the soul of the child, infidelity, atheism, and drives him from the Bible that his father and mother raised him by, which, as Mr. Bryan has so eloquently said, and drives man's sole hope of happiness and of religion and of freedom of thought, and worship, and Almighty God, from him.

"Bar the Door."

I say, bar the door, and not allow science to enter. That would deprive us of all the hope we have in the future to come. And I say it without any bitterness. I am not trying to say it in the spirit of bitterness to a man over there, it is my view, I am sincere about it. Mr. Darrow says he is an agnostic. He is the greatest criminal lawyer in America today. His courtesy is noticeable—his ability is known—and it is a shame, in my mind, in the sight of a great God, that a mentality like his has strayed so far from the natural goal that it should follow—great God, the good that a man of his ability could have done if he had aligned himself with the forces of right instead of aligning himself with that which strikes its fangs at the very bosom of Christianity.

Yes, discard that theory of the Bible—throw it away, and let scientific development progress beyond man's origin. And the next thing you know, there will be a legal battle staged within the corners of this state, that challenges even permitting anyone to believe that Jesus Christ was divinely born—that Jesus Christ was born of a virgin—challenge that, and the next step will be a battle staged denying the right to teach that there was a resurrection, until finally that precious book and its glorious teachings upon which this civilization has been built will be taken from us.

Religion in American History.

Yes, we have all studied the history of this country.

How many have read the story in history, when the Puritan fathers of this land went on Sunday to their church through the dense woods, no one perhaps except the father and mother and one or two little children, braved the dangers that lurked behind each tree in the forest for in those days the Indians killed the Puritans on frequent occasions. Why did they do these things? Going on Sunday to the religious worship and on other days to worship God according to the dictates of their own conscience?

We are taught that George Washington, on one occasion, before a battle he fought, led his army in prayer, and on another occasion that he secreted himself in a hiding place and prayed in private to the great God for victory. We are told that the great general of the southern Confederacy, Robert E. Lee, prayed to God before each battle and yet here we have a test by science that challenges the right to open the court with a prayer to God. I ask you again, who is it, and what is it, that comes here to attack this law and to say to this people that even though we are but a handful, you are a bunch of fools—who is it, I say—I do not know just who they might be, but they are in strange company. They come and say, "Ye shall not open your court with prayer, we protest"—they say we shall not teach our Bible to our children, because it conflicts with scientific investigation. I say scientific investigation is nothing but a theory and will never be anything but a theory. Show me some reasonable cause to believe it is not. They cannot do it.

Mr. Hays—Give us a chance.

Mr. Stewart—A chance to what?

Mr. Hays—To prove it, to show you what it is.

Not Entitled to a Chance.

Gen. Stewart—If your honor please, that charge strikes at the very vitals of civilization and of Christianity and is not entitled to a chance (applause and laughter throughout house) to prove by the word or mouth of man that man originated in the bottom of the sea. It is as absurd and as rediculous as to say that a man might be half monkey, half man. Who ever saw one—at what stage in development did he shed his tail—where did he acquire his immortality—at what stage in his development did he cross the line from monkeyhood to manhood. Yes, I confess, your honor, their purpose might be to show that to me, but not because they descended from a lower order of animals.

Now, if your honor please, this has been an unusual discussion. We have all gone beyond the pale of the law in saying these things, and I submit to your honor that in its analysis it must rest upon a construction of the statute and upon the law, as given by Mr. Bryan this morning, and is an invasion of the province of the court and jury, I submit, your honor, that under a correct construction of this statute that this scientific evidence would be inadmissible, and I ask your honor, and I say to your honor, to let us not make a blunder in the annals of the tribunals in Tennessee, by permitting such as this. It would be a never-ending controversy, it would be a babble of song, so if the court please, I ask your honor respectfully and earnestly, to disallow the admission of this testimony, and I ask it because I believe under the law of Tennessee, it is absolutely inadmissible.

Mr. Hays—May I ask you a question?

Gen. Stewart—Yes, sir.

Mr. Hays—You understand, do you not, that our scientists are going to state facts from which the court and jury can draw opinions. Does your same argument apply, assuming that our scientists will testify to facts?

Gen. Stewart—Personally, yes.

Mr. Hays—Does it, as a lawyer and attorney-general, not only personally?

Gen. Stewart — Within myself there is only one man.

Mr. Malone—He is a good talker.

The Court—The court will adjourn until 9 o'clock tomorrow morning.

CHAPTER VI.

SIXTH DAY'S PROCEEDINGS—FRIDAY, JULY 17, 1925.

The Bailiff—Raps for order. Everybody stand up, please. Is Rev. Mark in the house?—Rev. Rabbi Mark. Is Rev. Dr. C. G. Eastwood in the house?

The Court—Dr. Eastwood, open court with prayer.

Dr. Eastwood—Our Father and our God, we thank Thee for the privilege that is ours of living in this glorious land that Thou hast given to us through the sacrifice and heroism of those who have lived and gone. We thank Thee, Oh God, that Thou didst inspire them to press onward and upward in the building of a civilization that should last and we pray Thee that the same spirit that impelled them may grip our hearts and seize upon us that we may give to the generations that shall yet follow as rich a heritage as they have bequeathed unto us. And, our Heavenly Father, we thank Thee for the courts of justice in our land, where men can come and receive justice and this morning we pray that Thy blessings may rest upon the Court at this hour and upon this occasion. Wilt Thou give him clearness of vision and of mind for the solution of the problems that are before him? And, our Father, we pray that Thy blessings may rest upon the jury in its deliberations and upon the counsel and upon all those engaged in or participating in this case and, Oh God, we ask Thee that Thy blessings may rest upon those who are members of the press as they send out the messages to the waiting millions of the world. Now again we pray that Thy blessings may rest upon the Court and Thou wilt give Thy divine guidance in the things that shall be done and the decisions that shall be made. These things we ask in the name of our Lord and Master Jesus Christ. Amen.

The Court—Open court, Mr. Sheriff.

The Bailiff—Oyez, oyez, this honorable circuit court is now open, pursuant to adjournment. Sit down please.

TEXT OF JUDGE RAULSTON'S RULING IN EXCLUDING EXPERTS

State of Tennessee vs. John T. Scopes.

This case is now before the court upon a motion by the attorney-general to exclude from the consideration of the jury certain expert testimony offered by the defendant, the import of such testimony being an effort to explain the origin of man and life. The state insists that such evidence is wholly irrelevant, incompetent and impertinent to the issues pending, and that it should be excluded.

Upon the other hand the defendant insists that this evidence is highly competent and relevant to the issues involved, and should be admitted.

The first section of the statute involved in this case reads as follows:

"Be it enacted by the general assembly of the state of Tennessee, that it shall be unlawful for any teacher in any of the universities, normals and all other public schools of the state which are supported in whole or in part by the public school funds of the state, to teach any theory that denies the story of divine creation as taught in the Bible, and to teach instead that man has descended from a lower order of animals."

The state says that it is both proven and admitted that this defendant did teach in Rhea county, within the limits of the statute, that man descended from a lower order of animals; and that with these facts ascertained and proven, it has met the requirements of the statute, and has absolutely established the defendant's guilt; and with his guilt thus admitted and established, his ultimate conviction is unavoidable and inevitable, and that no amount of expert testimony can aid and enlighten the court and jury upon the real issues, or affect the final results. In other words, the state insists that by a fair and reasonable construction of the statute, the real offense provided against in the act is to teach that man descended from a lower order of animals, and that when this is accomplished by a fair interpretation and by legal implication, the whole offense is proven. That is, the state says that the latter clause interprets and explains what the legislature meant and intended by the use of the clause, "any theory that denies the story of divine creation as taught in the Bible."

But the defendant is not content to agree with the state in its theory, but takes issue and says that before there can be any conviction the state must prove two things:

First, that the defendant taught evolution in the sense used in the statute;

Second, that this teaching was contrary to the Bible.

That these are questions of fact, that the proof must show what evolution is, so that the jury may determine whether evolution as taught by the defendant conflicts with the Bible; that it is not merely what the defendant said, or what the book taught; and that they cannot do this without evidence. That is, that the defendant must have taught the descent of man from a lower order of animals, and a theory contrary to that of divine creation as taught by the Bible. That the teaching of either would not be a crime.

Now upon these issues as brought up, it becomes the duty of the court to determine the question of the admissibility of this expert testimony offered by the defendant.

It is not within the province of the court under these issues to decide and determine which is true, the story of divine creation as taught in the Bible, or the story of the creation of man as taught by evolution.

If the state is correct in its insistence, it is immaterial, so far as the results of this case are concerned, as to which theory is true; because it is within the province of the legislative branch, and not the judicial branch of the government to pass upon the policy of a statute; and the policy of this statute having been passed upon by that department of the government, this court is not further concerned as to its policy, but is interested only in its proper interpretation and, if valid, its enforcement.

Let us now inquire what is the true interpretation of this statute. Did the legislature mean that before an accused could be convicted, the state must prove two things:

First—That the accused taught a theory denying the story of divine creation as taught in the Bible;

Second — That man descended from a lower order of animals.

If the first must be specially proven, then we must have proof as to what the story of divine creation is, and that a theory was taught denying that story. But if the second clause is explanatory of the first, and speaks into the act the intention of the legislature and the meaning of the first clause, it would be otherwise.

To illustrate, when the legislature had provided that it shall be unlawful to teach a theory that denies the divine story as taught in the Bible; and, then, by the second clause, merely clarified their intention, and that the real intention as provided by the statute taken as a whole, was to make it unlawful to teach that man descended from a lower order of animals, then there would be no such ambiguity and uncertainty as to the meaning of the statute, and as to the offense provided against, as to justify the court in calling expert testimony to explain.

The court will seek the aid or opinion of expert evidence only when the issues involve facts of such complex nature that a man of ordinary understanding is not competent and qualified to form an opinion.

In Tennessee an act should be construed so as to make it carry out the purposes for which it was enacted.

The legislative intent will prevail over the strict letter, and in order to carry into effect its intent, general terms will be limited, and those which are narrow expanded.

In construing a statute we must look to the act as a whole, to the object with which it deals, and the reason and the spirit of the enactment, and thereby, if possible, discover its real purpose. The meaning must be determined, not from the special words in a single sentence or section, but from the act taken as a whole, comparing one section with another, and viewing the legislation in the light of its general purposes.

In the act involved in the case at bar, if it is found consistent to interpret the latter clause as explanatory of the legislative intent as to the offense provided against, then why call experts? The ordinary, non-expert mind can comprehend the simple language, "descended from a lower order of animals."

These are not ambiguous words or complex terms. But while discussing these words by way of parenthesis, I desire to suggest that I believe evolutionists should at least show man the consideration to substitute the word "ascend" for the word "descend."

In the final analysis this court, after a most earnest and careful consideration, has reached the conclusions that under the provisions of the act involved in this case, it is made unlawful thereby to teach in the public schools of the state of Tennessee the theory that man descended from a lower order of animals. If the court is correct in this, then the evidence of experts would shed no light on the issues.

Therefore, the court is content to sustain the motion of the attorney-general to exclude the expert testimony.

Mr. Hays—Your honor will permit me to take an exception? To state my grounds of exception. We say that it is a denial of justice not to permit the defense to make its case on its own theory.

The Court—You mean the state?

Mr. Hays—No, sir, not to permit the defense to makes its case on its own theory. I say further that it is contrary to every element of Anglo-Saxon procedure and jurisprudence to refuse to permit evidence as to what evolution is and what it means and what the Bible is and what it means. Take my exception on the further ground that for the court of Rhea county to try to determine whether or not this law is unreasonable without informing itself by evidence assumes plenary knowledge on a subject which has been the subject of study of scientists for generations and for these reasons and those placed on the record yesterday the defense most respectfully excepts.

The Court—Let the exception be entered on the record.

Gen. Stewart—I desire to except to exceptions stated in that manner. Such a procedure as that is unknown to the laws of Tennessee and I except to the manner in which the counsel for the defense excepts to the Court's ruling. I think it is a reflection upon the Court.

The Court—Well, it don't hurt this Court.

Gen. Stewart—I think there is no danger of it hurting the Court for that matter.

Mr. Darrow—There is no danger of it hurting us.

Gen. Stewart—No, you are already hurt as much as you can be hurt.

Darrow Is Sarcastic.

Mr. Darrow—Don't worry about us. The state of Tennessee don't rule the world yet. With the hope of enlightening the Court as a whole I want to say that the scientists probably will not correct the words "descent of man" and I want to explain what descent means, as starting with a low form of the life and finally reaching man.

Gen. Stewart—We all have dictionaries.

Mr. Darrow—I don't think the Court has one.

Gen. Stewart—I think the Court knows what "descent" means all right.

Mr. Darrow—We will submit your honor's request to the Association of Scientists.

The Court—I think the Court understands some things as well as the scientists.

Mr. Hays—May I respectfully move if the Court regards this question as one of law for the Court and if the Court believes that the question as to whether or not this law is unreasonable is wholly one for the Court, that the Court hear evidence in order to inform itself on that question in the presence of the Court only and in the absence of the jury.

Gen. Stewart—They are entitled to have entered on record the substance of what they expect to prove. We do not question that. I make no question as to that, but then, of course, they have no right to examine witnesses and conduct a long drawn-out examination and make a farce of your honor's opinion. They are entitled to have sufficient in the record to enable the supreme court to pass upon the proposition, and, in my opinion, a sufficient amount of which is already in the record. How many branches of science have you represented here by witnesses?

Mr. Hays—About six. As I interpret your opinion it does not cover this proposition. The court still has to charge the jury and the court still has to pass on questions of law. We wish to raise, not only before your honor, but before your higher court, our proposition that this law is unreasonable. If your honor will permit me to give an example. Suppose the legislature passed a law prohibiting workmen from working more than six hours in a paint factory. The court would declare that law unconstitutional. But in doing that the court would find out the effect of working more than six hours, and if the work was deleterious to the health of the workmen, then the court would hold such law constitutional.

Raulston Explains Stand.

The Court—Let me state what I have in mind. I think you are entitled to have in the record a sufficient amount of your proof to indicate to the appelate court, in case of conviction here, what your proof would have been. I think you have a right to introduce that proof that is under such limitations as the court may prescribe and let it be written in the record in the absence of the jury, and I meant all the time for you to do that.

Mr. Hays—I would like to state further—if I can prevail upon you to do so—I understand the rule is that we can put in the evidence in that fashion in order that we may make a record for the appellate court, but we not only want to do it for that reason, but we feel we have a right to argue before the court and the court will hear us upon the question of whether or not this law is reasonable. Gen Stewart says that that motion has been denied. That is true, but I hope the court will hear me with an open mind, and we want to introduce the evidence and ask that the court take that evidence and inform itself, and should the court come to a different conclusion, and we hope to persuade the court that this law is unreasonable—we ask the court to permit us to put in evidence for the sole purpose of informing the court so you can determine, after evidence, whether or not this law is unreasonable. I regard that as so important, if you will permit me again to refer to my Copernican illustration, which has seemed to be so humorous to the court in general—your honor knows there are people in the United States who would like to enforce on the people of the United States laws to the effect that nothing could be taught contrary to the theory that the planets moved around the earth and that the earth was the center of the universe, and I have learned of them in the hill country back of Dayton. When people, present the fact that science present the facts in court you would say that a law of that kind was unreasonable, and I state to your honor, in my judgment, if you permit us to come to the evidence your honor will come to the same conclusion on evolution that you have come to on the question of the Copernican theory, and I ask that it be put in as evidence in this case in order to inform this court and give us an opportunity to show whether that law is reasonable or not. Your honor told me yesterday that your honor would hear us with an open mind.

The Court—I am going to let you introduce evidence and I will sit here and hear it, and if that evidence were to convince me that I was in error I would, of course, reverse myself.

Mr. Hays—That is true. I know you would do that.

The Court—You can introduce evidence for the other purpose and I will hear it and I never hesitate to reverse myself if I find myself in error.

Mr. Hays—That being so I think your honor ought to permit us to enter the evidence for both purposes.

The Court—It looks like we are quibbling over a matter really without a difference.

Mr. Hays—If that is so won't your honor give me that privilege?

Mr. Malone—I want to ask Gen. Stewart whether he would mind withdrawing his remarks that the purpose of the defense in producing this evidence is to make a farce out of the judge's opinion. Certainly that is not our purpose and I don't think he meant that it is. We haven't really provided any low comedy here so far, so let us not—

Stewart Stands Ground.

Gen. Stewart—I will be glad to withdraw that and supplement it with this remark, which you will not deny. It is a known fact that the defense consider this a campaign of education to get before the people their ideas of evolution and scientific principles. This case has the aspect of novelty, and therefore has been sensationalized by the newspapers, and of course these gentlemen want to take advantage of the opportunity. I don't want to make any accusations that they are improperly taking advantage of it. They are lawyers and they have these ideas, and it is an opportunity to begin a campaign of education for their ideas and theories of evolution and of scientific principles, and I take it that that will not be disputed and all I ask, if the court please, is that we not go beyond the pale of the law in making this investigation and that we and that they not forget ourselves to the extent that we go beyond the pale of the law. Our practice, if the court please, has been in matters of this sort to let the substance of the evidence be stated by one of the attorneys and let it be placed in the record, in affidavit form, and I think that would be much better and would expedite the trial of this case, and I would much prefer that that course be taken. If witnesses are put on the stand, as your honor knows, a lawyer would ask a thousand questions that are not relevant, and if we do that we go beyond the pale of the investigation, and I respectfully ask your honor to confine this to the subject of that particular theory that is involved in the act and that no more be permitted. They say they have here six branches of science. I don't care how many branches they have, there is only one that is pertinent to this case—only one theory and that is that theory of evolution which teaches that man is descended from a lower order of animals, and if they want something for the higher courts to look at to support that theory—let that be put in substance.

Mr. Darrow—That is what I am willing to do.

Gen. Stewart—Let them put it in in substance—in affidavit form and not take up our time in the trial of the case. I don't object to your testimony or affidavits being printed.

Mr.Malone—I just want to make this statement for the purposes of the record, that the defense is not engaged in a campaign of education, although the way the defense has handled the case has probably been of educational value. We represent no organization nor organizations for the purpose of education. Your honor knows that everything the court says not only goes out to the world through the newspapers, but through the radio and it is difficult for a court these days to exclude a jury from what is going on in the courtroom, because it would be difficult for a juror to go anywhere in the utmost privacy and not hear what's going on, so the rules would have to be changed to meet the advance of science. If the defense is representing anything it is merely representing the attempt to meet the campaign of propaganda which has been begun by a distinguished member of the prosecution.

Bryan wants to Cross-examine Scientists

W. J. Bryan—May I ask if these witnesses are allowed to testify as experts, for the information of the judge, I presume they will be subject to cross-examination?

The Court—Well, Mr. Bryan, I will say, I think the court would make itself absurd after the court has passed upon the question to say he will hear testimony whether or not he was right in his former decision.

What I said was this: I want this proof put into record. I think they are entitled to some of it, under the limitations the court may prescribe. Now the court will be here to hear it and this court is always ready to correct any error it makes. If, after hearing this proof, I shall conclude my former decision was erroneous and unlawful, I would not hesitate to set it aside; but I am not inclined to set it aside in the beginning and say I will hear proof to determine whether or not I will set my opinion aside.

Mr. Bryan—I ask your honor: Will we be entitled to cross-examine their witnesses?

The Court—You will, if they go on the stand.

Darrow's Shot at Bryan.

Mr. Darrow—They have no more right to cross-examine than to bring in the jury to hear this issue. We want to submit what we want to prove. That is all we want to do. If that will not enlighten the court cross-examination of Mr. Bryan would not enlighten the court.

(Laughter in the courtroom).

Mr. Bryan—If I were to dispose—

Colloquy Which Caused Darrow to be Cited for Contempt

Mr. Darrow—What we are interested in, counsel well knows what the judgment and verdict in this case will be. We have a right to present our case to another court and that is all we are after. And they have no right whatever to cross-examine any witness when we are offering simply to show what we expect to prove.

The Court—Colonel, what is the purpose of cross-examination?

Mr. Darrow—The purpose of cross-examination is to be used on the trial.

The Court—Well, isn't it an effort to ascertain the truth?

Mr. Darrow—No, it is an effort to show predjudice.

(Laughter).

Nothing else.

Has there been any effort to ascertain the truth in this case? Why not bring the jury and let us prove it?

The Court— Courts are a mockery—

Mr. Darrow—They are often that, your honor.

The Court—When they permit cross-examination for the purpose of creating prejudce.

Mr. Darrow—I submit, your honor, there is no sort of question that they are not entitled to cross-examine, but all this evidence is to show what we expect to prove and nothing else, and can be nothing else.

The Court—I will say this: If the defense wants to put their proof in the record, in the form of affidavits, of course they can do that. If they put the witness on the stand and the state desires to cross-examine them, I shall expect them to do so.

Mr. Darrow—We except to it and take an exception.

The Court—Yes, sir; always expect this court to rule correctly.

Mr. Darrow—No, sir, we do not. (Laughter).

The Court—I suppose you anticipated it?

Mr. Darrow—Otherwise we should not be taking our exceptions here, your honor. We expect to protect our rights in some other court. Now, that is plain enough, isn't it?

Then, we will make statements of what we expect to prove. Can we have the rest of the day to draft them?

The Court—I would not say—

Mr. Darrow—If your honor takes a half day to write an opinion—

The Court—I have not taken—

Mr. Darrow—We want to make statements here of what we expect to prove. I do not understand why every request of the state and every suggestion of the prosecution should meet with an endless waste of time, and a bare suggestion of anything that is perfectly competent on our part should be immediately over-ruled.

The Court—I hope you do not mean to reflect upon the court?

Darrow Evidently Peeved.

Mr. Darrow—Well, your honor has the right to hope.

The Court— I have the right to do something else, perhaps.

Mr. Darrow—All right; all right.

Mr. Bryan—May it please the court. Do I understand that the defense has decided to put on no witness, but simply to present affidavits?

Mr. Darrow—That is it; to present statements.

Mr. Bryan—And no cross-examination. I understand they were to present witnesses and we were to have a right to cross-examine.

Mr. Darrow—You wouldn't have a right to cross-examine if we put on witnesses for the purpose of showing what we expect to prove.

Gen. Stewart—The court has held he has—we are conducting this case as the court directs.

Mr. Darrow—So far.

Gen. Stewart—So long as it continues, I hope.

Mr. Bryan—Your honor, then to be entitled to go in in the form of affidavits, would we have a right to produce any rebuttal?

Not for this court, but an upper court, is it to be a one-sided trial in the upper court, and will the upper court have nothing before it except the expert statements of the defendant? Or, will the plaintiff be entitled to put in, in the form of affidavits, its proof in rebuttal of what is promised or expected by the defendant?

Mr. Darrow—Mr. Bryan is naturally a little rusty in practice. Of course, the plaintiff has no such right. The question is, is it admissable now. After it has been heard, the state can introduce its rebuttal, but the question is, is this evidence which we offer admissable now? And, as long as the court has held it is not, we are expected to state what we will show.

The Court—I rather think, Col. Darrow is correct. The state's theory is that none of this proof is relevant to the issues, and I have excluded their evidence, holding that under the issues made up under the statute that it is not relevant. Now, the only purpose the court would have in allowing them to put their testimony in the record would be that the higher courts might properly determine whether this court was in error or not in excluding their testimony If the court there decides that evidence was admissable, then it would not be a question there to determine which theory was correct But the appelate court, independent of any number of affidavits, you would put in, would not attempt to pass upon the facts But, if they found that this court had erred in excluding this expert testimony, the case would be sent back. So, I think you would not be entitled to put in any rebuttal proof, would be my conception.

Mr. Hays—Doesn't that mean that they are not entitled to cross-examine?

The Court—That is another question.

Mr. Darrow—We will present it as I said.

The Court—Well, when it comes to taking the whole day, to prepare affidavits, I hate to lose the time. Col. Darrow is certainly laboring under a mistake when he says this court has ever taken a day to prepare an opinion. I read an opinion the other day. The court waited from 1:30 to 3—no,—the forenoon, about five hours, perhaps. It did take time, yes. I believe that is correct.

Gen. Stewart—Your honor needed that time.

Mr. Darrow—I want to ask if it is unreasonable for me to ask for the rest of the day to prepare the statements?

The Court—I don't know.

Mr. Darrow—I ought to know.

The Court—Do you think you need the time?

Mr. Darrow—I do need it, your honor.

The Court—You would know better than I.

Mr. Darrow—I will read them tomorrow.

Gen. Stewart—They wouldn't be read; just filed in the record.

The Court—Yes, they will be filed in the record; no occasion to read them.

Mr. Darrow—All right.

Mr. McKenzie—It has been held that they can go in any time in the world; why take the time of the jury? Put them in the record any time after the lawsuit is done.

The Court—You would dictate to the court stenographers what you expect to prove, and then let it be copied and filed later.

Mr. Darrow—No, I think it ought to be in the record.

Mr. Malone—We have these witnesses here who cannot stay here; we want to make use of them while they are here.

The Court—I mean right now, dictate it.

Mr. Darrow—No, we want to dictate it from our witnesses' statements.

Says He Wants to Be Fair.

The Court—Regardless of the opinion of counsel, I have no purpose except to be fair, but if it takes the day to do it, why of course, but I hate to lose the time, but justice is more important than time.

Mr. Darrow—Certainly, your honor. Your honor, we will come in tomorrow morning.

The Court—Have any of you gentlemen on the state's side any suggestions to make; do you want to be heard any further?

Gen. Stewart—I would like very much to have the afternoon, your honor. There is nothing left now except the argument of the case before the jury.

The Policeman—Order in the courtroom.

Gen. Stewart—We hate so much to lose this time. I do not want to be unreasonable. But, they have six men here.

The Court—Col. Malone, you think you could be ready by 1 or 1:30?

Mr. Malone—Your honor, we have these witnesses here, and they have summer assignments; we don't expect it is possible to make a statement in public here; we cannot do it in public, we have to concentrate upon it. (Consultation between counsel not heard by reporter).

Mr. McKenzie—Both counsels have agreed that a large number of counsel are worn out. These gentlemen want to try and prepare their affidavits; we know we cannot finish the case tomorrow, and there are many reasons why the jury should have a chance to go home and rest. This is the situation, and it is the unanimous agreement we made here, a minute or so ago, subject to your honor's agreement, to finish this case on Monday at 8 o'clock.

Mr. Malone—We think we can finish it up on Monday.

The Court—Today is Friday.

Mr. Malone—Yes, your honor.

The Court—That is agreeable to the court if it suits both sides.

Mr. McKenzie—Suits the attorneys on both sides.

Mr. Hays—Before we adjourn, we do not understand that we have agreed merely to file the affidavits, because if we make our offer of proof, we reserve the right to make it in open court.

The Court—You have made that, and the court has overruled it.

Mr. Hays—No. The suggestion of the general was that we file affidavits. Instead of filing affidavits we may wish to have the opportunity of stating our offer of proof in open court. We have not made up our minds on that.

Gen. Stewart—You have no right.

Mr. Hays—Are not trials public in Tennessee? Isn't it a part of the trial when we state what we expect to prove?

Mr. Neal—As I understand—

The Court—I have passed upon that when you presented it to me.

Mr. McKenzie—It is not part of the trial.

Gen. Stewart—We cannot meet here Monday morning and spend the whole day in statements—the statements are in affidavit form, and placed in the record.

The Court—I will tell you what has been a practice in my court, for the man whose evidence is excluded, is to step to the court reporter and give the proof, so that the jury does not hear it, and proceed with the trial. That is the way we have been doing. But, they say they cannot do that in this case intelligently.

Mr. Darrow—It is too elaborate.

The Court—But, if the statements are put in, in open court, why not make them today?

Mr. Hays—We are not prepared to do that. As you say, when that question comes up, we want to discuss it, but the General wants to discuss it before it comes up.

Gen. Stewart—I don't want to spend all next week—

Mr. Hays—Pardon me.

Gen. Stewart—I understand, if your honor please, they do not have a right under our procedure and practice to state in open court what their witnesses will testify to. What would be the purpose of a statement in open court, for the enlightenment of the crowd present? If they want it for the record—

The Court—If the court excluded a statement Monday morning, I could not give them time then to to prepare it.

Mr. Hays—I ask that your honor hear that question Monday morning.

The Court—I will hear it Monday morning. Let the court take a recess until Monday morning.

Mr. Malone—Until 8 o'clock.

The Court—Nine o'clock. Nine o'clock Monday morning.

Thereupon at the hour of 10:30 o'clock a.m., of Friday, July 17, A. D., 1925, a recess was taken to the hour of 9 o'clock a.m., of Monday, July 20, 1925.

CHAPTER VII.

SEVENTH DAY OF DAYTON EVOLUTION TRIAL—MONDAY, JULY 20, 1925.

Court met pursuant to adjournment. Present as before.

Prayer by Rev. Standefer:

Almighty God, our Father in Heaven, we thank Thee for all the kindly influences Thou hast surrounded our lives with. Thou hast been constantly seeking to invite us to contemplate higher and better and richer creations of Thine, and sometimes we have been stupid enough to match our human minds with revelations of the infinite and eternal. May we, as a nation, have Thy guiding and directing presence with us in all ultimate things, and wilt Thou this morning be the directing presence that supplements human limitations and enables each individual in his respective position to meet the full requirements of this position. Do Thou grant to all of us Thy presence and Thy direction in all things, we ask for Christ sake. Amen.

The Court—Mr. Sheriff, open court.

(Court was then opened.)

Judge Cites Darrow for Contempt of Court.

The Court—If there is any member of the jury in the courtroom, let him at once retire. Any member of the jury anywhere about the courtroom, let him at once retire. You gentlemen have seats in the bar. No member of the jury in the courtroom?

The Court—In the trial of a case there are two things that the court should always endeavor to avoid:

First—The doing of anything that will excite the passions of the jury, and thereby predjudice the rights of either party.

Second—The court should always avoid writing passion into his own decrees.

On last Friday, July 17, contempt and insult were expressed in this court, for the court and its orders and decrees, when the following colloquy occured between the court and one of the attorneys interested in the trial of the case:

Mr. Darrow—What we are interested in, counsel well knows what the judgment and verdict in this case will be. We have a right to present our case to another court and that is all we are after. And they have no right whatever to cross-examine any witness when we are offering simply what we expect to prove.

Court—Colonel, what is the purpose of a cross-examination?

Mr. Darrow—The purpose of cross-examination is to be used on the trial.

Court—Well, isn't it an effort to ascertain the truth?

What Darrow Said.

Mr. Darrow—No, it is an effort to show predjudice. Nothing else. Has there been any effort to ascertain the truth in this case? Why not bring in the jury and let us prove it?

Court—Courts are a mockery—

Mr. Darrow—They are often that, your honor.

The Court—When they permit cross-examination for the purpose of creating prejudice.

Mr. Darrow—I submit, your honor, there is no sort of question that they are not entitled to cross-examine, that all this evidence is to show what we expect to prove and nothing else, and can be nothing else.

The Court—I will say this: If the defense wants to put their proof in the record in the form of affidavits, of course they can do that. If they put witnesses on the stand, and the state desires to cross-examine them, I shall expect them to do so.

Mr. Darrow—We except to it, and take an exception.

The Court—Yes, sir, and always expect the court to rule correctly.

Mr. Darrow—No, sir we do not.

The Court—I suppose you anticipated it?

Mr. Darrow—Otherwise we would not be taking our exceptions here, your honor. We expect to protect our rights in some other court. Now, that is plain enough, isn't it? Then we will make statements of what we expect to prove. Can we have the rest of the day to draft them?

The Court—I would not say—

Mr. Darrow—If your honor takes half a day to write an opinion.

The Court—I have not taken—Yes, I did take five hours.

Mr. Darrow—We want to make a statement here of what we expect to prove. I do not understand why every request of the state and every suggestion of the prosecution would meet with an endless loss of time; and a bare suggestion of anything that is perfectly competent, on our part, should be immediately overruled.

The Court—I hope you do not mean to reflect upon the court?

Mr. Darrow—Well, your honor has the right to hope.

The Court—I have a right to do something else, perhaps.

Mr. Darrow—All right, all right.

The Citation.

The court has withheld any action until passion had time to subdue, and it could be arranged that the jury would be kept separate and apart from proceedings so as not to know of the matters concerning which the court is now about to speak. And these matters having been arranged, the court feels that it is now time for him to speak:

Both the state and federal governments maintain courts, that those who cannot agree may have their differences properly adjudicated. If the courts are not kept above reproach their usefulness will be destroyed. He who would unlawfully and wrongfully show contempt for a court of justice, sows the seeds of discord and breeds contempt for both the law and the courts, and thereby does an injustice both to the courts and good society.

Men may become prominent, but they should never feel themselves superior to the law or to justice.

The criticism of individual conduct of a man who happens to be judge may be of small consequence, but to criticise him while on the bench is unwarranted and shows disrespect for the official, and also shows disrespect for the state or the commonwealth in which the court is maintained.

It is my policy to show the same courtesy to the lawyers of sister states that I show the lawyers of my own state, but I think this courtesy should be reciprocated; those to whom it is extended should at least be respectful to the court over which I preside.

He who would hurl contempt into the records of my court insults and outrages the good people of one of the greatest states of the Union—a state which, on account of its loyalty, has justly won for itself the title of the Volunteer State.

It has been my policy on the bench to be cautious and to endeavor to avoid hastily and rashly rushing to conclusions. But in the face of what I consider an unjustified expression of contempt for this court and its decrees, made by Clarence Darrow, on July 17, 1925, I feel that further forbearance would cease to be a virtue, and in an effort to protect the good name of my state, and to protect the dignity of the court over which I preside, I am constrained and impelled to call upon the said Darrow, to know what he has to say why he should not be dealt with for contempt.

Therefore, I hereby order that instanter citation from this court be served upon the said Clarence Darrow, requiring him to appear in this court, at 9 o'clock a. m., Tuesday, July 21, 1925, and make answer to this citation.

I also direct that upon the serving of the said citation that he be required to make and execute a good and lawful bond for $5,000 for his appearance from day to day upon said citation and not depart the court without leave.

JOHN T. RAULSTON.

Mr. Darrow—What is the bond, your honor?

The Court—$5,000.

Mr. Darrow—That is, I do not have to put it up this morning.

The Court—Not until the papers are served upon you.

Mr. Darrow—Now, I do not know whether I could get anybody, your honor.

Mr. Neal—There will be no trouble.

(Frank Spurlock, of Chattanooga, thereupon volunteered his services in the matter.)

The Court Officer—Let us have order in this courtroom. If you people come up here to hear the trial, this is not a circus. Let us have order.

Mr. Spurlock—Do you want a signed bond, judge?

The Court—I reckon not, Mr. Spurlock. Oh, Mr. Spurlock.

(The court and Mr. Spurlock thereupon held a whispered consultation.)

The Court—Are you ready to proceed with the case on trial, gentlemen?

Mr. Hays—Yes, sir, if your honor please; shall we proceed?

The Court—Yes.

The Governor's Message.

Mr. Hays—Before coming to the evidence that we wish to read into the record, the defense wishes to introduce in evidence a certified copy of the message from the governor approving this bill, on the ground that the message of the governor approving the bill has a bearing on the public policy of this state. Is there any objection?

Gen. Stewart—Yes, we except to that.

The Court—All right, I will hear you.

Gen. Stewart—That is the message that the governor sent to the legislature at the time this bill was being considered by that body. It is not competent in this case.

Mr. Hays—Oh, no, this is the governor's message approving the bill.

Gen. Stewart—That message has no bearing on this case and I object to it.

Mr. Hays—He said, "having these views, I do not hesitate to approve this bill." This is the message approving the bill.

Gen. Stewart—Well, sent to the legislature? Who is the message to?

Mr. Hays—A message from the governor.

Gen. Stewart—To whom?

Mr. Hays—To the senate and house of representatives, approving the bill.

Gen. Stewart—We except to that.

Mr. Hays—I presume the signature is important also. He signs the bill.

Gen. Stewart—We except to that being put in the record.

Mr. Hays—May I read a part of it?

Gen. Stewart—I except to your reading any of it.

The Court—I will hear it.

Gen. Stewart—Why not get what some of the representatives said and introduce it in evidence?

Mr. Hays—I have not yet come to it. You don't give me time.

Gen. Stewart—I will not be surprised if you undertake to do it.

The Court—That would be a matter addressing itself to the powers of the legislature on the question of public policy. I think I will hear you.

Mr. Hays—The governor said, among other things: "It will be seen that this bill does not require any particular theory or interpretation of the Bible regarding man's creation to be taught in the public schools. We know that creeds and religions are commonly founded in the different refinements and interpretations of the Bible. * * * It seems to me that the two laws are entirely consistent. The widest latitude of interpretation will remain as to the time and manner of God's processes in His creation of man."

Another part says: "After careful examination I can find nothing of consequence in the books now being taught in our schools with which this bill will interfere in the slightest manner. Therefore, it will not put our teachers in any jeopardy. Probably the law will never be applied. It may not be sufficiently definite to permit of any specific application or enstatute."

Now, your honor, I believe that that statement is important on the question of the public policy of the state, and has a bearing upon the question of whether this statute is reasonable or within the police powers of the state.

The Court—That is the governor's opinion about it.

Mr. Hays—But when it is a statement made in approval of a bill, your honor will agree with me that his signature is important in the approval of bills.

The Court—Yes.

Mr. Hays—Why not his statement?

Gen. Stewart—His signature on the act is all that could be important, to put the law in force.

Mr. Hays—These are our reasons—

Gen. Stewart—If your honor please, it is absurd to put that into the record.

The Court—He has the floor.

Gen. Stewart—I want to except to his reading any more. If he wants to put that into the record it might be proper to put it in without reading. I say it is not competent for any purpose.

The Court—I will tell you, gentlemen, without further argument, as I have said before, the state government is divided into three branches, one, executive, the other legislative, the other judicial. Of course, the legislative branch has nothing to do with interpreting the law, the courts do that. Gov. Peay—with all deference to Gov. Peay—does not belong to the interpreting branch of the government. His opinion of what the law means, whether or not it would be enforced, is of no consequence at all in the court, and could not have any bearing, and I exclude the statement.

Mr. Hays—I take an exception.

The Court—Yes.

Mr. Hays—May I ask that it be marked for identification?

The Court—Let the record show it was offered and excluded.

Mr. Hays—Will you mark this for identification?

(Passing to court reporter.)

The Court—Yes, mark it for identification. Are you ready to proceed with the case?

Gen. Stewart—Yes, we are ready.

The Court—Do you want the jury?

Gen. Stewart—The state is ready to proceed with the argument.

Mr. Hays—I do not think your honor wants the jury yet.

The Court—No.

Offers New Text Book.

Mr. Hays—We offer in evidence the message of the governor of the state approving the bill; first, for showing the public policy of the state; second, for the purpose of showing that the law providing for the school work that was taught in the schools is not necessarily inconsistent with the teaching of which Scopes is accused, in other words, that the two laws are not necessarily inconsistent; and third, I ask the court to take judicial notice, without the presence of the jury, of the message of the governor as an indication of the public policy of the state. Motion denied. Excepted to. Statement is ruled out. Exception taken by defendant. Now, your honor, at the same line I wish to introduce in evidence the textbook that has now been adopted by your state commission, these parts—

Gen. Stewart—Your honor—

Mr. Hays—Why should I not be permitted to state what I wish to offer in evidence without interruption. May I make my statement without interruption?

Gen. Stewart—No, sir, you cannot do that.

The Court—The court can exclude that if it is improper. It cannot hurt you.

Gen. Stewart—Your honor, the exception I am making is, he is not entitled to read it, that is from a textbook that has just been adopted, since this trial began.

Mr. Hays—Again on the question of public policy, at any rate I wish to state what I wish to offer in evidence.

The Court—Well, state the substance.

Gen. Stewart—If he wants to get it into the record let it be treated as read. I do not see the benefit of reading it.

Mr. Malone—The jury is not here.

Gen. Stewart—What is the reason of reading it to the court?

Mr. Malone—No, he is reading to the stenographer.

The Court—I think the court should hear what he wants to read, if it is not proper.

Mr. Hays—Referring to page 6 (reading): "Charles Darwin, to whom the world owes a great part of its modern progress in biology, spent twenty years in getting answers to puzzling questions as to how plants and animals came to resemble and to differ from each other. He then published one of the epoch-making books of all time, on "The Origin of the Species." Even if we cannot hope to be Pasteurs or Darwins, we can at least keep our eyes and ears open; we can be constantly learning new and interesting facts, and we may be able to contribute something of real value to the sum total of human knowledge."

Referring to page 463 (reading): "The highest order of mammals, the primates." You will remember we referred to that the other day—"There remains one other group of mammals of which we shall speak, namely, the highest, that to which man belongs. This group also includes the monkey, the baboon and the ape. To the latter group belong the orangoutang, the chimpanzee and the gorilla. Because these animals excel the rest of the animal kingdom in brain development and in intelligence, this order of mammals is known as the primates (from the Latin, meaning first). Some of these animals, 'while resembling the human species in many characteristics, must, of course, be recognized as having evolved (developed) along special lines of their own, and none of them are to be thought of as the source or origin of the human species. It is futile, therefore, to look for the primitive stock of the human species in any existing animals.'"

Then, there are questions following for the student to take up. The tenth is, "What animals belong to the order of mammals known as primates? Why are they so called?"

Also, this book, of course, contains a picture of Darwin, and on the order of primates contains a picture of the gorilla, and this is the book we are prepared to prove was recently adopted by your textbook commission. I want to offer these parts in evidence.

The Court—I will hear you, gentlemen.

Gen. Stewart—What is the purpose of offering that?

Mr. Hays—To show the public policy of the state, trying to prove to the court that this law is unreasonable.

Gen. Stewart—The public policy now, or at the time the law was passed?

Mr. Hays—Both.

Gen. Stewart—When do you claim that book was adopted?

The Court—Both?

Gen. Stewart—What?

Mr. Hays—I am going to ask you to tell me or to call witnesses. I thought you would concede our statement.

Gen. Stewart—I mean since the passage of the law.

Mr. Hays—Oh, yes.

Gen. Stewart—We do not think that is competent. The book involved in this case, along with the other evidence, is Hunter's Civic Biology. The book he reads from now is not the same book that was taught last year in the public schools. What is the name of that book?

Mr. Hays—That is true. I do not claim that it is. The testimony as to the other is here. I claim this book was adopted in substitution for that. I claim it indicates you cannot teach biology without teaching something about Darwin and evolution. If this law is unreasonable, it is, of course, unconstitutional. That shows how unreasonable that law is, that is one of the questions we have to make on the constitutionality of the statute. I take it that the court takes judicial notice of every fact that bears upon that question.

The Court—I have already passed upon it, but you have the right to have my action reviewed, of course.

Mr. Hays—No, I think we have a right to have your honor to pass upon it. We want to be heard in order to have it before the court should we desire to make a motion in arrest of judgment or the direction of a verdict.

The Court—Mr. Stewart, for the present I will let this book be filed. If I see proper I will exclude it later. It might be competent, I am not sure.

Gen. Stewart—We except.

Mr. Hays—I am referring to page 6 of "Biology and Human Welfare," by Peabody & Hunt, and page 463. I am quite ready to suggest, if the prosecution wants to use any other part of the book on appeal, or if we want to use any other part, that this same ruling be adopted as to the other parts. (Said book was thereupon received and marked defendant's exhibit, No. 2.)

Mr. Hays—If your honor please, we next desire to make another offer. I have, since the last hearing, looked up the law and inquired from prominent members and jurists of your bar as to the practice in your courts. I understand, of course, that the offer of proof must be made in the absence of the jury. I understand, further, that it is done in any one of three ways. Either you call your witness and first bring out the testimony by question and answer, so as to make your record; or, secondly, you state to the court what you intend to prove; or, thirdly, you make an affidavit, first handing it to opposing counsel. I believe all three ways are properly recognized and used. I am told that an attorney, so long as the jury is not present, is seldom, if ever, denied the right to make this offer of proof in his own way. We are anxious, your honor, to state what our offer of proof is, and we are particularly anxious to state it in reference to a statement that your honor made in the discussion on last Friday. You will remember that I suggested to your honor that it might be, after hearing some of the statements, you would change your ruling, at least as to some of it. For instance, we are prepared to prove what evolution is by a witness, and by the same witness what the Bible is, qualifying him as an expert on both subjects, and show according to a proper interpretation or translation of the Bible, or translation, these two parts of the act are not conflicting and Scope's act has not conflicted with the first part. I don't say that will be convincing to your honor, but I suggest we want to prove it on that ground and also on the ground that after hearing the evidence your honor might change your opinion as to the reasonableness of this law. In the discussion I said: "I asked to be given an opportunity to show whether or not that law is reasonable or not." Your honor then told me this: That your honor would hear us. Here is what happened. I asked that it be put in evidence in this case in order to inform the court and give us an opportunity to show whether that law is reasonable or not. Your honor told me yesterday that your honor would hear us with an open mind.

Your honor said: "I am going to let you introduce evidence and I will sit here and hear it, and if that evidence were to convince me that I was in error, I would, of course, reverse myself."

Mr. Hays—That is true. I know you would do that.

The Court—You can introduce evidence for the other purpose and I will hear it, and I never hesitate to reverse myself if I find myself in error.

Mr. Hays—That being so, I think your honor ought to permit us to enter the evidence for both purposes.

So I suppose we may assume in this offer to proof that we can make it in our own way.

The Court—I said, further, that I did not know about it. Of course all this discussion we have now we did not have before the court, the question as to how that would be introduced.

Mr. Hays—Yes, but your honor said you will hear us.

Gen. Stewart—Permit me to interrupt for just a minute.

Mr. Hays—Yes.

Gen. Stewart—On Friday we adjourned until Monday to give them an opportunity to prepare statements or affidavits of witnesses to be filed. I stated I would not agree to a continuance if we were to meet on Monday morning to spend a whole day in a harangue. I stated it expressly, and the record shows that. We adjourned with the express understanding that they would be permitted to prepare the affidavits for the record, and the only thing left for the court to determine, according to the court's own statement in response to an inquiry by Mr. Hays, as to whether or not the affidavits would be read.

Mr. Hays—That is what I wanted to do. I want to present my offer of evidence, stating that we would prove such and such a thing by such and such a witness.

The Court—You mean you will make the statement yourself, Mr. Hays?

Mr. Hays—Oh, yes; yes, sir.

The Court—Gen. Stewart, do you object to that?

Gen. Stewart—How is that?

The Court—Do you object to his statement that he hopes to prove so and so by certain witnesses?

Gen. Stewart—Your honor, the statement could be made for the record and we would except to it being stated in open court.

The Court—Of course. I want to hear it or want to read it, one.

Gen. Stewart—Your honor could read it. About how long would it take to make these statments?

Mr. Hays—I do not know. If I presume on the patience of the court, the court will stop me. But this is in the absence of the jury.

The Court—I know, of course, you would not expect to read all of the statements, but would merely summarize it.

Mr. Hays—I will merely state what I expect to prove. Some I would summarize, others I would read.

Gen. Stewart—That is not the correct way to put it into the record. If it is to go in in affidavit form, it must have the effect of testimony.

Mr. Hays—I do not understand. I understand statements are allowed to go in.

Gen. Stewart—They must be sworn to.

Mr. Hays—I do not understand the practice. That is not the practice in your state, Cruso vs. State, 95 Tennessee, appears the statement that where you do file a statement, "one way is for counsel to write it out at that time, what is expected to be proved, hand it to opposing counsel, so there is no dispute." I cannot find anything in that case that requires any affidavits being sworn to whatever.

The Court—What about the agreement when we adjourned Friday?

Gen. Stewart—It was to be prepared. We made no such statement. You said you would present the statements to be read in court. That you would have to prepare these, because your scientists were going away.

Mr. Hays—Exactly. We could not prepare these in their absence.

Gen. Stewart—I suggest, your honor, the record will show the word "affidavit" was used, that is the understanding.

Mr. Hays—It was used in one place and not another. "Statement" was used, statements, of course.

Gen. Stewart—We adjourned on Friday to give them the opportunity to prepare them.

Mr. Hays—After our discussion on Friday the court said: "If the statements are to be put in open court, why not put them in today?" That was at the very end. Do you now insist we not put in affidavits?

Gen. Stewart—Read what was said when you asked them for a continuance to Monday?

Mr. Hays (Reading)—"I don't want to spend all next week." That was the reason, probably.

Gen. Stewart—That is the reason; a pretty good reason.

Mr. Hays (Reading) from Friday's minutes of the record.)

Gen. Stewart—The record further shows his honor stated you better have your affidavits prepared.

Mr. Hays—Statements, your honor said. I would like to make my reasons a little clearer, but Gen. Stewart perhaps will agree—

The Court—Yes.

Mr. Hays—First, I offer the proof in open court, while we are all here, so that, if the state's attorney desired to they have a right to deny that the witness would so testify. Secondly, the court should hear read the statements in order to properly certify them as part of the bill of exceptions on appeal. Thirdly, to consider whether the court erred in excluding the testimony from the jury; next, holding the statute unconstitutional, and to consider whether the testimony is not properly before the jury in that it tends to show that the theory of the divine creation of man, as set forth in the Bible, and that the science Scopes taught merely portrayed the manner of man's creation——. There are manifold reasons why the court should read these, and if we are wrong the court could point out to us, and if we are right we should have the benefit of reading them.

The Court—Why not read a synopsis?

Gen. Stewart—Why do you object to preparing them in written form and handing them to state's counsel?

Mr. Hays—It may be a habit or custom of mine, but I like to try my case in open court.

Gen. Stewart—I stated on Friday—

Mr. Hays—I believe under your practice I have a right to make my offer of proof in the form I want to?

Gen. Stewart—No.

Mr. Hays—I believe the prosecution should not insist we make our proof or prepare our record, but in a proper way as long as we are right.

Gen. Stewart—In the discretion of the court.

Mr. Hays—Of course it is, if we ask for anything unreasonable.

Gen. Stewart—It is anyway.

Mr. Hays—Is it unreasonable to state what we mean to prove by certain witnesses, when they can do no harm?

Gen. Stewart—I stated that the primary purpose of the defense is to go ahead with this lawsuit for the purpose of conducting an educational campaign and say to the publice through the press their idea of their theory. And I think that this thoroughly demonstrates that that statement was more than correct.

Mr. Hays—You see the prosecution not only attempts to state our theory of the case, but also to tell our purpose to the court. Why not do it?

The Court—Let us hear from the attorney-general.

Gen. Stewart—There could be no purpose in reading the statement, or making the statement in open court to this crowd, the people here, except for the purpose of furthering its educational campaign as they call it, or spreading propaganda as I call it. That is the only purpose. Put them in the record for the supreme court that it may review the statement when the statement reaches the supreme court. The record is being made up by the stenographers here and they can take the statements prepared and write them into the record, and we can proceed to the disposition of this case without the necessity and the time of arguing this matter out here before the court. Of course, these statements will have to be submitted to the court and counsel here.

The Court—Of course it would relieve the court of a great amount of work, instead of sitting here and reading them.

Gen. Stewart—I think we are fixing to lose two or three days on these statements, right now if that is permitted.

Mr. Hays—Justice is more important than time.

Gen. Stewart—The crowd is not going to try the lawsuit.

Mr. Malone—We are not talking to the crowd. We are talking to his honor.

Gen. Stewart—Put them in—

Mr. Malone—Let his honor dismiss the crowd, and have your honor and the attorneys—

Gen. Stewart—Why not put them in written form?

Mr. Hays—We feel the prosecution has been allowed to state what the theory of our case is, and we insist at any rate to state our purpose. I am afraid perhaps my methods of explanation may be somewhat confused because after ten minutes of explaining why I prefer to present my offer of proof, and the general explanation, and I have a reason, and I am entitled to do it, as long as my procedure is right according to your state practice. The attorney-general may not like our methods, and suspect our purposes, but we have a right to state them to your honor for ourselves.

Mr. Stewart—It is my desire that these proceedings retain what legal aspect they may. It may be contended that it is going to be a Sunday school class or a chautauqua, if so it is time to adjourn.

Mr. Hays—I take exception. No one on this side of the table talks on the chautauqua.

The Court—It is not my purpose to withhold anything from the crowd, or give anything to the people who happen to be here. It is purely a legal question for me.

I would like to see the holdings in Tennessee on this point, if there are any. In my practice we have not had a big case like this. This is my first case of this kind, perhaps, the first the court has had. Ordinarily, when we offer proof that has been ruled out, counsel for the party against whom the court rules just steps over to the shorthand reporter and gives it to him quietly, with the jury in their seats. Now that is the way we ordinarily do it.

As to whether it shall be put in affidavit or in statement form, I am not prepared to rule.

Mr. Hays—Your honor, affidavits have never been required. I can state from the cases I examined yesterday.

Gen. Stewart—You might be right on that.

Mr. Hays—We already insisted it should be done out of the presence of the jury?

Gen. Stewart—I would object as vigorously as I know how as to the statements being made in open court. It is unnecessary. They are being made for the appellate court and whether verbal or written makes no difference to the appellate court.

Mr. Hays—It may make some difference to this court. He still stated he had an open mind. What is the fear?

Gen. Stewart—My objection is to making a Sunday school out of this at the expense of Rhea county, of the courthouse.

Mr. Hays—It may lead to intelligent thought and that can do no harm.

Gen. Stewart—The fact that it may lead to unintelligent thought may do harm.

Mr. Hays—I don't think intelligent thought ever does any harm. I have a right to make my offer of proof in my own way.

Mr. Bryan—If the court pleases—

The Court—Gentlemen—

Mr. Stewart—Mr. Bryan, if your honor pleases—

The Court—If there is no theory why this should not be—to prevent the court from sitting down and reading this, the court and counsel going to a private room and having this put into the record—I do not mean to intimate that it should be done this way.

The Court—Col. Bryan, I will hear you.

Must Not Be a One-Sided Case.

Mr. Bryan—If your honor please, if the object of the defense is to make a record for the higher court, that can be done by affidavits and we will not be allowed any affidavits, on the other side.

If the purpose of the defense is to present an argument with the purpose of pursuading your honor that he was wrong, and in order to induce him to reverse himself, if that is the purpose of this, it cannot be a purely ex parte matter.

If they are allowed to present argument to the court that it should be wrong, and should reverse itself, we certainly should be allowed to present an argument on the other side. As long as it goes into the record for the other side, we are excluded, but so long as the defense is attempting to persuade this court and to secure action in this matter, it cannot be on side ex parte, it seems to me. We must be allowed to present our side to the court so that the court, when it comes to consider whether it should reverse itself, should have both sides of the case and not only the other side.

Mr. Malone—Mr. Bryan is guilty of the same fallacy in his statement now that he was guilty of the other day when he asked the right to cross-examine our witnesses who might be called merely for the purpose for which these statements are offered. The prosecution and the court sustained that objection to the admissibility of the testimony of our witnesses who were here. If the prosecution had not objected and your honor had admitted our witnesses, then Mr. Bryan would have the right he now wishes to claim to cross-examine witnesses. But after limiting us to witnesses to testifying to mere points in synopsis, he wanted to maintain all the broad rights which he would have had if our testimony had been admitted even without the prosecution having objected to our testimony being limited. Now, this morning he claims the right when limited not to witnesses, but statements, and I have the same right to answer that he cannot have the issue limited as to our offer of proof, the court having ruled upon it, and then claim all the broad rights which he would have if the proof had been admitted that we wish to offer.

Mr. Bryan—The point which the gentleman makes is not the point in this case. He says I object to the introduction of witnesses without the right to cross-examine. Now, even if the court had held that we had no right to cross-examine these witnesses on the ground that the testimony was not for the court to consider, but for the higher court, even if the court had so held and we had been permitted to cross-examine witnesses, I submit that this morning is in for an entirely different purpose. The argument to be made by the gentleman from New York is not for the higher court, but for this court, to persuade this court that it was wrong, to secure from this court a decision in this trial, and surely we are not to be banned from presenting our side, whenever they try to persuade this court to take an action that vitally affects this cáse. This is an entirely different point this morning, Mr. Malone, and had the other side been right when they objected to cross-examination, they cannot be right now, because if they had been right then, it would be simply because the evidence was for the upper court, which could not render a decision, but only remand the case for a new trial. Had they been right then, they cannot be right now, when their purpose is not to make a record for the higher court, but to persuade this court and secure a decision from this court for the acquittal of the defendant at this time.

Mr. Darrow—May I say a word in reference to Mr. Bryan's statement, if your honor please?

The Court—Judge MacKenzie, couldn't you furnish me some authorities on this question?

Mr. Hays—I have the authorities.

The Court—Just a minute, Col. MacKenzie.

Mr. MacKenzie—Your honor—

Mr. Darrow—I don't suppose there is any dispute between us lawyers on it, but you may differ, Mr. Bryan. If there is I suggest, your honor, that is a good way to send out for them, but I do not believe there is any dispute between the lawyers on this method.

The Court—I will hear you.

Mr. Darrow—I only want a moment. I agree exactly what the practice is, not only here, but elsewhere. We offer certain evidence; the court refused it. We offer to call witnesses and the court said it was not competent. Now, we cannot predicate error on that unless we put in the record what we expected to prove by witnesses.

Your honor is quite right, and that is ordinarily right, too, stepping up to the shorthand reporter and stating what we expect to prove, telling the court and the shorthand reporter taking it down. That is exactly what we want to do here. We want to state it to the court and have it taken down by the shorthand reporter, or else pass them the statements we have already prepared to be used in the record in lieu of that.

We want to state to the court exactly what we expect these witnesses to swear to. How can there be any question?

The Court—Have you the affidavits?

Mr. Darrow—They are not affidavits, but statements.

Gen. Stewart—Have you the statements prepared?

Mr. Darrow—Yes.

Gen. Stewart—Then simply place them—let them turn them into the record, and proceed with the case.

Mr. Darrow—We think we have the right—

Gen. Stewart—To make a speech? That is what you are talking about!

Mr. Darrow—To choose our own way of protecting the record.

Gen. Stewart—I think not.

Mr. Darrow—We have a right, if we choose, to state in open court we expect to prove, for instance, Dr. Osborn—

The Court—How long will it take?

Mr. Darrow—I think, your honor, we will not need to read all of them; I think we could read all we wanted to in an hour, and then adopt their method on the rest.

The Court—What do you say?

Mr. MacKenzie—What do you say?

The Court—Let me ask you a question; they ask to file statements; they want an hour to briefly review what is in the statements.

Mr. MacKenzie—Can we see the statements?

Mr. Darrow—Certainly.

Mr. McKenzie—After he speaks an hour and tells us what he expects to prove, this is excluded testimony for the supreme court to review, how much closer to the facts than you are right now? Is your honor going to say under your statement, as judge of the Eighteenth Judicial Circuit of the state of Tennessee, if these gentlemen could prove that—have either of the witnesses fainted? Have they run off and has it gotten down to the point where these distinguished gentlemen have to take the statements, too? Not even concerned to? What does the statement of an hour mean in this record? Of course, they are entitled to preserve their exceptions.

Mr. Darrow—That is all?

Mr. McKenzie—Not what Mr. Hays of New York thought he hoped to prove? This is not an application for a continuance?

Mr. Hays—Why of New York?

Mr. McKenzie—I noticed you don't want to be of Tennessee, and hence I thought I would place you. We want you to have the respect of your own wishes, Brother Hays, and we have no objections to your living in New York.

Mr. Hays—(Not heard in the noise and continued talking of counsel.)

Mr. McKenzie—Please do not interrupt me. I am talking to the court, if you please. I will answer anything you want to ask, and write you a letter to boot.

Mr. Hays—You cannot—

Mr. McKenzie—Your honor, I was proud to see as a friend of these distinguished gentlemen among the many able Chattanooga lawyers up here, my distinguished friend, Frank Spurlock, one of the best lawyers in Tennessee, standing by Col. Darrow. If you want to get the Tennessee laws as to how to get this in the record, let him make a statement in the record.

Mr. Hays—He told me.

Mr. McKenzie—We are perfectly willing for you to have it, but we don't want to give you three hours over it; your honor is not going to let you prove that unless you could show some symptoms you could prove that.

Mr. Hays—I want to see the symptoms.

Mr. McKenzie—We have had several of them. I think we have heard the speeches of my good friends, Hays and Malone. We kind of enjoy Brother Darrow speaking, but we heard their speeches and sitting around for two hours, on every exception. Now it is a mere matter of law and procedure what shall go into the record in this case.

Mr. Hays—That is right.

Mr. McKenzie—It has been excluded. Now, will your honor put it in the record in this case? In the first place, this honorable court must be satisfied that they could have proved by these witnesses this excluded testimony.)

The Court—How can I be satisfied?

Mr. McKenzie—The onus is on them.

Mr. Darrow—Let me ask you a simple question.

Mr. McKenzie—All right, Col. Darrow.

Mr. Darrow—If you were asking for the admission of some evidence or John Smith here to testify, and you told the court what it was and the court said it was not competent and ruled you could not give it, isn't that simple statement of what you expect to prove by John Smith enough to preserve the record?

Mr. McKenzie—No, sir; not unless it is agreed to by the other side.

Mr. Darrow—What?

Mr. McKenzie—Never has been in Tennessee. If that is the way of it we ought to just practice law on the statements of the lawyers on each side as to what they want to prove and dispense with the witnesses, and argue the case.

Mr. Darrow—I don't like to dispute on Tennessee law, but I am sure I am right.

Mr. McKenzie—Let Col. Stewart look at your statements there, and if he will agree your witnesses will swear to them—that trouble is all over.

Mr. Hays—Why not read them in open court?

Mr. McKenzie—I don't want to read them and nobody else wants to read them.

Mr. Darrow—It won't take over an hour, and take the statements of the rest of them.

Gen. Stewart—I don't think we will have any trouble about what goes into the record; the only thing is the reading of these in open court.

The Court—They do not purpose to read them.

Gen. Stewart—Read them and make speeches on them.

Mr. Hays—No.

Mr. Darrow—We expect to show that a certain professor will say so and so and read the statement; read two or three of them, and let it go at that.

Mr. Hicks—As I understand and remember, they made the statement the other day to this effect: What they intended to prove, that evolution does not conflict with the Bible, or they want to interpret the Bible or show evolution does not contradict the Bible. Now, your honor has ruled that line of evidence is not admissable in this case. Now, will your honor rule time and again on it? Is there an end to that?

The Court—Didn't they say what they intended to prove; didn't Mr. Hays say that he wanted to offer proof they wanted to show what was meant by it?

Mr. Hicks—That is true, your honor has already ruled on that; what is their purpose?

The Court—I thought that the defendant admitted that he taught that man descended from a lower order of animals.

Mr. Hicks—If they exclude everything else but the only evidence on this merely to save time.

The Court—The higher courts may differ with me.

Mr. Hicks—What is the use of reading them in open court?

Mr. Darrow—We are just trying to make the record, nothing else.

Mr. Hays—We are entitled to make this record in our own way as long as it is in accord with the practice of Tennessee and the constitution. Aren't we entitled to make it as long as it is in accord with your practice or the way you gentlemen say?

Gen. Stewart—Like the court says.

Mr. Hays—If it is in accord with your practice.

Gen. Stewart—Why state them in open court?

Mr. Hays—I understand that three times.

Gen. Stewart—I would not be able to understand you.

Mr. Hays—That is not my fault; beg pardon.

Mr. McKenzie—As I understand these gentlemen, the other day, they offered this scientific testimony, and your honor held that was not competent; am I right?

The Court—I held it was immaterial and incompetent because it would not reflect upon the issues involved in the case.

Mr. McKenzie—Now, as I understand, if your honor please, the only purpose in their offering these statements now is to make up the record, and in the event the case goes to the appellate court to convince the court your honor is in error in refusing to admit this particular testimony. Now, if this is true, may it please this honorable court, what right would they have to come into this court by reading these statements and then after that as indicated by Mr. Hays, make an argument on the very statements of these scientists that you have held their testimony was not competent?

The Court—Just an hour to make up the record.

Mr. Hicks—To make up the record, not to make an argument.

(Thereupon after a further colloquy between counsel, the court said):

The Court—I give you an hour, gentlemen, to go over that, I will then hear you on both sides. I will let you have a chance to see the statements offered as proof.

(Thereupon Mr. Hays proceeded to read):

Mr. Malone—We can finish it in an hour, your honor.

The Court—I am very much inclined to give them an hour, general. I believe I will give the defense an hour to make up their record for the appellate court. I want to be fair to both sides and it occurs to me that that is fair.

Mr. Hays—We expect to prove by the Rev. Walter C. Whitaker—

The Court—I wish you wouid stand over here near the stenographer, Mr. Hays.

Mr. Hays—We expect to prove by the Rev. Walter C. Whitaker, rector of—

Gen. McKenzie—I just want to ask for information. Is he first going to state what he expects to prove here by each of these witnesses and then read the affidavit covering the some thing?

The Court—No, I don't think he wants to do that. Mr. Hays, what do you want to read, your statement of summaries? I don't understand he wants to argue the statements.

Bible Not to Be Taken Literally.

Mr. Hays—I will offer a portion of them. I have two here which have not been prepared and I will state what they are and then I will offer just one where the witness is in court and I want to read from that what we offer to prove. The others will state in general and we will save time if we can. We expect to prove by the Rev. Walter C. Whitaker, rector of St. John's Episcopal church, Knoxville, Tenn., and chairman of the committee which passes on the competency of new ministers for the United States that a man can be a Christian and an evolutionist at the same time. He says "As one who for thirty years has preached Jesus Christ as the Son of God and as 'the express image of the Father' I am unable to see any contradiction between evolution and Christianity.

"And also a man can be a Christian without taking every word of the Bible literally. Not only so, but the man has never lived who took every word of the Bible literally. When St. Paul said: 'I am crucified with Christ,' and when David said, 'The little hills skipped like rams,' neither expected that what he wrote would be taken literally. The sense of Scripture is Scripture. That sense is conveyed to us sometimes in a story and sometimes in a poem. The higher and truer meaning would often be lost if we held ourselves exclusively to the letter and rejected that which it suggests or figures. The story of Abraham's two sons, as contained in Genesis, is interesting and valuable: but in his epistle to the Galatians, St. Paul does not hesitate to say that it is an allegory, and that its true value is its teaching as to the two convenants or testaments.

"I am thoroughly convinced that God created the heavens and the earth, but I do not know how he proceeded. I am sure that He made man in His own image, but I find nothing in the Scriptures that tells me His method. Since God is not subject to the categories of time and space a thousand years being in His sight as a single day, I am unable to see that there is any incompatibility between evolution and religion. Some evolutionists are irreligious, but so are some who are not evolutionists. I myself hold with the writer of the Epistle to the Hebrews that 'God, who at sundry times and in divers manners, spoke in time past unto the fathers by the prophets hath in these last days spoken unto us by His Son, by whom He made the world.' " That would be the testimony of Dr. Whitaker.

Would Use Shailer Matthews.

We expect to prove by Dr. Shailer Mathews, dean of the Divinity school of the University of Chicago, and one of the leading American authorities on the Bible, author of the book on "Contribution of Science to Religion," that "a correct understanding of Genesis shows that its account of creation is no more denied by evolution than it is by the laws of light, electricity and gravitation. The Bible deals with religion.

Two Accounts of Creation.

There are two accounts in Genesis of the creation of man. They are not identical and at points differ widely. It would be difficult to say which is the teaching of the Bible. The aim of both, however, is clear and wonderfully inspired. Each shows how God created man and how man differs from beasts.

In the first account in Genesis, Chapter 1 to Chapter 2, Verse 3, it is said that God made beasts, cattle and all creeping things by having the earth bring them forth as living creatures. The Hebrew expression here used to quote Nephesh Shayah is the same as that used in Genesis, Chapter 2, Verse 7, to describe man when created. The first story then continues with the creation by God of man in the divine image, male and female being created on the sixth day. In the second account Genesis, Chapter 2, Verses 2 to 24, God is said to have formed man from the dust of the ground and to have breathed into him the breath of life. Man thus became a living soul. In the Hebrew the same word is used as that previously used to describe the animals which the earth brought forth.

This living creature, Adam, is placed by God in a garden, which he is to till. He is forbidden to eat of the tree of the knowledge of good and evil. He, however, disobeys and eats the fruit. God then declares that man has become "one of us knowing good and evil." Genesis thus says that an animal life, produced by God from the earth by his spirit, came to be like God through a developmnet born of experience. Thus so far from opposing the Genesis account of the creation of man, the theory of evolution in some degree resembles it.

But the book of Genesis is not intended to teach science, but to teach the activity of God in nature and the spiritual value of man. It is a religious interpretation, its writers use the best of the then current knowledge of the universe to show how God was in the creative process, and how that process culminated in man possessed of both animal and divine elements.

The theory of evolution is an attempt to explain the process in detail. It does not take place in a vacuum, but in an environment in which is God. Genesis and evolution are complementary to each other, Genesis emphasizing the divine first cause and science the details of the process through which God works. This view that evolution is not contrary to Genesis is held by many conservative evangelical theologians, such as Strong, Hall, Micou, Harris and Johnson, Mullins also holds to a theistic evolution."

Mr. Hays then read the statement of Dr. Fay Cooper Cole, anthropologist, University of Chicago; the statement of Kirtley F. Mather, chairman of department of geology, of Harvard university, and the statement of Dr. Winterton C. Curtis, zoologist, University of Missouri.

At 11:40 a. m., during the reading of Dr. Curtis' statement, the further hearing of this case was adjourned to 1:30 p. m., when the following proceedings were had:

Darrow Apologizes to the Judge.

Gen. Stewart—This morning the court read a citation to one of the counsel for the defense, referring to a certain matter which occurred here on Friday and during the noon hour I conferred with some of the gentlement for the defense, particularly the gentleman involved, Mr. Darrow, and Mr. Darrow has a statement that he wants to make at this time and I think it is proper that your honor hear him and I want to ask the court to hear the statement.

The Court—All right, I will hear you, Col. Darrow.

Mr. Darrow—Your honor, quite apart from any question of what is right or wrong in this matter which your honor mentioned and which I will discuss in a moment—quite apart from that, and on my own account if nothing else was involved, I would feel that I ought to say what I am going to say. Of course, your honor will remember that whatever took place was hurried, one thing followed another and the truth is I did not know just how it looked until I read over the minutes as your honor did and when I read them over I was sorry that I had said it. This is not all I am going to say—I am just going to preface it. So on Friday I determined immediately on reading it over I would tell the court just what I thought about it this morning. In the meantime, I had seen the paper which stated that the court thought that I was trying to get in position where I would be held in contempt and they thought so and the like and I was at loss what to do, but I knew your honor wanted to be heard first. Now I want to say that what I say is in good faith, regardless of what your honor may think, it is right for you to do. But I say it because I think I ought to say it for myself. I have been practicing law for forty-seven years and I have been pretty busy and most of the time in court I have had many a case where I have had to do what I have been doing here—fighting the public opinion of the people, in the community where I was trying the case—even in my own town and I never yet have in all my time had any criticism by the court for anything I have done in court. That is, I have tried to treat the court fairly and a little more than fairly because when I recognize the odds against me, I try to lean the other way the best I can and I don't think any such occasion ever arose before in my practice. I am not saying this, your honor, to influence you, but to put myself right. I do think, however, your honor, that I went further than I should have gone. So far as its having been premeditated or made for the purpose of insult to the court I had not the slightest thought of that. I had not the slightest thought of that. One thing snapped out after another, as other lawyers have done in this case, not, however, where the judge was involved, and apologized for it afterwards, and so far as the people of Tennessee are concerned, your honor suggested that in your opinion—I don't know as I was ever in a community in my life where my religious ideas differed as widely from the great mass as I have found them since I have been in Tennessee. Yet I came here a perfect stranger and I can say what I have said before that I have not found upon anybody's part—any citizen here in this town or outside, the slightest discourtesy. I have been treated better, kindlier and more hospitably than I fancied would have been the case in the north, and that is due largely to the ideas that southern people have and they are, perhaps, more hospitable than we are up north. Now I certainly meant nothing as against the state of Tennessee, whom I don't think is any way involved, as your honor knows that these things came up in court time and time again and that it is not unusual perhaps in a case where there is a feeling that grows out of proceedings like this that some lawyers will overstep the bounds. I am quite certain that I did that. I do not see how your honor could have helped taking notice of it and I have regretted it ever since on my own account and on account of the profession that I am in, where I have tried to conform to all rules and think I have done it remarkably well and I don't want this court, or any of my brethren down here in Tennessee, to think that I am not mindful of the rules of court, which I am, and mean to be, and I haven't the slightest fault to find with the court. Personally, I don't think it constitutes a contempt, but I am quite certain that the remark should not have been made and the court could not help taking notice of it and I am sorry that I made it ever since I got time to read it and I want to apologize to the court for it. (Applause.)

The Judge Forgives Darrow.

The Court—Anyone else have anything to say? In behalf of Col. Darrow in anyway? (No response.) If this little incident had been personal between Col. Darrow and myself, it would have been passed by as unnoticed, but when a judge speaks from the bench, or acts from the bench, his acts are not personal but are part of the machine that is part of the great state where he lives. I could not afford to pass those words by without notice, because to do so would not do justice to the great state for which I speak when I speak from the bench. I am proud of Tennessee, I think Tennessee is a great state. It has produced such men as the Jacksons, such men as James K. Polk and such men as Andy Johnson and such men as the great judge that recently went from our neighborhood to the supreme bench of the United States—Judge Sanford—so I feel that we must preserve the good name of this great state that has produced such great men—such great characters as these that I have mentioned. We have had another man who lived in Tennessee—I believe he is dead now—he was a poet and he wrote these words:

"Dost thou behold thy lost youth, all aghast,
Or dost thou feel from retributions' righteous blow
Then turn from the blotted archieves of the past
And find the future pages white as snow.
Art thou a mourner? Rouse thee from thy spell;
Art thou a sinner? Sin may be forgiven.
Each day gives thee light to lead thy feet from hell.
Each night a star to lead thy feet to heaven."

Raulston Acts on Christian Principles.

My friends, and Col. Darrow, the Man that I believe came into the world to save man from sin, the Man that died on the cross that man might be redeemed, taught that it was godly to forgive and were it not for the forgiving nature of Himself I would fear for man. The Savior died on the cross pleading with God for the men who crucified Him. I believe in that Christ. I believe in these principles. I accept Col. Darrow's apology. I am sure his remarks were not premediated.

Listening to Darrow Apologize for Contempt of Court

Judge Raulston Forgiving Him

I am sure that if he had had time to have thought and deliberated he would not have spoken those words. He spoke those words, perhaps, just at a moment when he felt that he had suffered perhaps one of the greatest dissapointments of his life when the court had held against him. Taking that view of it, I feel that I am justified in speaking for the people of the great state that I represent when I speak as I do to say to him that we forgive him and we forget it and we commend him to go back home and learn in his heart the words of the Man who said: "If you thrist come unto Me and I will give thee life." (Applause.)

I think the court should adjourn downstairs. I am afraid of the building. The court will convene down in the yard.

(Court thereupon adjourned to the stand in the courthouse lawn and upon reconvening the following proceedings occurred:)

Mr. Hays—If your honor please, I will not take very much time. I have condensed these statements considerably.

The Court—Where is my officer? Announce to the jury if any are present they must retire.

Officer Kelso Rice—Now, if any of the jurors are present please retire, by orders of the court.

Mr. Hays—Your honor, as to the next order of proof which the defense would offer I should like to say that the defense, as lawyers, take no position of this. It has to do wholly with the question of what the Bible means, and what we would be able to prove from witnesses—we wish to state that we should be able to prove from learned Biblical scholars:

(The statement of defense counsel was thereupon read, which has heretofore been multigraphed and delivered to the press.)

Rabbi Rosenwasser's Statement.

Mr. Hays—Next, your honor, we come to the question of what we would like to prove on the questions of translation that occur in the King James version from the original.

(The statement of Dr. Herman Rosenwasser was thereupon read, which follows.)

Dr. Herman Rosenwasser is a rabbi whose qualifications are vouched for by Dr. Kaufman Kohler, president emeritus of the Hebrew Union college of Cincinnati, and the leading Hebrew Scholar of America, who says:

"I consider Rabbi Rosenwasser well qualified to interpret Genesis scientifically and fully agree with him in his endeavor to reconcile evolution with the Bible, as I did in all my teachings."

(Biography—Dr. Herman Rosenwasser resides at 180 Commonwealth avenue, San Francisco, Cal. He is 46 years of age, was born in Hungary and came to the United States in 1893. He studied in the West High school of Cleveland, O. Upon graduation he went to the Hebrew Union college, where he was a pupil of Dr. Isaac M. Wise. After two years, and before graduation, he was called to the rabbinate of the congregation at Springfield, Mo., and while there, in addition to his religious duties, he taught in the public high school. He left Springfield for Cleveland in 1903 to continue his academic studies at the Western Reserve university of Cleveland. He specialized there in semitics and philosophy. In the year 1905, he received a degree of master of arts from the Western Reserve university. In 1906 he continued his rabbinical studies at the Hebrew Union college in Cincinnati, and in 1908 was there ordained a rabbi. His first charge was Lake Charles, La., two years; then Baton Rouge, three years. While there he was a member of the Protestant Ministerial alliance. Then he went to San Francisco, where he occupied for ten years the rabbinate of Temple Sholem, leaving there two years ago to devote himself to research.

During all this time he was a student and teacher of the Bible and has contributed largely to theological papers.

He speaks fluently, English, German, Yiddish, Hungarian and Hebrew. He reads and translates the above languages and in addition, Latin, Greek, Chaldaic, French and Italian. On the Bible he has done original research work for years.)

The defense counsel, of course, disclaims any knowledge on the subject and knows there are any number of translations, but this witness would testify that the King James version is not an accurate translation; not true to texts vitally teaching creation, man, life and soul.

In 1611, when the King James revision was made, little was known of the Hebrew language. The scholarly study did not begin until 1753. 142 years after the King James version. From that time on, great strides have been made. To understand the Bible one must know Hebrew.

The original Bible was without vocalization (that is, the vowels were missing), and without punctuation, and the five books of Moses are read in Hebrew synagogues from unvocalized or unvowelized and unpunctuated texts.

Mistakes in Bible Translation.

In the translation of the Hebrew Bible, from which the King James Protestant version is derived, there are many errors, none of them basic. The word "create" purports to be a translation of "bara." This word, "bara" is used with reference to both inorganic and organic creation, man as well as animals and plants. The word "bara" is used to represent the whole cosmic scheme. The correct translation is "to set in motion." From the incorrect translation into English in the King James version great confusion has resulted.

In Verse 2 of the King James version of the Protestant Bible appears the following: "The spirit of God moved upon the face of the waters." That is not a correct translation of the Hebrew. A correct translation of the Hebrew word "marachefeth" is, "And God animated, imparted life, vivified." The words, "The face of the waters" are "alpenai humayin," which means "to animate the face of the fluid mass."

In Psalms cxlviii:6, the King James version says: "He hath made a decree which shall not pass." That is not a correct translation. The word "chak" in Hebrew means "natural law" or "law of nature." Here it is translated "decree." The words "which shall not pass" do not represent a correct translation, either. The words should be "which He doth not transgress." The proper English translation of the whole would be as follows: "He hath made a law of nature, which He doth not transgress." In other words, the laws of nature are unchanging.

In the Bible there are four distinct terms for man: Adam, Enoch, Gever and Ish. Some of these are used as meaning animals.

In the Book of Ecclesiastes 3:19: "Adam (the physical man) and animals are declared to be subject to the same laws. The original, properly translated, is "There is no pre-eminence of the 'Adam' (of the natural man) over the animal, for all is unstable." The word "eucsh" also refers to the physical man, because that man turns to dust. (Psalms xc: 3). These two words, "Adam" and "Eaosh" refer to the physical man only and identify him with the physical creation.

In the first chapter of Genesis, the word "Adam" is used. The word Adam means a living organism containing blood. If we are descended from Adam we are descended from a lower order—a living, purely organism containing blood. If that is a lower order of animal, then Genesis itself teaches that man is descended from a lower order of animals.

The terms "Gever" and "Ish" refer to the intellectual and spiritual man.

Wherever the higher attributes of man are referred to, such as love, mercy, justice, righteousness, purity, etc., or any ethical attribute, the words used are "Gever" and "Ish." Every translation of a term here is a literal translation. The Hebrew dictionary will bear out every translation referred to.

If the Hebrew Bible were properly translated and understood, one would not find any conflict with the theory of evolution which would prevent him from accepting both.

Mr. Hays—The defense counsel, of course, disclaims any knowledge on the subject, but knows there is a number of translations, and this witness would testify to them.

What Dr. H. E. Murkett Would Say.

We would also be able to prove that the Bible, properly interpreted, does not conflict with the theory of evolution by Dr. Herbert E. Murkett, pastor of First Methodist Church, Chattanooga.

There is nothing whatever in the belief in evolution that denies the divine story of creation. The non-Calvanistic churches have never believed, through their leaders, in divine fiat or determined and fixed processes as acts of God.

The divine story does not tell how man was made. It says that he was made out of the dust—that is, the material—it tells what God did with him when made—breathed into him his spirit. The process is not mentioned anywhere.

If the second chapter of Genesis is taken literally then the creation of man was progressive. First man is formed and he is put to sleep and through another process that no man can interpret, woman was made, and then through another process children were made and this process has been going on for centuries.

Take the statement that God said, "Let us make man in our image." This is open to interpretation. Was man already made? The story does not say, "Let us make another creature and call him man and let us make him after image." No, let us make man—already known, already a part of the animal life—let us make him after our image. He was then endowed with the spirit of God, possessing his moral, spiritual and intelligent nature.

Again note the story in the second chapter of Genesis. Man is introduced as perfectly naked, and does not know it; he is ignorant of right and wrong. This is a story of a man awakening to the consciousness of right and wrong, of the consequences of such a knowledge, and he begins the only process known to allay the pangs of conscience and lack of harmony with his Creator.

To science and not to the Bible must man look for the answer to the question as to the process of man's creation. To the Bible and not science must men look for the answer to the cause of man's intelligence, his moral and spiritual being.

Man is here and must be accounted for from two standpoints. He is a physical being and lives the life of all other physical beings and is a study for material science. He is spiritual and lives in the realm of spirit and for understanding of that spiritual side one must study the science of theology. When these two shall be harmonized then will we have an understanding of this dual personality that follows after God rather than the animal existence, who plays with God's laws and learns how. He operates them, who sees in spirit and then transforms the vision into locomotives, airplanes, telegraph instruments, radio, and by many inventions overcomes time and space.

Students have a right to be taught the truth about the whole man rather than a half truth. The future of human progress demands it.

Mr. Hays—Our next witness would be Donald F. Metcalf and I think I stated his qualification the other day.

The Court—His testimony is in the record?

Mr. Hays—I have a few statements I will read.

The Court—All right.

(Excerpts of the statement of Mr. Metcalf were thereupon read.)

Mr. Darrow—I take it you want all of the testimony incorporated in the record?

Mr. Hays—Yes, of course, the whole statement will go into the record.

The Court—Yes, let the whole statement go into the record.

Mr. Hays—The next is Herbert A. Nelson, who, as your attorney knows, is state geologist of Tennessee. (Reading statement by Nelson.)

Mr. Hays—I will next read Dr. Jacob Lipman, as your honor no doubt knows, Mr. Lipman is a very eminent scientist.

(The statement of Mr. Lipman was thereupon read.)

Letter From Luther Burbank.

Mr. Hays—While we are on that subject I will say Mr. Luther Burbank makes a full statement.

The Court—Is he here?

Mr. Hays—He is not here, your honor. This is his letter. We will take his deposition if you will let it in as evidence.

Mr. Hicks—Will you let us cross-examine him?

Mr. Hays—Do you want to cross-examine Mr. Burbank, Mr. Hicks?

Mr. Hicks—We would cross-examine him if you put him on.

Mr. Hays—I would like to hear you cross-examine Mr. Burbank.

Mr. Hicks—I would like to hear you too. (Applause and rapping for order by Policeman Kelso Rice.)

(Letter by Mr. Burbank was then read.)

Gen. Stewart—That is just a letter from Mr. Burbank?

Mr. Hays—That is what we would be able to prove and if the scientific witnesses went on the stand, I assume we could take his deposition and prove it if we could get it here in time. Dr. Charles Hubbard Judd would testify:

(The statement of Mr. Judd was thereupon read.)

Mr. Hays—And the last statement which I would read to your honor, showing what I could prove, is from Dr. Horatio Hackett Newman, zoologist of the University of Chicago.

(Reading.)

The Court—Have you had the statements marked filed, Mr. Hays?

Mr. Hays—Yes, sir, I will.

STATEMENTS ARE FILED.
By Defense Counsel.

Of course, the defense, as lawyers, take no position on the truth of the stories of the Bible, but we wish to state that we should be able to prove from learned Biblical scholars that the Bible is both a literal and figurative document, that God speaks by parables, allegories, sometimes literally and sometimes spiritually.

We should be able to prove:

First—That the entire Bible teaches the fact of the fundamental difference between the soul and the body. This is clearly shown by the following passages: Ecclesiastes vii:8; Luke viii:55, xxiii:46, xxiv:39; John vi:63; I Corinthians vi:17,20; Hebrews iv: 12, xii 22, 23; James ii:26—all of which show the Bible attitude on the question of the nature of the soul.

Typical examples of the teaching of the Bible in reference to the body or flesh are given below:

"My substance was not hid from thee, when I was made in secret, and curiously wrought in the lowest parts of the earth. Thine eyes did see my substance yet being unperfect: and in thy book all my members were written, which in continuance were fashioned, when as yet there was none of them." (Psalm cxxxxix:15-16.)

Here there is a distinct statement that the human body was created by the process of evolution.

Also Roman viii:22 says "For we know that the whole creation groaneth and travaileth in pain together until now."

Second—That the entire Bible teaches that God is a spirit and "the father of spirits," and not the father of flesh. (See Numbers xxvii:16; John iv:23-24; Hebrews xii:9.)

Third—Therefore, it is man's soul or spirit, and not his body, that is the Son of God, and which consequently is in the image of God.

Fourth—That the Bible is concerned with the ethical and spiritual side of life, and not with the body, or chest of tools, which is the means of self-development or self-expression of that soul.

Fifth—That natural science is concerned with the developmental history, the structure and the functions of all living bodies, and not with any religious or any ethical questions.

Sixth—That the Bible simply states that God created the human body and the material he used in doing it, and not how he did so. There are at least four separate accounts of the creation of the human body in Genesis, and they can only be harmonized in accordance with this viewpoint.

Science Discovers Method.

Science has discovered the developmental history (evolution) of that body—i. e., the method by which God has brought it into being.

Another theory of some Biblical scholars is that the Bible interprets itself. In Roman iv:17 appears the statement that God "calleth things that be not as though they were."

For instance, some scholars would say, where the Bible states that man was made in the image of God, it refers only to Christ and His body, and in the Bible are found passages to uphold this. As an instance, in Philippians iii:21 is the statement concerning Christ, "Who shall change our vile body that it may be fashioned like unto His glorious body?"

We can merely give illustrations. Genesis said, "Let there be 'light' and there was 'light'." According to some scholars, the word should be law. According to others, as appears in Psalms cxix:105, "Thy word is a lamp unto my feet and a light unto my path,"—the word light should be construed in a different sense. In Psalms cxix:130, the statement is "The entrance of thy words giveth light, it giveth understanding unto the simple.' In Psalms xliii:3 appears, "Send out thy light and thy truth." "Let there be light" should be interpreted, these men say, as "let there be understanding," according to those other statements in the Bible. So, within the Bible itself, can be found many interpretations. Even those who do not choose to go outside the Book, interpret from within the Book. Innumerable illustrations might be given bearing upon almost every word in the Bible.

In other words, we should prove that the Bible is subject to various interpretations depending upon the learning and understanding of the individual, and that, if this is true, there is nothing necessarily inconsistent between one's understanding of the Bible and evolution. Many accept these statements in the Bible as legends or parables. They may accept them as legends or parables, and thus not find them inconsistent with any scientific theory.

In II Timothy iv:4 appears the following, according to the translation from the Greek of Prof. Goodspeed, of the University of Chicago:

"For the time will come when they will not listen to wholesome instruction, but will overwhelm their whims and tickle their fancies, themselves with teachers to suit and they will turn from listening to the truth and wander off after fiction."

Statements of Noted Scientists as Filed Into Record by Defense Counsel.

By Charles Hubbard Judd

Director of the School of Education, University of Chicago.

(Biography.—Director of the School of Education and head of the Department of Education at the University of Chicago; has been in this position sixteen years. Prior to that was professor of psychology at Yale University. He was educated in Connecticut Wesleyan, a Methodist college, where the doctrine of evolution is taught by all of the instructors in the Science Department. He received the degree Ph. D. at Leipsig University, where he took comparative anatomy as a minor subject, with psychology as a major. In 1909 he was president of the American Psychological Association; was twice president of the Society of College Teachers of Education, president of the National Society for the Study of Education, president of the North Central Association of Colleges and Secondary Schools, vice-president of the section of psychology of the American Association for the Advancement of Science. He is author of seven books and of numerous articles of psychology and education.)

In the normal schools of the state of Tennessee it will, I think, be impossible to obey the law without seriously depriving teachers in training of a proper view of the facts of human mental development. Every psychologist recognizes the fact that the human organs of sense, such as the eye and the ear, are similar in structure and action to the organs of sense of the animals. The fundamental pattern of the human brain is the same as that of the higher animals. The laws of learning, which have been studied in psychological and educational laboratories, are shown to be in many respects identical and always similar for animals and man. It is quite impossible to make any adequate study of the mental development of children without taking into account the facts that have been learned from the study of comparative or animal psychology.

Would Handicap Teaching.

It will be impossible, in my judgment, in the state university, as well as in the normal schools, to teach adequately psychology or the science of education without making constant reference to all the facts of mental development which are included in the general doctrine of evolution. The only dispute in the field of psychology that has ever arisen among psychologists so far as I know has to do with the methods of evolution. There is general agreement that evolution in some form or other must be accepted as the explanation of human mental life.

Elaborate studies have been made in the field of human psychology dealing with such matters as the evolution of tools, the evolution of language and the evolution of customs and laws. All of these studies are based on definitely ascertainable facts and show without exception that a long process of evolution has been going on in the life of man as it is definitely know through historical record and prehistoric remains. In my judgment it will be quite impossible to carry on the work in most of the departments in the higher institutions of the state of Tennessee without teaching the doctrine of evolution as the fundamental basis for the understanding of all human institutions.

Whatever may be the constitutional rights of legislatures to prescribe the general course of study of public schools it will, in my judgment, be a serious national disaster if the attempt is successful to determine the details to be taught in the schools through the vote of legislatures rather than as a result of scientific investigation.

By Jacob G. Lipman,

Dean of the College of Agriculture and Director of the New Jersey Agricultural Experiment Station, State University of New Jersey, New Brunswick, New Jersey.

Dr. Jacob G. Lipman, of Rutgers and the state university of New Jersey, is a specialist in the field of soil science. He received his bachelor's degree at Rutger's in 1894, his master's degree at Cornell in 1900, and the degree of doctor of philosophy also at Cornell in 1903. His alma mater gave him the honorary degree of doctor of science in 1923. He has been soil chemist and baceteriologist of the New Jersey Experiment stations since 1901; director of the stations in 1911, dean of the college of agriculture, State university of New Jersey since 1915. Since 1902 he has been a member of the faculty of Rutgers.

He is editor-in-chief of Soil Science, associate editor of the Journal of Agricultural Research, International Mitteilungan fur Bodenkunde and of Annales Sciences Agronomiques. He is also editor of the Wiley Agricultural Series, and associate editor of the Pennsylvania Farmer.

He is a member of the National Research council, the American association for the Advancement of Science, the American Chemical society, the American Society of Bacteriologists, the American Society of Agronomy, the American Academy of Science, the Washington Academy of Sciences and a number of other American scientific societies. He is president of the International Society of Soil Science and corresponding member of the Swedish Royal Society of Agriculture and Veterinary Medicine.

Organic Evolution from the Point of View of the Soil Investigation.

The student of soils is obliged to consider the materials from which they are made. These materials are represented by rocks and minerals, and by the remains of plants, animals, insects, bacteria and other micro-organisms. The change of rocks into soils is a slow and gradual process. In the older geological ages the mantle of soil covering the rocks was not as thick as it is today. Going back for enough, we come to the time when the depth of soil was not great enough to support plants of any but very primitive forms. Like plants and animals, our soils had to pass through a long period of change to support the varied forms of life on the earth. A direct relation may be traced between soil, plants and animals in the evolution of organic life.

Among the early forms of life there were bacteria capable of developing in a purely mineral medium. Such forms are still found in the sea, in mineral springs and in soils. Some of them can obtain the energy for their life processes by oxidizing hydrogen gas, mothane (marsh gas), carbon monoxide, sulphur, sulphurated hydrogen, iron and even carbon. In the primitive seas, and on the rock surfaces, these simple forms of life prepared the way for the more highly organized beings. Some bacteria are able to manufacture nitrogen compounds out of the simple nitrogen gas of the air. They thus supply material out of which the protoplasm of plant and animals cells is made. Other bacteria convert the nitrogen of the plant and animal substances into ammonia and nitrates. Mineral acids, like nitrous, nitric, sulphuric and phosphoric, are partly, if not entirely, the products of bacterial activity. Carbon dioxide is generated in enormous quantities through the activities of nitro-organisms. In the course of ages the by-products of microbial activity served to dissolve enormous quantities of rock material, and this dissolved material started on its way to the sea. Silicates, phosphates, nitrates, sulphates and carbonates, went to supply the building stones for the bodies of marine organisms. Some of the salts dissolved from the rocks ultimately became the source of salt deposits, such as rock salt, gypsum, potash, salts, limestone, etc. Bacteria are thus recognized as the primary or secondary cause of extensive mineral deposits, in other words, as geological agents of importance. By way of example, mention may be made of the potash deposits of certain European countries, estimated to be 20,000,000 years old. The green sand formation of New Jersey and states further south originated in the sea about 10,000,000 years ago. The phosphate deposits of Central Tennessee are derived from limestone rock 50,000,000 years old at the very lowest estimate. The extensive deposits of coal represent the remains of the ancient vegetation. We are now burning coal derived from plants that grew at least 20,000,000 years ago. The coal deposits contain nitrogen which today is the source of fertilizer. In making coke, illuminating gas and other products from coal, a large part of the nitrogen is saved and converted into ammonia for refrigeration and fertilizer purposes. We know of extensive deposits of sulphur which originated millions of years ago and which today are used for industrial and agricultural purposes. In a smaller way, mention may be made of deposits of iron ore, gypsum, or limestone, in the formation of which bacteria played an important part.

Today, like many millions of years ago, bacteria are busy creating conditions necessary for the growth of plants and animals. Bacteria are responsible for the circulation of carbon and nitrogen in nature. The material of plant and animal bodies is used over and over again, and processes of decay must go on in order that the carbon, nitrogen, sulphur phosphorus, lime and other elements locked up in the bodies of plants and animals may be released for the development of countless generations of living things. It has been truly said that we may have in our bodies today the carbon or the nitrogen which were once in the bodies of the kings of Egypt or of living organisms of whose origin and history we know nothing.

After the lowly bacteria and other miscroscopic forms of life had lived and produced extensive changes on land and in the sea, conditions became more favorable for the growth of plants. The primitive forms of plant life gradually developed into more perfect organisms, until the mosses, ferns, cycads gave way to flowering plants, perhaps 10,000,000 years ago at a very conservative estimate. In some way bacteria learned to establish a partnership with some kinds of plants, such as clover, alfalfa, soy beans, etc. These plants, together with the bacteria, are the important factors in our agriculture as regards the maintenance of a supply of nitrogen in our soils.

Thus plants had to develop both as to quantity and quality in order that there might be sufficient food for the advancing forms of animal life. One may properly speak of the genesis and evolution of soil as one would speak of the genesis and evolution of plants and animals. Man has learned to use this knowledge to improve his condition, and in following the laws laid down by the divine Creator, he has been able to fashion more perfect forms of plant and animal life. The story of Genetics, which deals with the principles of plant and animal breeding, is full of interest. It has to its credit more perfect flowers, fruit of higher yielding qualities and better flavor, fibre crops of superior fibre, sugar crops with a higher content of sugar, crops resistant to plant diseases, crops suitable for dry climates and wet climates, for sour soils and sweet soils, and, in general, for a wide range of soil and climatic conditions. In the same way, genetics has made it possible for us to improve on the types of animals of economic importance in our farming industry.

We are indebted to science for a clearer vision of the great laws of nature, and of the methods of the divine Creator. The men of science, carrying on their labors in a spirit of reverence and humility, try to interpret the great book of knowledge in order that the paths of man may fall in more pleasant places, and the ways of human society may be in better keeping with the divine purpose.

With these facts an interpretations of organic evolution left out, the agricultural colleges and experimental service to our great agricultural instations could not render effective dustry.

By Dr. Fay-Cooper Cole,

Anthropologist, University of Chicago.

(Biography—Dr. Fay-Cooper Cole received the degree bachelor of science at Northwestern university. After work as a graduate student at Rush Medical college and the University of Berlin, he took the degree doctor of philosophy at Columbia university. He is now anthropologist at the university of Chicago. Before that he was connected with the Field Museum of natural History at Chicago, one of the three chief museums in America, for nineteen years, for the greater part of that period he was in charge of the museum's work in physical anthropology and Malayan ethnology. He conducted three expeditions covering a period of five and one-half years in the Philippine Islands, Borneo, Java, Madura, Nias, Sumatra and the Malay peninsula, making a particular study of the origin and the migration of the pygmies and Malays, and of their social organizations. He was a member of various expeditions to the American southwest, excavating the ruined cities of cliff dwellers in the southwest and carried on investigations among the Pueblo and Navajo Indians. From 1907 to 1912 he was special investigator for the Philippine Bureau of Science, codifying the laws and making a study of the social, economic and mental life of the uncivilized tribesmen. During the last three years of connection with the field museum he was also lecturer in anthropology at Northwestern university. He is a fellow of the American Association for the Advancement of Science, fellow of the American Geographical society, member of the council of this association and now one of the vice-presidents of the American Anthropological association, and member of the Social Research council of this association. He is author of four monographs and various scientific papers dealing with the folk lore, physical types, social, religious and economic life of the primitive tribes of the Philippine islands.)

Anthropologists accept evolution as the most satisfactory explanation of the observed facts relating to the universe, to our world and all life on it. They hold that evidence abundantly justifies us in believing that development has been from the simple to the complex and that present forms of life, including man, have been produced from earlier existing forms, but through immense periods of time.

The field of the anthropologist is man, man's body, and man's society, and in this study he finds himself working side by side with the biologist and the geologist. For the study of man's body he has worked out a set of instruments and has selected a series of points for observation, by means of which he can accurately describe each individual of a group, the length, breadth and height of head, the facial proportions, the length of limbs and so on.

In this way the anthropologist determines the average of a group or tribe or race, and to determine its normal variation. Anything strikingly beyond the normal at once becomes the subject of inquiry to determine its cause. In addition to the mathematical description there are added observations—color of skin, shape of teeth, the form of the hair, and many others.

On man's skeleton these observations are even more exact and are so definite that given a single skull or skeleton it is possible to tell with considerable certainty the age, sex and race of the individual, while for a series of skeletons the results are definite. The skeletons tell much of man's history, for the articulation of the bones and the lines of attachment of the muscles reveal how he walked, how he held his head and many other details of his life. It also reveals the fact that man presents many variations difficult to explain without referring to similar conditions found in the animal world. To gain further light on these variations the anthropologist works with the anatomist and comparative anatomist and he quickly finds that every human being of today possesses many muscles for which there is no apparent use, such muscles are those behind the ears, those going to the tail, the platysma—a muscle going from the chin to the Clavicle. These are but a few among many which today are functionless in man, but are still in use by certain animals. Going to the human embryo we find these vestiges of an earlier condition much more developed while others appear for a time and then vanish before birth. Such a case is the free tail possessed by every human embryo, a few weeks before its birth.

Man's Useless Organs.

It is difficult to explain the presence of these useless organs in man unless we believe that sometime in his development they were in use.

This study also reveals the fact that man closely resembles certain members of the animal world in every bone and organ of his body. There are differences, but they are differences of degree rather than of kind. The animals most closely resembling man are the anthropoid apes. A careful study shows that they have specialized in their way quite as much as man has in his, so that while they are very similar, yet it is evident that man's line of descent is not through any of these anthropoids. It does appear, however, that both man and the other primates have a common precursor, but that the anthropoids must have branched off from the common stock in very remote times. If this is true, then we might hope to find in ancient strata of the rocks some evidences of earlier forms of men, who might perhaps more closely approach the common ancestor. This is exactly the case. The geologists have established the relative age of the strata of the rocks, while the palaentologists have made plain the forms of life which lived in the epochs when these strata were deposited.

In the strata laid down at the end of the Pliocene period, at least 500,000 years ago, there has been found the bones of a being which appears to be an attempt of nature toward man. In the year 1891 on the island of Java, there was found the bones of an animal which in many ways seems to be intermediate between man and the anthropoids. These bones were found in undisturbed strata, forty feet below the surface, at a point where a river had cut through the mountainside. There can be no doubt that these bones were laid down at the time that stratum was deposited and by studying the associated fauna, consisting of many extinct animals, the age of these rocks was established. These bones were not lying together, but had been scattered over a distance of about forty-five feet by the action of the ancient river which deposited them.

These semihuman bones consisted of a skull cap, a femur, and two molar teeth. The skull was very low with narrow receding forehead and heavy ridges of bone above the eye-sockets, while a bony ridge extended from between the eye-brows to the top of the head approaching a condition found in the cranium of the anthropoids. The brain capacity of this individual was between 850 and 900 cubic centimeters, or a little more than half of that of modern man. On the other hand it is half as much again as that of the adult gorilla, and the special development has taken place in these regions whose high development is typical of the brain of man. Hence in this respect this being seems to stand midway betwen man and the highest anthropoids. The teeth approach the human type and indicate the peculiar rotary mode of mastication of the human, which is impossible in animals having their interlocking canine teeth. The thigh bone is straight, indicating an upright posture and ability to run and walk, as in man. And the muscle attachments show he was a terrestial and not an arboreal form. If, as seems probable, these four bones belonged to the same individual, he must have been more man-like than any living ape and at the same time, more ape-like than any human known to us. He is known as Pithecanthropus erectus or the erect ape-man.

Another find of somewhat similar nature was made only a few months ago in Bechuanaland of South Africa by Prof. Dart, of the University at Johannesburg. This find consisted of the skull of an animal well developed beyond modern anthropoids in just those characters, facial and cerebral which are to be expected in a form intermediate between man and the anthropoids. Neither of these two beings are of certainty, directly ancestral to man, but they do seem to indicate that nature at a very early period was making experiments toward man.

Two other fossil beings, found in the early strata of the rocks, also seem to indicate a development toward man. In the strata of the second interglacial period, probably at least 250,000 years ago, there lived a being with a massive jaw, a jaw human in every respect, except that it had no chin and the ramus or upright portion toward the socket was very broad, as in the anthropoids. This jaw is so narrow behind that it is thought the tongue could not have sufficient play to allow of articulate speech. The teeth, although very large, are essentially human with even tops, as in man, while the canines lacked the tusk-like character which they still retain in the apes. This jaw was found in the year 1907 in a sand pit working near Heidelberg, Germany. It was discovered in place at a depth of nearly eighty feet and lay in association with fossil remains of extinct animals which make possible its dating in geologic time. It is difficult to picture a man from the jaw alone, but this much we can say the mouth must have projected more than in modern man, but less than in the chimpanzee or gorilla. He had a heavy protruding face, high muscles of mastication, essentially human teeth, and he was already far removed from his primatic ancestors with large canines. He was nearer to man than to the apes; he was further along the line of evolutionary development than Pithecanthropus erectus, the Java ape-man, and he lived at a much later period. This being is known as the Heidelberg man.

The second of these two finds which we have mentioned occurred near Piltdown in Sussex, England. This consisted of the crushed skull of a woman and a jaw which can scarcely be distinguished from that of a chimpanzee. For a time there was much question if the two could possibly belong together, but a more recent find, which occurred about three miles distant from the first, again showed portions of the same type of skull and jaw. The skull is exceedingly thick and its capacity much less than that of modern man, but it is distinctly human, while, as indicated, the jaw approaches that of an anthropoid. Here again we seem to have an approach toward man in very ancient strata.

Toward the end of the second interglacial period in Europe at least 225,000 years ago we begin to find stone implements which give indication of having been intentionally formed and used by intelligent beings. By the third interglacial period, more than 150,000 years ago these utensils have taken on definite form and we find thousands of stone axes of crude type scattered over a large portion of central and southern Europe. We have no fossil remains of man during this third interglacial period, for he then lived in the open and it would only be by the merest chance that his skeletons might be preserved to us. But when the fourth glacial epoch spread over Europe these men were compelled to make their homes in the shelters and caves of the rocks, and here in the debris around their ancient hearths we can read the record of their home life, and from this period on for a period of at least 50,000 years, we can read the record of man's occupancy of Europe as clearly as though we were reading from the pages of a book. Fortunately for the scientists, these people buried their dead and we have preserved for us a considerable number, ranging from children to adult men and women, so there is no guessing as to the sort of man who occupied Europe at this time.

They were massively built, with long arms and short legs, in height they averaged about five feet three for the men, and four feet two, for the women, or about the same as the modern Japanese. The head was long and narrow, above the eyes was a heavy bony ridge, back of which the forehead retreated abruptly, indicating rather little development of the fore brain. The nose was low and broad, the upper lip projecting, but the jaw was weak and retreating. The head hung forward on a massive chest, this we know because the foramen magnum, the opening by which the spinal cord enters the cranium, was situated further back than is the case in modern man, and the points of articulation with the bones of the neck also show conclusively that the head hung habitually forward. In all cases we find the thigh bone to be curved and this, together with the points of articulation, show that the knee was habitually bent and that this man walked in a semierect position. Those people known as the Neanderthal race spread out over the western half of Europe and we now know and have excavated very large numbers of the stations in which they lived. They were men—they were human—but they were much more like the anthropoids in many respects than is modern man. They lived in Europe for a period of at least 25,000 years, probably much longer, when they were displaced by newcomers who pushed in from around the eastern end of the Mediterranian and from Asia. The newcomers known as Cromagnon, are a much finer physical type, but so closely related to modern man that it is not necessary to describe their physical type; but it is of interest that we can study his home life, his art and his life among certain animals now extinct, for a period beginning about 20,000 years ago and extending down to the coming of modern races.

Only a few points relating to man and his history have been reviewed, but enough has been said to indicate that the testimony of man's body, of his embryological life, of his fossil remains strongly points to the fact that he is closely related to the other members of the animal world, and that his development to his present form has taken place through immense periods of time.

From the above it seems conclusive that it is impossible to teach anthropology or the prehistory of man without teaching evolution.

By Wilbur A. Nelson,
State Geologist of Tennessee.

(Biography—Wilbur A. Nelson is state geologist of Tennessee, president of the American Association of State Geologists, past president of the Tennessee Academy of Science, chairman executive committee, Southern Appalachian Power conference, 1923, member of the executive committee of the division of states relations of the Natural Research council; member of the council of the American Engineering council, and president of the Monteagle Sunday School assembly, of Monteagle, Tenn., the leading interdenominational chautauga and summer resort in the south, founded forty-three years ago, and after Sept. 1st, Corcoran professor of geology and head of the department of geology, University of Virginia, and state geologist of Virginia. He received the degree bachelor of science at Vanderbilt university and the degree master of arts at Leland Stanford university. He has held responsible posiions with commercial firms as well as in the service of the state. He is a fellow of the American Association for the Advancement of Science, a fellow of the Geological Society of America, member American Institute of Mining and Metalurgical Engineers, American Association Petroleum Geologists, Seismological Society of America and other organizations. He has published a number of papers on geological and related subjects, both scientific and of a popular nature.)

The different layers of rock which form the surface of the earth unfold a remarkable story of evolution. These rock layers may be read as clearly as the leaves of a book, and they are the book which tells the true history of the earth; and the buried remains of animal and plant life which they contain likewise show the rise of life and its development on this earth. All forms of life have changed and developed to meet the conditions which have existed on the earth, as it has developed to meet the conditions which have been developing from the beginning of geological time.

Tennessee is an ideal place in which to study and learn the story of the rock layers which have been laid down, from the earliest times in which any life existed up to the present. Life forms suitable for one period of the earth's history, proved unsuitable for another period, and so new forms, therefore, evolved through natural causes.

This is not a new study in Tennessee, as geology and its study of buried animal and plant remains has been taught in this state since 1828, at which time Gerard Troost, one of the founders of the Philadelphia Academy of Science, was elected professor of geology at the University of Nashville, and three years later was elected state geologist of Tennessee. From that date to the present time, this science, dealing with the age and study of the earth, and its rocks and the buried life which they contain, has been continuously taught in Tennessee.

Such teaching could not have been carried on through ninety-seven years of time, unless the teaching of evolution had been permitted as it was permitted by our religious ancestors who formed this state.

We know that streams and rivers carry sediment; that muddy waters are full of the soil of some field, washed into a nearby stream by a hard rain, and some such soil, when it once gets into a stream, starts on a long journey to the ocean. Most of the streams in this section are muddy for many months in each year, and this mud, which is the soil washed from our gullied hillsides, in this particular case goes down the Tennessee river, into the Mississippi river and to the Gulf of Mexico.

We know that at the mouth of the Mississippi river the sediments brought down by this river are deposited so rapidly that land is formed which is extending into the Gulf of Mexico at the rate of many feet a year. As a rule, these processes of weathering of rocks to produce soil, of erosion of this soil, and of deposition of this transported soil through rivers into some nearby sea or ocean, takes place so slowly, as time is generally measured, that we can only see through detailed and scientific observation the results within our own lifetime. But at the delta of the Mississippi river this very process is taking place so rapidly that anyone can easily measure it year by year and can understand that these same processes have been taking place all through all geologic time, and in each and every part of the world.

We also know that practically all of the earth has at some time or other been covered by water, and in these ancient seas life has existed, which has left its record to us in fossil form. It must, however, also be understood that large parts of our present water areas were at some period in past geologic time also land areas. These seas have come and gone over limited areas of the earth's surface many times during the geologic history of the earth.

We know that originally the mouth of the Mississippi river was near Cairo, Ill., and that all of the Mississippi valley, as we now know, it was at that time (which was the close of the Cretacious period) a part of a much larger Gulf of Mexico than the one that now exists. All of West Tennessee, during this time, was in a northern extension of the Gulf of Mexico, and the fine China clay deposits of that section were laid down in shallow water at the time tropical plants flourished in that section.

East Tennessee.

East Tennessee is made up of many layers of rocks, limestone, shale and sandstone, all of which were likewise laid down under water, and many of these layers contain the remains of animal and plant life. Some of the oldest rocks which contain animal life are found in East Tennessee. They are known as Cambrian rocks, and in these rocks occur the first abundant remains of sea form of life. This was the age of the early invertebrates. These rocks are well exposed to the east of Dayton in the East Tennessee valley region.

Then came the time interval which the geologist calls the Ordovican, the time when primitive fishes, corals, and land plants came into existence. Some of these first corals in fossil form have been found in the western edge of Dayton. This time interval was followed by another series of rocks which, in East Tennessee, contain the red iron ore deposits which are used by the iron furnaces of this section. The rocks of this age are known as the Silurian, and during this time life further developed and scorpions and lung fishes came into existence.

The series goes on. Layer after layer of rocks were laid down, each series of which has been given a name by geologists so that they can be easily referred to. Next came the great age of fishes, and their remains are found in the rocks which the geologists call the Devonian and Mississippian series. The black slate, which crops out at the foot of Walden's ridge, as well as the limestone lying above it, which form the side of the mountain to the west of Dayton, are layers belonging to these series. These rocks are full of the remains of animal life.

Then came the period in which the ancient plants flourished and produced great coal deposits, the age which has ben called the carboniferous. The extensive coal deposits of the Tennessee coal field, the edge of which caps the mountain a few miles west of Dayton, are of this age, and wonderfully preserved plant remains are found in the slates which lie on top of the different coal seams. This is a fact well known by the coal miners of this section. And what has been stated above as to Tennessee is but one illustration of how the different geologic periods passed and life developed over the earth.

And even when this carboniferous period in the development of the earth has been reached, we are still many millions of years back from the age of man; we must still pass through many geological time periods, through that age known as the Permian, when land vertebrates first arose; through the Triassic, when reptillian mammals arose; through the Jurassic, when flying reptiles were in existence. This was the age of reptiles. Then into the Cretaceous when flowering plants came into existence, and a great group of the reptiles known as dinosaurs, became extinct.

And then we came to that period in the earth's history, at the beginning of which the ancient mammals and birds were first known to exist. Fossil remains show clearly that birds evolved from flying reptiles. This is the great age of mammals. Thru this period, the modern life forms developed. A period of glacial activity took place, during which five distinct glacial stages existed, one after the other, with four interglacial intervals, and man-like beings came into being at least the beginning of this time. Such, very briefly, is an account of the evolution of the earth from Cambrian time to the present, with a brief outline of the life forms which existed during these different periods. We know that this took many millions of years, and yet we also know that the earth existed untold millions of years before Cambrian time.

For the formation of the earth and its early stages we must turn to the science of astronomy. The relations of the earth to the stars and the planets are shown in the depths of the leavens, and there must exist in the heavens those cosmic conditions which gave rise to our world and the other planets of our system. Through the telescope and spectroscope, the astronomers have solved many of these secrets.

But what of the age of the earth measured in years as we measure other happenings. From the brief outline just given one can see that it has been in existence unknown millions of years, but just how many it is impossible to say.

We can, however, measure back to the more recent events in geological time to the last ice age, before which we know man existed, and get a fairly accurate result, in terms of years.

One of the most accurate ways in which to measure such time intervals, is by measuring and counting the light colored and dark colored bands of clay, deposited by the melting of the ice sheet in the fresh water lakes which existed on the edges of those continental glaciers, as it retreated to its present position in the north polar regions. Each dark layer of clay was laid down during one winter and each light layer during one summer. By such detailed studies, it has been determined that it has taken, approximately, 5,000 years for the glaciers of Sweden to melt back 270 miles, and it is further known that this melting took place 8,500 years ago. We know that the glaciers in North America extended into the northern part of the United States and reached as far south as the Ohio river. We know that now their southern edge lies far to the north in northern Canada over a thousand miles away. We know that it took approximately 4,000 years for the continental glacier which last covered the New England states to melt back from Hartford, Conn., to St. Johnsburg, Vt. This is only one way of measuring in years some of the more recent happenings. There are many more methods that could be given if it were necessary.

In connection with evolution, it is especially of interest to note that the relative ages of the rocks correspond closely to the degrees of complexity of organization shown by the fossils in those rocks. The simpler organizations being found in the more ancient rocks, each type of organism becoming more and more complex as we come nearer to the present day, man and his fossil and cultural remains being no exception.

It, therefore, appears that it would be impossible to study or teach geology in Tennessee or elsewhere, without using the theory of evolution.

By Kirtley F. Mather,

Chairman of the Department of Geology of Harvard University.

(Biography—Kirtley F. Mather graduated in 1909 from Denison university, a Baptist college at Granville, O., in which evolution has for years been taught by every science teacher. In 1915 he received the degree Ph. D. from the University of Chicago. He taught geology at the University of Arkansas for three years, at Queens university, a Presbyterian institution at Kingston, Ontario, for three years, and from 1918 to 1924 he was head of the department of geology at Denison university. In 1923 he was appointed professor of geology at Harvard and has recently been made chairman of the department of geology at Harvard. He has been a geologist of the United States geological survey for many years, and has made geological examinations for various oil companies in Bolivia, Peru, Mexico, Canada, etc. He was for several years a trustee of the Baptist church at Granville, O., and chairman of the Baptist church at Newton Centre, Mass., and teacher of the "Mather class" in Bible school of that church. He is a fellow or member of such scientific organizations as the Geological Society of America, the American Academy of Arts and Sciences, and the American Institute of Mining and Metallurgical Engineers. In 1923-24 he was president of the Ohio Academy of Science. He is the author of numerous scientific publications and bulletins of the United States Geological Survey, dealing with the petroleum resources of Kentucky, Oklahoma, Alaska and Colorado; technical papers on geology, paleontology and evolution in scientific journals; "Christian Fundamentals in the Light of Modern Science," etc. In 1919 he prepared a bulletin of the Tennessee Geological survey, dealing with the geology and oil resources of Summer county, Tennessee.)

The facts of life development are so numerously displayed and so evident in the rocks of the earth's crust that every geologist with whom I am acquainted has accepted the evolutionary principle as demonstrated. Much of the exposed part of the earth's crust is composed of rocks deposited in layers as sand, mud, gravel or limestone in the seas, lakes, or ponds of past time, or upon the surface of the dry land. These are in many places broken through by masses of rock which has formed by solidification of molten lava. The successive ages of the various kinds and formations of rock are determined by their physical relations.

Where not greatly disturbed by crumpling or upheaval of the earth's crust, the rocks formed in layers are obviously still in their original order, the oldest underneath and the younger layers in order one upon the other, just as they may now be observed in the hills overlooking Dayton, Tenn. Where cut through by rocks which were once in a fluid state, it is apparent that each body of rock is younger than the youngest rock through which it broke and older than the oldest rocks deposited upon its surface after it was solid. Thus the succession of physical events in the history of the earth may be determined by patient and careful scrutiny of the earth's surface as it now is visible, either in natural or artificial exposure such as canyon walls, valley slopes, mines and wells.

In many of these rocks there are found entombed the fossil remains of the animals and plants which were alive at the time the rocks were formed. Some of these are the shells or bones of animals that lived in the seas or lakes, some are the harder parts of animals that lived on the land and were buried beneath the mud of river flats or the ashes blown out of volcanic vents. Discovering these fossil remains and knowing by their physical relations the successive ages of the rocks in which they are found, the geologist is able to sketch the history of animal and plant life upon the earth.

At Least 100 Million Years.

In the very oldest rocks which have yet been discovered, which are at least 100,000,000 years old there are absolutely no traces whatsoever of any animal or plant life. In somewhat younger rocks, but rocks, also referred to the oldest era of geological history, the archeozoic era, there are remains of one-celled plants of the type known as albae. The next era of earth history has been named the proterozoic. In rocks formed during it, there are a very few fossils of lowly types of shell-bearing animals and some rather obscure markings which are probably in part due to the presence of worms and in part represent the remains of sea-weeds. The rocks of these two oldest eras are nearly everywhere much distorted and broken by volcanic activity and crustal upheavals.

Upon these ancient formations there rest in orderly succession the layers deposited during the several periods of time which geologists group into what is called the paleozoic era, which began at least 50,000,000 years ago. Most of the rocks of Tennessee were laid down during that long space of time. In this state, as elsewhere, these strata are known at many places to contain a great abundance of fossils. In the oldest rocks of that era, the fossils are of many and various invertebrate animals, many of which are of kinds not now known to exist anywhere on the face of the earth today. There are no fossils of animals which had a backbone of any sort in any of these rocks. In somewhat younger beds, referred to the second period of the paleozoic era, there are, however, very scanty and fragmentary remains of primitive fishes, the first known animals which possessed a backbone. The oldest known forest, composed of trees of fern-like rather than of seed-bearing types, was found a few years ago in New York in rocks formed about at the middle of this paleozoic era. That was the time when fishes ruled the waters, for remains of sharks and lungfishes are present in great numbers in the rocks formed in the seas, but in the rocks laid down on the land or in swamps there is not a trace of animals with a backbone, although insects and land snails have left their fossil remains in them. Toward the end of the paleozoic era, however, the rocks formed of desert sands and swamps contain the footprints and petrified bones of amphibians and reptiles, the first animals with a backbone which could breathe air by means of lungs. This part of the paleozoic system of rocks includes the coal seams of the eastern states, and associated with the coal are many beautiful specimens of ferns and primitive evergreen trees, but none of the modern types of flowering plants. About at the close of the paleozoic era the Appalachian mountains were formed by the crumpling of the earth's crust in this region.

That episode of crustal crumpling is taken as the milestone to mark the end of the paleozoic and the beginning of the mesozoic era, which began at least 25,000,000 years ago. Since that time, Tennessee and neighboring states have, with minor exceptions, remained continually above sea level, so that we have to transfer our search to other localities to find the continuation of the fossil record. The mesozoic era, the fourth great era of earth's history, is frequently referred to as the age of reptiles. In practically all the stratified rocks of this era there are petrified bones and footprints which tell that cold-blooded, scaley animals with backbones and four limbs lived in great numbers on land, in the sea, and in the air. The largest and most ferocious animals that ever inhabited the lands left their bones among the fossils of that era. Animals with enough feathers to enable them to fly, yet with claws on their forelimbs and teeth in their jaws, lived then and indicate the transition forms between reptiles and birds. In the same rocks with those reptiles, most of which have long since vanished from the face of the earth, a very few fragments of quite primitive mammals have been found.

These are small and insignificant creatures, most of whom laid eggs as do a couple species of small mammals today, but who suckled their young, were warm-blooded and presumably had no scales as surface covering. For the most part the reptiles were small-brained and large-bodied; they placed their trust in strength of talon and claw, rather than in mentality and agility. Observing the earth at that time, one could not help but feel that no good could possibly come from that welter of blood-thirstiness and cruelty. Yet the small minority of puny mammals, present then, was so endowed with instinct, such as parental love for offspring, that at the end of Mesozoic time it became the dominant form of life on land, while the few reptiles which did not become extinct were for the most part banished to the swamps and deserts or other out-of-the-way places. The close of Mesozoic time, the age of reptiles, was marked by the upheaval of the Rocky mountains. In a small fraction of the time that has elapsed since then, the entire Grand Canyon of the Colorado river has been carved by the ceaseless wear of running water. For this, and many other reasons, geologists believe that each of these eras of time should be measured in terms of tens of millions of years.

The Cenozoic era, which began 5,000,000 or 10,000,000 years ago, began as the Rocky mountains were formed. Most of the rocks of that era are still unconsolidated layers of silt or sand or volcanic ash, although some are firmly cemented into sandstone, limestone, etc. In the earliest beds deposited around the flanks of the new-born mountains of the western states, the bones of a great variety of mammals have been found. They are evidently the improved offspring of the puny mammals which had lived in constant fear of the ponderous reptiles during the preceding era. Not until about this time had there been any large quantity of the kinds or vegetation upon which modern mammals feed, and this presumably explains in part the slowness of the mammalian minority in throwing off the yoke of the reptilian majority during the age of reptiles. The first flowering plants had left their leaves and seed pods in the rocks formed during the middle of the Mesozoic era, but grasses and herbs, fruit-and-nut-bearing trees were not numerous until the beginning of the Cenozoic era.

With an abundance of the right kind of plant food and freed from reptile dominion, the mammals increased rapidly in numbers, and their bones in great variety may today be seen in the rocks of the Rocky mountains and other regions. Among those of the earliest Cenozoic strata may be mentioned the five-toed and four-toed ancestral horses, the trunkless and small-tusked ancestral elephant, the cat-like forerunner of the modern seal. At that time, too, we find the first record of a primate, that order of mammals to which the zoologists refer man. This was evidently a small quadruped with toes terminated neither in hoofs nor claws, but with rather horny nails, and with teeth adapted neither for grinding grain like those of a horse nor for tearing flesh like those of a tiger nor for gnawing nuts like those of a squirrel, but like those of a man for eating herbs, fruits and eggs. But in general appearance this creature resembled a rat much more closely than a monkey, ape or man. Bones of that lowly type of primate have been found in North America, Asia and North Africa.

Somewhat higher in the series of Cenozoic strata of India, there were recently found a fragment of jaw which had teeth totally different from those of any nonprimate, somewhat different from those of a monkey, and closely resembling those of the great apes and of man. That animal lived somewhere between 2,000,000 and 10,000,000 years ago. He is believed to have been ancestral to the apes, chimpanzees, gorillas and mankind, all of which had by that time become completely differentiated from the monkey strain. If that be true, man has become distinct from the other anthropoids since that creature left his bones on the banks of an Indian stream. Narrowing our attention now to the strain that leads to man, the next fossil of significant interest is that known as the ape-man of Java. Some thirty years or so ago there was found on the island of Java a partially cemented layer of gravel and sand containing fossil bones and fossil plant remains. The plants were of the same sort as found elsewhere in rocks known to have been formed rather late in the Cenozoic era just before the first glaciers of the great ice age were accumulating, therefore, it must be that the associated animal bones are also of that age. The skull of this animal had brain capacity somewhat greater than that of the most brainy apes now living and somewhat less than that of the smallest-brained human tribe. He had a receding forehead and a heavy ridge of bone above his eyes like an adult chimpanzce; yet his leg-bones show unmistakably that he stood and walked erect upon his hind limbs. The name ape-man describes him exactly; he was truly intermediate in body structure between the apes and man. He lived 1,000,000 or 2,000,000 years ago. In rocks of just about that same age in England there have been found crudely fashioned flint implements, unmistakably shaped by some intelligent creature with hands so developed as to be capable of holding a stone and striking it with another stone. Modern apes have been observed to hold clubs in their clumsy hands, but none of them can at will touch his thumb against the tip of each finger on the same hand. Presumably the creature who chipped the flints found in those rocks near Foxhall, England, could do so.

Then came the first of the great glacial advances of the ice age about 1,000,000 years ago. Five times the northern lands were buried beneath a mantle of moving ice. Five times the ice melted until the glaciers were at least as small as those now remaining on Greenland and in the valleys of Alaska. In the gravels deposited in Germany by the rivers flowing from the melting ice of either the first or the second of these interglacial intervals, there has been found the jaw of the so-called Heidelberg man. The jaw resembles that of a modern man; its sides are nearly parallel, the canine teeth are only a little higher than the incisors and molars. But it has no chin at all, and the portion of the jawbone which articulates with the skull just in front of the ears looks considerably like the equivalent portion of an ape's jaw. Scientists classify that creature as a member of the same genus to which modern man belongs, but as a different species.

Gravels of later interglacial stages have revealed the bones of still another extinct species even closer to modern man. More than a score of practically complete skeletons and hundreds of fragmentary bones of this the Neanderthal man have been found in France, Spain and Germany. It is chiefly in the characters of the skull rather than in the other bones of the skeleton that he differs from modern man. His forehead was very receding, his brain capacity was just a little less than that of the most primitive of existing savage tribes; his brow ridges were more prominent than those of the negro, his chin was approximately half way between the chinless profile of the Heidelberg man and the clearly defined chin of the white race of today. With his petrified bones there are frequently found the stone spearheads and the bone knives which he fashioned. To this array of facts concerning him, I want to add just one inference. Many skulls of Neanderthal type were broken when found, as though struck with a hammer on top of the head either at the moment of death or very shortly thereafter. Several tribes of aborigines in recent years break the skulls of their dead in order, as they say, to permit the spirit to start on its journey to the happy hunting ground. The inference is that the Neanderthal man, a couple of hundred thousand years ago, had the same thought that man was immortal.

During the last of the glacial stages, about the same time that the ice pushed southward across Ohio and Indiana to the Ohio river, 40,000 or 50,000 years ago, there lived in southern Europe a race of men known as the Cro-Magnons. They were stalwart highbrows with prominent chins and large brain capacity, and eyebrow ridges no more protruding than those of the existing white race, but with massive cheekbones like the North American Indian. Clearly they belonged to the same species as that which today includes the white, yellow, brown and red races, but they cannot be included in any of these races. Their implements were much better manufactured than those of their predecessors, the Neanderthals, and they had a remarkable artistic ability as shown by the pictures they engraved or painted on the walls of caves in southern France. For thousands of years they maintained their life in Europe, but about 10,000 years ago they were displaced by the first members of the races of mankind which are today in existence.

During all this time no known record of the presence of man or man-like creatures was left in either North or South America. Not until the ice sheets of the latest glacial episode had dwindled nearly to disappearance was any clear indication of man's presence left in the New World. The oldest human inhabitants of North America were members of the existing races of mankind. They reached this continent not more than 10,000 or 12,000 years ago.

The facts stated in the foregoing paragraphs have been discovered by many different individuals. Probably no one man could be found who could testify to all of them as having been personally observed by himself. Knowledge of them is the common property of countless scientists. I can, however, affirm the truth of many of these facts from personal observations; the others I believe to be true because of my confidence in the technical ability and integrity of those who have seen the actual evidence. I have also studied many of the specimens collected by those fellow-workers and now on exhibition in various museums. In 1916 and 1917 I examined the oldest known rocks of the Archeozoic era in eastern Ontario and was unable to discover any fossil remains in them. The presence of these rocks had already been made known by a Canadian geologic survey party. I was accompanied by four or five of my students. In this bleak and windswept waste of rounded rock hills and impassable swamps, these ancient rocks are cleanly displayed. On the same trip I saw in slightly younger rocks of the same era in that locality the evidence of the presence of primitive organisms, but no record of any of the higher forms of life. In 1906 I collected fossil shells of lowly invertebrates from the early Paleozoic rocks of Wisconsin. During the spring of 1916 I found the remains of somewhat higher types of invertebrates in slightly younger rocks of the same era in eastern Ontario and later described these fossils in publications of the Ontario bureau of mines and in the Ottawa Naturalist. Other invertebrate fossils of about the same age and about the same kinds were observed when I was in Bolivia in 1919 and 1920. Accompanied by half-bred guides and camp hands I, together with K. C. Heald, formerly chief of the oil and gas section of the United States geological survey, pushed far beyond the outposts of civilization into the rocky fastnesses of the eastern Andies and there we found these fossil remains.

Saw Evidences in Rocks.

I have seen the fossil remains of primitive fishes of middle Paleozoic age on a number of occasions near Columbus, Ohio; in 1917, in Allen County, Kentucky, and in 1919, in Sumner County, Tennessee. I observed the foot prints of large reptiles in rocks formed shortly after the upheaval of the Appalachian Mountains at several places in the Connecticut valley during 1921. While exploring in Alaska during the summer of 1923, I searched for fossils in rocks of middle Mesozoic age, but found in them only the remains of shellfish and corals. There was a party of six dispatched by the United States geological survey to search for mineral resources in a previously unknown and altogether uninhabited portion of the Alaska peninsula, not far from the famed valley of Ten Thousand Smokes, so named because of the countless vents from which steam roared heavenward. We had to cut steps with our geologic hammers across glaciers and snow fields in traversing the almost inaccessible mountains of that bleak, barren and rugged land. In Colorado, during the summer of 1924, I had occasion to study the petrified bones of mammals imbedded in flat-lying rocks of Cenozoic age directly overlying tilted strata of late Mesozoic age, in which were the fossil bones of reptiles. The tilting of those beds was a part of the crustal movement which formed the Rocky Mountains; the flat layers on top of them were deposited while those mountains were being eroded.

To this summary of known facts concerning the life of the past, there might be added a multitude of other facts concerning the body structures of the various animals, the life history of the individual animal from its start as a single fertilized cell until its attainment of adult stature, etc. I have, however, personal knowledge of only a few of the facts in these fields in which I am not a specialist. While exploring the headwaters of the Amazon in Bolivia and Peru in 1919 and 1920, I lived for some time among quite uncivilized peoples, many of whom had never before seen a white man. At the same time I watched the habits and examined the bodies of several different kinds of South American monkeys. I have studied with care the skeletons of many of the Asiatic apes and Old World monkeys, as they were available in various university laboratories and museums. From these studies and from the studies of others, I can affirm the following generalized statements: Comparing the body structure of monkeys, apes and man, it is apparent that they are all constructed upon the same plan; with only trivial exceptions every bone in the body of one has its counterpart in the body of the others. Only in details of shape, in relative size and in method and angle of articulation with their neighbors do these bones differ in the different creatures just mentioned. Monkeys have long tails; some apes have long and some have short tails; man has a vestigial tail composed generally of about four vertebrae so small and so short as to be entirely concealed in the flesh and muscles at the base of the spine. In relation to the total dimensions of the body, the brain of monkeys is quite small, that of the apes is much larger, while that of man is largest of all. This determines in large degree the contour of the head; thus the face of the monkey occupies more space than the top and back of its head, that of the apes is comparatively smaller, while the face of man is smallest of all in relation to the total area of head surface. No one would be surprised or shocked to learn that apes and monkeys had a common ancestor, nor would he regard it as a startling scientific theory, yet in general there are more differences between the modern monkeys and the modern apes, such as the chimpanzee, the gorilla, the gibbon and the orangoutang than there are between the apes and man. Yet in general there are more differences between the apes and man than there are between the existing races of men. The gaps between these various groups are, however, largely filled by the fossils, some of which I have already described. There are in truth no missing links in the record which connects man with the other members of the order of primates.

Such facts as I have stated above can be explained only by the conclusion that man has been formed through long processes of progressive development, which when traced backward through successively simpler types of life, each living in more remote antiquity, lead unerringly to a single primordial cell. The facts ascertained by natural science are obviously incomplete; the record of the rocks by no means tells the whole story. Man not only has an efficient and readily adaptable body, he also possesses a knowledge of moral law, a sense of rightness, a confidence that his reasoning mind finds response in a rational universe, and a hope that his spiritual aspirations will find increasing answer in a spiritual universe. Such things as these cannot be preserved in the fossil record, yet their presence must be accounted for. Nor have we a direct record of whence came the first living cells. The inference is unmistakable that material substances from which living cells were first constructed were previously present among the rocks and minerals of the earth. All the necessary ingredients were certainly present in the outer shell of the youthful earth of even pre-Archeozoic time. But life is something more than matter. Living creatures are characterized by vital energy, something about which we really know very little, but something which is absolutely indispensable to every living creature. T. C. Chamberlin, the dean of American geologists, closes his volume on the origin of the earth with the following sentence: "It is our personal view that what we conveniently regard as merely material is at the same time spiritual, that what we try to reduce to the mechanistic is at the same time volitical, but whether this be so or not, the emergence of what we call the living from the inorganic, and the emergence of what we call the psychic from the physiologic, were at once the transcendent and the transcendental features of the earth's evolution." With this conclusion I am in hearty accord. I believe that life as we know it is but one manifestation of the mysterious spiritual powers which permeate the universe. The geologic factors assembled in the primitive earth provided an environment within which the spiritual could manifest itself in the material. The form which it should assume may have been largely determined by that environment; the primitive cell was the result. Thus, in truth, was man made from the dust of the ground.

Again, the record of the rocks tells nothing except by inference of the previous state of the mineral matter of which the earth is made. Several theories, varying from one another in greater or less detail, are now under consideration by geologists and astronomers in their attempt to understand the actual beginnings and the antecedents of the earth and its fellow planets in the solor system. So far as we now know all the planets, suns and stars within range of our telescopes are composed of the same sort of matter, reducible upon analysis to about eight different elements, nearly all of which are present in the earth. In other words, it is a fair sample of the material substances of the entire universe. Science has not even a guess as to the original source or sources of matter. It deals with immediate causes and effects, not at all with ultimate causes and effects. For science there is no beginning and no ending; all acceptable theories of earth origin are theories of rejuvenation rather than of creation—from nothing. Indeed, there is some evidence for the prevalent view that our sun had had at least one earlier generation of planets in its train before the disturbing effect of the close approach of another star caused the reorganization of part of its matter into our present solar system. Conversely, it is probable that at some remotely distant date in the future this group of planets, on one of which we live, will be similarly destroyed by another rejuvenating disturbance and still another cycle of planetary organization may take place.

But none of these facts is really in any way disturbing to the adherent to Christianity. Not one contradicts any teaching of Jesus Christ known to me. None of them could for his teachings deal with moral law and spiritual realities. Natural science deals with physical laws and material realities. When men are offered their choice between science, with its confident and unanimous acceptance of the evolutionary principle, on the one hand, and religion, with its necessary appeal to things unseen and unprovable, on the other, they are much more likely to abandon religion than to abandon science. If such a choice is forced upon us, the churches will lose many of their best educated young people, the very ones upon whom they must depend for leadership in coming years. Fortunately, such a choice is absolutely unnecessary. To say that one must choose between evolution and Christianity is exactly like telling the child as he starts for school that he must choose between spelling and artithmetic. Thorough knowledge of each is essential to success —both individual and racial—in life.

Although it is possible to construct a mechanistic, evolutionary hypothesis which rules God out of the world, the theories of theistic evolution held by millions of scientifically trained Christian men and women lead inevtiably to a better knowledge of God and a firmer faith in his effective presence in the world. For religion is founded on facts, even as is the evolutionary principle. A true religion faces the facts fearlessly, regardless of where or how the facts may be found. The theories of evolution commonly accepted in the scientific world do not deny any reasonable interpretation of the stories of divine creation as recorded in the Bible, rather they affirm that story and give it larger and more profound meaning. This, of course, depends upon what the Bible is and what the meaning and interpretation of the stories are to each individual. I have been a Bible student all of my life and ever since my college days I have been intensely concerned with the relations between science and the Bible. I have made many addresses and have written several articles upon this subject. I have many times lectured to Biblical students, such as those in the Boston University School of Religious Education.

It is obvious to any careful and intelligent reader of the book of Genesis that some interpretation of its account must be made by each individual. Very evidently it is not intended to be a scientific statement of the order and method of creation. In the first chapter of Genesis we are told that man was made after the plants and the other animals had been formed, and that man and woman were both created on the same day; in the second chapter of Genesis we read that man was formed from the dust of the ground before plants and other animals were made, that trees grew until fruit was upon them that all the animals passed in review before man to be named, and then after these events woman was made. There is obvious lack of harmony between these two Biblical accounts of creation so far as details of process and order of events are concerned; they are, however, in perfect accord in presenting the spiritual truth that God is the author and the administrator of the universe. And that is the sort of truth which we find in the Bible. It is a textbook of religion, not a textbook of biology or astronomy or geology. Moreover, it is just exactly the Biblical spiritual truth concerning God which rings clearly and unmistakably through every theory of theistic evolution. With it modern science is in perfect accord.

Reasons for Distrust.

There are a number of reasons why sincere and honest Christians have recently come to distrust evolution. These reasons must be understood and discussed frankly, before the world will believe that science and religion are not in conflict. Some of the opposition to evolutionary science results from failure to read the Bible. Too many people who loudly proclaim their allegiance to the book know very little about what it really contains. The Bible does not state that the world was made about 6,000 years ago. The date 4004 B. C. set opposite Genesis i:1 in many versions of the Bible was placed there by Archbishop Ussher only a few centuries ago. It is a man's interpretation of the Bible; it is in the footnotes added recently: it is not a part of the book itself. Concerning the length of earth history and of human story, the Bible is absolutely silent. Science may conclude that the earth is 100,000,000 or 100,000,000,000 years old; the conclusion does not affect the Bible in the slightest degree. Or if one is worried over the progressive appearance of land, plants, animals and man on the successive six days of a "creation week," there is well known Biblical support for the scientists' contention that eons rather than hours elapsed while these things were taking place. "A day in the sight of the Lord is as a thousand years, and a thousand years as a day." Taking the Bible itself as an authority dissipates many of the difficulties which threaten to make a gulf between religion and science. The fact that the seventh day was stated to be a day of rest has no bearing upon the length of the other days. I have no doubt that the man who made that chapter of Genesis had in his mind days of twenty-four hours each, but I reserve for myself the right to make my own interpretation of the meaning of words, as does every Christian, be he literalist, trivialist or modernist.

Another of the reasons for the modern distrust of science in the religious world is the idea that evolution displaces God. Many seem to think that when the scientist enthrones evolution as the guiding principle in nature he dethrones God, that the two words are somehow synonymous, that there is not room for both and one must go. But the facts are as follows: Evolution is not a power, nor a force; it is a process, a method. God is a power, a force; he necessarily uses processes and methods in displaying His power and exerting force. Many of us believe that science is truly discovering in evolution the processes and the methods which God, the spiritual power and eternal force, has used and is using now to effect His will in nature. We believe we have a more accurate and a more deeply significant knowledge of our Maker today than had the Hebrew patriarchs who thought a man could hide from God in a garden, or who believed that God could tell man an untruth. (Genesis ii:17 states that God told man he would surely die if he ate the fruit of the tree of knowledge; man ate, he did not die, God knew he would not die therefor.)

Again there is the widespread misconception that if one accepts the evolutionary process as the method which God uses he will find himself in a moral dilemma. Regardless of sect or creed, all followers of Christ must accept his teaching that the law of life is love, that service to others is the true guiding principle, that self-sacrifice even to death is the best trait a man can display. To many, evolution means the survival of the fittest in the struggle for existence; and that is taken to imply that the selfish triumph, the most cruel and blood-thirsty are exalted, those who disregard others win. Obviously, this is the very anthithesis of Christianity; both principles cannot be true; one must be false. The Christian needs not to be told which of the two it is. Here is a real reason for opposition to evolution; men are not driven from it by the fear of discovering that their bodies are structurally like those of apes and monkeys; it doesn't bother us to discover that we are mammals, even odorous mammals—"by the sweat of your brow must man earn food" states the Bible. It does bother us to find the implication that the law of progress has thus apparently been opposed to the love of Christ. But here are the facts. It has been my privilege as a geologist to read the record in the rocks; knowing the ages of the rocks has led to better knowledge of the Rock of Ages; I have watched the procession of life on the long road from the one-celled bit of primitive protoplasm to the present assemblage of varied creatures, including man. At times of crisis in the past it was rarely selfishness or cruelty or strength of talon and of claw that determined success or failure. Survival values at different times have been measured in different terms. Ability to breathe air by means of lungs rather than to purify the blood by means of gills meant success in escaping from the water to the land. Love of offspring and tender care for the young gave the weak and puny mammals of long ago the ability to triumph over much stronger and more powerful reptiles like the dinosaur. Especially in the strain that leads to man can we note the increasing spread of habits of co-operation, of unselfishness of love. The survival of the "fit" does not necessarily mean either the survival of the "fittest" or of the "fightingest." It has meant in the past, and I believe it means today and tomorrow, the survival of those who serve others most unselfishly. Even in evolution is it true that he who would save his life must lose it. Here, if nowhere else, do the facts of evolution lead the man of science to stand shoulder to shoulder with the man of religion.

Another difficulty arises from our present limitations of knowledge. If man has evolved from other forms of animal life by the continuous process of evolution it is asked how can there be any difference between him and them, how can we believe that he has an immortal soul. Again, the appeal to facts makes it clear that somehow out of the continuity of process real differences have emerged. When the cow pauses on the hillside to admire the view, when the dog ceases to bay at the moon in order to construct a system of astronomy then and not till then will we believe that there are no differences between man and other animals. Even though we may not understand how these differences arose, the facts are there; knowledge and mystery exist side by side; mystery does not invalidate the fact. Men of science are working on those very problems. They have not learned—and may never learn how God breathed a living soul into man's body. If they discover that process, and the method used, God will still be just as great a power. In the image of God cannot refer to hands or feet, heart, stomach, lungs. That may have been the conception of Moses; it certainly was not the conception of Christ who said that God is spirit, and proclaimed that man must worship Him in truth. It is man's soul, his spirit, which is patterned after God the Spirit.

Soul Theologian's Business.

It is the business of the theologian not the scientist to state just when and how man gained a soul. The man of science is keenly interested in the matter, but he should not be blamed if he cannot answer questions here. The theologian must tell when the individual gets his soul, whether at the moment of conception, or when the unborn babe first stirs within the womb, or at the moment of birth, or at the first gleam of intelligent appraisal of his environment and how he knows this.

Men of science have as their aim the discovery of facts. They seek with open eyes, willing to recognize it, as Huxley said, even if "it sears the eyeballs." After they have discovered truth, and not till then, do they consider what its moral implications may be. Thus far, and presumably always, truth when found is also found to be right, in the moral sense of the word. Men of religion seek righteousness; finding it they also find truth. The farther along the two avenues of investigation the scientists and the theologian go, the closer together they discover themselves to be. Already many of them are marching shoulder to shoulder in their endeavor to combine a trained and reasoning mind with a faithful and loving heart in every human individual and thus to develop more perfectly in mankind the image of God. Neither the right kind of mind nor the right kind of heart will suffice without the other. Both are needed if civilization is to be saved.

As Henry Ward Beecher said, forty years ago, "If to reject God's revelation of the book is infidelity, what is it to reject God's revelation of himself in the structure of the whole globe?" With that learned preacher, men of science agree when he stated that "the theory of evolution is the working theory of every department of physical science all over the world. Withdraw this theory, and every department of physical research would fall back into heaps of hopelessly dislocated facts, with no more order or reason or philosophical coherence than exists in a basket of marbles, or in the juxtaposition of the multitudinous sands of the seashore. We should go back into chaos if we took out of the laboratories, out of the dissecting rooms, out of the field of investigation, this great doctrine of evolution." Chaos would inevitably destroy the whole moral fabric of society as well as impeded the physical progress of humankind.

By Dr. Maynard M. Metcalf.

Biography—Dr. Maynard M. Metcalf is engaged in private research work at the Johns Hopkins university, specializing in zoology. From 1893 to 1914 he taught college zoology, first at Goucher, then at Oberlin college, at Oberlin, O. He received his bachelor's degree at Oberlin, the degree or doctor of philosophy at the Johns Hopkins university, and the degree of science at Oberlin. He has memberships and has held offices in the American Association for the Advancement of Science, the American Society of Zoologists and numerous other scientific and economic societies. During the past year he has been chairman of the committee on biology and agriculture of the National Research council. He is author of numerous books and articles on zoology and evolution.)

Intelligent teaching of biology or intelligent approach to any biological science is impossible if the established fact of evolution is omitted. Discussion of the methods by which evolution has been brought about is less essential but the fact of evolution must be appreciated and the evolutionary point of view must be emphasized for any understanding of the growth of the universe, of the earth of plants or animals; for any proper grasp of the facts of structure or function of living bodies as involved in medicine and in animal and plant husbandry; psychology, whether of normal or diseased minds, must constantly remember the processes of evolution; human societies, with their diverse customs, are unintelligible without the facts of their origins and changes their evolution. God's growing revelation of Himself to the human soul cannot be realized without recognition of the evolutionary method he has chosen. Teaching in any field that deals with living things is disgracefully, yes, criminally, inadequate if it omits emphasis upon evolution. An intelligent teacher could omit such emphasis only at the expense of his self-respect and of his moral integrity. Such teaching would be criminal malpractice just as truly as would be a physician's failure to follow established sound methods of treatment because of fear of persecution by ignorant neighbors. For a teacher to fail to bear testimony is as essentially sinful as for a man to fail to stand by his religion. Truth is one, whether scientific truth or religious truth, and it calls for loyalty from every worthy man. The fact of evolution—of man, of all living things, of the earth, of the sun, of the stars—is as fully established as the fact that the earth revolves around the sun. Change, growth evolution, is a fundamental, a pivotal truth in all nature. Those familiar with the phenomena of nature testify with unanimity to this. The great mass of evidence of different sorts from different sources, when once seen, is overwhelmingly convincing to any normal, human mind. It can be only the uninformed who fail to accept evolution as a fact established beyond doubt. On the other hand, there is great uncertainty as to the method by which evolution has been brought about. Many different factors have been in operation, among them probably the chief has been the mysterious intimate activities of the living substance itself about which as yet we know so little. As to the numerous "causes" of evolution and their relative importance, there are about as many varieties of opinion as there are students of evolution. I am somewhat acquainted personally with nearly all the zoologists in America who have contributed extensively to the growth of knowledge in this field and I know many of the botanists and a goodly number of the geologists and I doubt if any two of these put exactly the same relative emphasis upon all the numerous interacting "causes" of evolution. But of all these hundreds of men not one fails to believe, as a matter of course, in view of the evidence, that evolution has occurred.

None of this, of course, has any bearing upon the question of God as the creator of the universe. It is only a matter of the method He has chosen in creation—whether immediate fiat or gradual growth accompanied by divergence. The evidence is overwhelming that the latter was and is His method. God is just as truly and just as intimately acting in the gradual growth of a plant from a seed or of a man from a fertilized egg as He would be in creating the full grown plant or man all at once in a thousandth part of a second of time.

No Contest Between Bible and Fact of Evolution.

There is no conflict, no least degree of conflict, between the Bible and the fact of evolution, but the literalist intepretation of the words of the Bible is not only puerile; it is insulting, both to God and to human intelligence.

But the fundamentalist would do much worse than insult God. He is in reality, although he doesn't realize this, trying to shut man's mind to God's ever-growing revelation of Himself to the human soul. He teaches, in effect, that God's revelation of Himself was completed long ago, that He long ago ceased to unfold His mind to men in new revelation. This is evil influence, criminal, damnable. Truth is sacred and to hinder men's approach to truth is as evil a thing, as unChristian a thing, as one can do. The thought that God is at odds with Himself, that his revelation of Himself to men of old is at variance with His works in nature is as blasphemous as it was for the Jewish leaders to say of Jesus that "He casts out devils through Beelzebub, the prince of the devils." Jesus made short work of this attack upon him.

No, the thing is not to attempt to guide God's self-revelation into channels of our own ignorant choosing, but is, rather, humbly and in a wholly teachable spirit to seek His thought and Himself in nature, in history, in the vision of Himself He has given to men of old and is still giving to the humble minded today. One of the greatest of God's revelations of Himself to men has come through His showing us His habit of producing results, by gradual growth, by evolution, rather than by immediate fiat.

Not only has evolution occurred; it is occurring today and occurring even under man's control. If one wishes a new vegetable or a new flower it is, within limits, true that he can order it from the plant breeder and in a few years he will produce it. Hundreds of new plants and animals have been and are being produced in this way. This is evolution of just the sort that has always occurred, only it is influenced by man's purpose. We can see evolution occurring in our experiment stations and our laboratories and we can control and modify the conditions of the experiments and can thus modify the resultant product to suit ourselves. Evolution is a present observable phenomenon as well as an established fact of past occurring. The organisms produced by this present day controlled evolution in our experiments are as divergent from one another and from the original stock as are animals and plants in nature. The different kinds of domestic horses, produced by human experiment, differ far more than do the different kind of horses found in nature. Domestic fowl under man's control have evolved into a large number of kinds far more widely divergent than are the wild kinds in the genus Gallus, from which our domestic chickens came. The genus Brassica, plants belonging to the mustard family, include a number of different sorts of plants. One of these, Brassica Cleracoa, is the ancestor, the form from which man has evolved the cabbage, the cauliflower, kale, Brussels sprouts, kohl rabi and the Swedish turnip, which differ among themselves far more than do the wild members of the genus Brassica. The same sort of thing is seen in hundreds of domestic animals and plants, dogs, cattle, sheep, pigeons, cucumbers, radishes, lettuce, dahlias, roses, wheats, corns, strawberries, peaches, apples, pears, etc., etc., etc. This is all true evolution and is going forward today with ever-increasing strides. To describe adequately the tremendous mass of phenomena which establish the fact of past and continuing evolution would require not a book or a series of books, but a library. In the main, these evidences may be arranged in four chief groups: (1) The phenomena of comparative anatomy; (2) the phenomena of comparative embryology; (3) the phenomena of paleontology and geology, and (4) the phenomena of geographical distribution. Much in the fields of physiology, psychology and human cultures has very important bearing upon evolution.

First—We can arrange plants and animals in a double, parallel series, showing increasing complexity of organization.

Second—In the development of an individual from egg to adult this individual passes through a series of stages of increasing complexity and this individual series in one of the higher organisms strangely parallels and agrees with the racial series first mentioned.

Third—In the fossiliferous rocks we find actual bodily remains of organisms of the past and these form a series showing increasing complexity within each taxonomic group, the animals and plants in the older rocks being more and more simple, while the successively younger rocks show more and more complex organisms in each group under observation.

Fourth—The distribution of animals and plants over the earth is such as to suggest strongly the origin of each group of animals or plants at some one place, and their gradual spread from that center, divergent evolution occurring while they are spreading. No other suggestion even plausible, let alone convincing, has been made to explain these phenomena. Evolution is the only key we can find.

In each of the four groups of phenomena mentioned there are many very striking things. One set of these things, in the first, morphological group, is that of the vestigal organs in animals and plants. There are in man, for example, very many structures of no conceivable present use, but showing resemblance to organs in other animals which are useful. The appendix vermiformis is one such structure, a mere vestige of an organ of great great importance in some lower mammals. The human tail—bony coccyx with its rudimentary muscles—is another. The wisdom teeth of man are approaching a vestigial condition.

It is interesting to observe that an organ in one kind of animal may have a different use from the similar organ in a related animal. There are very few, if any, structures in man, for example, which do not show clear indications of relationship to, descent from, an organ of different use in some related animal. The lungs of man correspond to the swim bladder of fishes; hair has apparently been derived from tactile sense organs in the skin of aquatic vertebrates; certain bones connected with the human larynx were derived from the supporting arches in the bars between the gill slits of our aquatic ancestors; our teeth were once scales in the skin and so on and so on. Probably there is no structure in the human body which was not at some time used for a different purpose. As the use of an organ changes, in evolution, its struture correspondingly changes and we see most complete series of intergrades between the earlier and the later conditions.

In all this discussion I have not used the word "species." There are no such things as species in nature. In nature we find different kinds of animals and plants. The words "species," "genus" "family," etc., are terms used to describe the fact that animals and plants differ among themselves and differ to different degrees. Those that are closely similar, that is, closely related, we class in one species; those less closely related, but still not too different, we place in different species, putting the related species together in one genus and so on. Species, genera and so forth, are man-made pigeon holes in which to classify the real animals and plants seen in nature. I have recently made about 150 species of protozoa, but I have never made an animal. The word species is indefinable, and is used by biologists merely as a convenience, and it has wholly different meanings when applied to different groups of animals and plants. There are many genera of animals and plants in which most of all the species completely intergrade so that specific distinctions are purely artificial. This is true to large degree among the protozoan forms I have been studying recently. I have made species among them on the basis of distinctions far too minute to be considered for a moment as of "specific" value among, say insects or mammals.

By Dr. Winterton C. Curtis

Zoologist, University of Missouri.

Biography—Dr. Winterton C. Curtis received the degree of doctor of philosophy at Johns Hopkins in 1901. He has served the University of Missouri since the latter date, and is now chairman of the department of zoology in this institution. He has also been associated with the Marine Biological laboratory at Woods Hole, Mass., for many years, being at the present time one of its trustees. At various times he has acted as an investigator for the United States Fisheries bureau, notably in studies upon the pearl-button mussels. His numerous technical papers have been along the general lines of invertebrate zoology, regeneration and parasitology. His recent work entitled, "Science and Human Affairs," undertakes a discussion, from the standpoint of biological science, of the relationships between the advancement of scientific knowledge and our civilization. Dr. Curtis is particularly qualified to speak in the matters under consideration, because in this volume he has emphasized the spiritual rather than the material influences of science. He is a member and past secretary of the American Society of Zoologists, of the American Society of Ecologists, the American Naturalists, and a fellow of the American Association for the Advancement of Science.

Nature and Current Aspects of the Doctrine of Evolution.

Definitions are wearisome. But we may ask ourselves, by way of limitation, what is evolution in general and organic evolution in particular. The answer can best be given by means of illustrations. The term evolution, as today used in science, means the historical process of change. When we speak of the evolution of man-made products, like automobiles and steam engines, of social institutions like democratic government, of the crust of the earth, or solar system, of animals and plants, we mean a gradual coming into existence of what is now before us, in contrast to its sudden and miraculous creation. Such an idea is of recent origin. Our intellectual forbears of a few centuries ago thought in terms of a world created in its present form. The evolutionary point of view marked an advance from the concept of a static universe to one that is dynamic. In the phraseology of the street, the world is a going concern, historically as well as in its present aspects.

Evolution is, therefore, the doctrine of how things have changed in the past and how they are changing in the present. It may be naturally divided into its cosmic, geologic, and organic aspects, as represented by the sciences of astronomy, geology and biology.

Cosmic Evolution.

Cosmic evolution includes all other forms, for by the cosmic we mean the entire visible universe, our very bodies, as well as the farthest star. But in practice, one thinks of the cosmos as remote. And what we have in mind under cosmic evolution is the changes that are postulated by the science of astronomy. It is believed by astronomers that our solar system with its central sun, its planets and lesser bodies, has not always possessed its present form, although it has been in exitence from a remote period of time. Our earth seems to have been more molten, and before that perhaps gaseous. Although the famous nebular hypothesis of La Place has been in part replaced by other theories, the belief of modern astronomers is that our solar system and perhaps countless others have arisen by an evolutionary process whose extent is infinite in both time and space. I take it that few will combat the concepts of astronomy regarding the nature of our sun and planets. Even when some of us were children the ideas of cosmic evolution, as set forth by the nebular hypothesis, the planetesimal hypothesis, or the like is correct, but that the astronomer regards the heavenly bodies as having reached their present state by an evolutionary stage continuous through an unfathomable past and presumably to be continued into a limitless future. There is no longer talk among intelligent or educated men—or there should not be—of "heaven, earth, center and circumference, created all together, in the same instant, and clouds full of water, on October 23, in the year 4004 B. C., at 9 o'clock in the morning," as was determined by the chronology of Dr. John Lightfoot in the seventeenth century. The astronomical evidence for the development of such a dynamic universe in space and time is of course limited. But it all points in the direction of evolution.

Geologic Evolution.

Geologic evolution overlaps with cosmic, since the geologist takes the evolutionary problem where the astronomer leaves it. Geology deals with the history of our earth, how it originated and how it has assumed its present form. Astronomy deals with the origin of the earth as a planet of our solar system. Geology finds evidence that the earth was once a molten mass which has since been cooler. What may be called the "countenance" of the earth is the subject matter of geology, how the land lies at the present day, how rocks and soil are being produced, and what these facts imply regarding historical origins. The evolutionary evidence of astronomy is vague and remote, although generally accepted by the layman. The evidence from geology is written in the ground beneath our feet. The geologist's belief in a vast lapse of time and stupendous changes rests upon evidence that is everywhere at hand. Leonardo da Vinci, in the fifteenth century, grasped the significance of important geological facts, when he wrote concerning the saltness of the sea and the marine shells found as fossils in the high mountains. Since the publication of James Hotton's "Theory of the Earth," in 1795, it has been the cardinal principle of geological science that past changes of the earth's surface are explicable in terms of changes now in operation. For example, such a vast chasm as the Grand Canyon is explained not as produced by miraculous creation or by sudden catastrophe, but by running water acting upon the rocks throughout innumerable centuries. The process may be observed in miniature in the wash of the soil in Tennessee fields. The weathering of rock into soil, erosion with its transportation of the products of weathering, deposition of the material in the oceans or in large bodies of fresh water, uplift of the ocean's floors and its hardening into rock may all be seen in slow but certain progress in various parts of the world at the present day, and their occurrence in the past is recorded in the rocks. The subtitle of Charles Lyell's famous book, the "Principles of Geology," published in 1830, runs as follows: "An attempt to explain the former changes of the earth's surface by reference to causes now in operation." Lyell established the idea of evoluion as the only reasonable interpretation of geological facts and his elaboration of Hutton's doctrines still constitute the very foundation of geologic science. Today, geology without an evolution of the earth's surface, from a molten mass to its present form, and extending over millions of years, would be on a par with a science of geography postulaing a flat earth. The conclusions of modern astronomy and geology, therefore, point to an evolutionary process—involving many millions of years and still in progress—to an earth hoary with age and still growing old.

Astronomy and geology, despite their practical importance, are remote from human concern, insofar as their evolutionary doctrines are concerned. To borrow from the phraseology of a distinguished anti-evolutionist, the age of the rocks is of no particular consequence insofar as the Rock of Ages is concerned. Cosmic evolution and geologic evolution are readily accepted by the laity on the authority of science, because they do not seriously interfere with doctrines that are deemed vital. But the evolution of plant and animal life, and hence human evolution, is inseparable from that of inorganic matter as described by astronomy and geology, because of the fossils in the rocks.

Organic Evolution.

Organic evolution resembles the cosmic and geologic evolution above described, since it concludes that the living bodies, which are the objects of its investigation, have not always existed as they are today, but have undergone a process of change. As with the evidence of geologic change, the evidence for an evolution of animals and plants rests upon facts that are immediatly before us, for example, the structure and development of animals, their distribution over the earth, the fossils in the rocks. Our time will permit of only enumeration and brief characterization of the recognized lines of evidence for organic evolution, which are as follows:

First—Evidence from structure is derived from:

Comparative anatomy.
Comparative embryology.
Classification.

Second—Evidence from distribution, past and present, is derived from:

Palaeontology.
Zoogeography.

Third—Evidence from physiology is derived from:

Fundamental resemblances in vital processes.

Specific chemical resemblances of closely related forms; e. g., blood tests.

Fourth—Evidence from experimentation rests upon:

Unconscious experimentation upon animals and plants since their domestication.

Conscious experimentation of breeders and of scientific investigators.

The nature of these lines of evidence may now be indicated.

Evidence From Comparative Anatomy—In the animal kingdom as a whole and in every group of animals, whether large or small, we find facts that may be interpreted most reasonably in terms of evolution. The vertebrates or backboned animals will serve as an illustration. We find here a certain plan of structure, for example, backbone, two pairs of limbs, body, head and various internal organs, all laid down according to a similar general plan, but with endless modifications to suit the mode of life. The flipper of a whale, the wing of a bird or a bat, the forefoot of a horse, the arm of a man, and the like, all show the same plan of structure. One of the pre-Darwin ideas was that each animal, while created separately, was nevertheless formed in accordance with a certain type that the Creator had in mind, hence the resemblance. Such an idea is a theoretical possibility, provided there is any evidence to show that animals were created all at once and separately. But there is not a shred of such evidence that will appeal to one who approaches the matter with an open mind and uninfluenced by preconceived notions.

On the other hand, the biological explanation of this anatomical resemblance is that the present vertebrates (fishes, amphibia, reptiles, birds and mammals) have all descended from a primitive race, somewhat like the present fishes. All vertebrates are now alike, because they have never lost the underlying plan of structure inherited from their common ancestry. They have come honestly and naturally by present organization.

The Evidence From Fossils—(Paleontology) Interlocks with the above, since the first vertebrates known to appear were primitive fish-like forms. These were succeeded by amphibians, reptiles, mammals and birds in the order named, the last two having connecting links with the reptiles. The invertebrate groups tell a similar story.

Turning to the Facts of Comparative Embryology—The kind of evidence everywhere discoverable may be illustrated by the gill-slits in the embryos of higher vertebrates like reptiles, birds and mammals. All these forms exhibit in their early stages of development a fish-like plan of structure, particularly in the neck region where the gill-slits are located. The reasonable interpretation of the existence of such structures in the embryo of a human being, or any land-living vertebrate, is that we have never lost these telltale evidences of our ancestry. The later stages of our development are modified so that they lead to the adult human body. The earlier stages still show the primitive conditions of a fish-like organization. Modern fishes have survived to the present day without a fundamental departure from the ancestral condition. Modern amphibia (frogs, toads and salamanders) have survived in the halfway state between an aquatic and a terrestrial existence, through which higher vertebrates have passed as indicated by the fossil record and by the above fish-like stages in their development.

The facts of classification are commonly cited as evidence for evolution. Since classification is based on structure (anatomy), this is but an aspect of the general evidence from comparative anatomy and embryology. While the facts cannot be detailed here, they are striking and bear out the doctrine.

Another line of evidence is that of geological geographical distribution. The facts in this connection are utterly senseless and insulting to an intelligent Creator, if viewed as a result of special creation. One can simply say, "God did it," and not ask why. But such explanations do not satisfy modern minds. On the other hand, their explanation in terms of evolution give reasonableness and consistency to a large body of facts. The fossils appear in such an order in time as to constitute evidence for evolution. Existing animals are distributed over the surface of the earth in a manner that confirms their geological origins.

The facts of physiology tell a similar story. Life and the living stuff is the same sort of thing wherever we find it, thus lending support to the idea that it has all descended from the same primitive source from which it has inherited its resemblances. A more striking line of physiological evidence is the recently discovered chemical resemblance between the blood of animals previously supposed to be closely related on grounds of their anatomical similarities, for example, apes and monkeys, birds and reptiles and the like. Two entirely independent lines of evidence are here found to interlock to such an extent that evolution is the one reasonable interpretation.

Experimental Evidence.

Finally there is the evidence from experimentation: Evolution has taken place before the eyes of men, during the period since animals and plants were first domesticated. The changes have not been profound, because the ten or twenty thousand years since the first animals and plants seem to have been brought under domestication is a brief span of time for evolutionary modification. But it is clear that such modification has occurred and is today occurring under the direction of skillful breeders. The modern science of genetics is beginning to solve the problem of how evolution takes place, although this question is one of extreme difficulty.

The foregoing summary of the various lines of evidence is hopelessly inadequate, since books could be written on each. The point to be appreciated is that all the multitudinous facts of biology hang together in a consistent fashion when viewed in terms of evolution, while they are meaningless when considered as the arbitrary acts of a creator who brought them into existence all at once a few thousand years in the past. Modern biology has developed around two major generalizations, the cell doctrine, and the doctrine of organic evolution. Modern evolutionism dates not from Darwin's "Origin of Species," published in 1859, but from the historic Naturello of Buffen, the first volume of which appeared in 1749, and from the work of the other philosopher-naturalists of the eighteenth century. It is a sad comment upon the state of popular information that the practical facts of biological science are everywhere acknowledged, while the status of its greatest philosophical generalization remains so commonly unknown. In view of its implications and applications, the doctrine of evolution is second to none in modern thought, it has been established by a gradual but irresistible accumulation of facts.

At this point we may examine a common misunderstanding with reference to evolution and the work of Charles Darwin. Suppose we begin with an analogy, illustrating what many be termed the fact, the course and the causes in a progressive series of events. A ship leaves a European port and sails across the Atlantic to New York harbor. We may distinguish between: (1) The fact that the ship actually crossed the ocean, instead of being "created" in the harbor of New York; (2) the course the ship may have pursued, whether direct or indirect, and the like; and (3) the causes that made the ship go, whether an internal propelling force like steam or electricity, an external force like wind or current or even direction by wireless. Compared with the doctrine of evolution, we have: (1) the fact of evolution, as representing the historical series of events; (2) The course followed in evolution, for instance, whether the land vertebrates arose from the fish-like ancestors, birds the causes of evolution or what made from reptiles, or the like; and (3) and makes it happen. These three aspects, like those in the voyage of a ship, are separate though related items. They must be constantly distinguished if there is to be any clear thinking on this matter by one who is not a scientist.

It is now possible to explain the misunderstanding above cited. The historical fact of evolution seems attested by overwhelming evidence. Science has nothing to conceal, it stands "strong in the strength of demonstrable facts," and invites you to view the evidence. The course pursued by evolution is known broadly in many instances, but in the nature of the case the evidence is limited and many of the steps will always remain uncertain, without, however, a calling in queston of the historic fact. The causes of evolution present the most difficult problem of all and the one regarding which we know the least. The recent strictures of Prof. Bateson, which have been exploited by anti-evolutionists were directed wholly at current explanations of evolutionary causation and the course of evolution. He affirmed his belief in the historic fact when he said "our faith in evolution is unshaken"—meaning by "faith," of course, a reasonable belief resting upon evidence.

That such an interpretation of Prof. Bateson's views is the correct one, appears from the following communication:

"11 December, 1922,
"The Manor House,
"Merton,
"London, S. W. 20.

"Dear Prof. Curtis:

"The papers you have sent me relating to the case of Mr. ——— give a curious picture of life under democracy. We may count ourselves happy if we are not all hanged like the Clerk of Chatham, with our pens and ink horns about our necks!

"I have looked through my Toronto address again. I see nothing in it which can be construed as expressing doubt as to the main fact of evolution. In the last paragraph (copy enclosed) you will find a statement in the most explicit words I could find giving the opinion which appears to me forced upon us by the facts—an opinion shared, I suppose, by every man of science in the world.

"At Toronto I was addressing an audience, mainly professional. I took occasion to call the attention of my colleagues to the loose thinking and unproven assumptions which pass current as to the actual processes of evolution. We do know that the plants and animals, including most certainly man, have been evolved from other and very different forms of life. As to the nature of this process of evolution, we have many conjectures, but little positive knowledge. That is as much of the matter as can be made clear without special study, as you and I very well know.

"The campaign against the teaching of evolution is a terrible example of the way in which truth can be perverted by the ignorant. You may use as much of this letter as you like and I hope it may be of service.

Yours truly,
W. BATESON."

The paragraph to which Prof. Bateson refers above is the concluding one of his address and runs as follows:

"I have put before you very frankly the considerations which have made us agnostic as to the actual mode and processes of evolution. When such confessions are made the enemies of science see their chance. If we cannot declare here and now how species arose, they will obligingly offer us the solutions with which obscurantism is satisfied. Let us then proclaim in precise and unmistakable language that our faith in evolution is unshaken. Every available line of argument converges on this inevitable conclusion. The obscurantist has nothing to suggest which is worth a moment's attention. The difficulties which weigh upon the professional biologist need not trouble the layman. Our doubts are not as to the reality or truth of evolution, but as to the origin of species, a technical, almost domestic, problem. Any day that mystery may be solved. The discoveries of the last twenty-five years enable us for the first time to discuss these questions intelligently and on a basis of fact. That synthesis will follow on an analysis we do not and cannot doubt."

With this distinction between fact, course and causes clearly in mind, the significance of Darwin's work in the history of biological thought can be understood. Darwin's accomplishment was two-fold. In the first place he established organic evolution as the only reasonable explanation of the past history of living things. Secondly, he offered, in natural selection, what then appeared an adequate explanation for the origin of species, and, hence, for the causes of evolution. Darwin's evolutionary argument in his "Origin of Species" was that one species could give rise to another "by means" as he believed, "of natural selection or the preservation of favored races in the struggle for life." If one species could be shown to give rise to another, the same process could be continued. No limit could be set. The types thus produced could depart indefinitely from the parent form. Once the mutability of species be admitted the only reasonable conclusion is that evolution has taken place. His argument was supported by an immense collection of facts along observational and experimental lines. The total result was overwhelming, coming as it did, more than 100 years after setting forth of transmutation, and its repeated rejection by the main body of naturalists. Evolution was accepted so quickly by scientists that the world was startled. This sudden conversion gave rise to the impression, even among scientific workers, that no serious contribution to evolutionary theory had been made before the work of Darwin. Such an impression does not represent the facts and it does grave injustice to the pioneer thinkers of the eighteenth century, to whom we have alluded.

Darwin's second accomplishment. natural selection, was accepted by science as a causo-mechanical explanation of evolutionary change. The cogent statement and the simplicity of the principle of selection were of great importance for its acceptance as the cause of evolution, along with the broader theory of evolution as the historic fact. Extended exposition of the selection process will not be attempted. It may be found in numerous elementary books, and in the early chapters of the "Origin of Species." The tabulation known as Wallace's chart, which is an admirable outline of the argument, may be cited in this connection:

Wallace's Chart of Natural Selection.

Proved Facts—(a) Rapid increase of numbers; (b) total numbers stationary; (c) struggle for existence; (d) variation and heredity; (e) survival of the fittest; (f) change of environment.

Consequences—Struggle for existence; survival of the fittest (natural selection); structural modifications.

The importance of Darwin's work in the history of scientific thought is that it convinced science of the truth of organic evolution and proposed a then plausible theory of evolutionary causation. Since Darwin's time evolution as the historic fact has received confirmation on every hand. It is now regarded by competent scientists as the only rational explanation of an overwhelming mass of facts. Its strength lies in the extent to which it gives meaning to so many phenomena that would be meaningless without such an hypothesis.

But the case of natural selection is far different. Of recent years this theory of the causes of evolution has suffered a decline. No other hypothesis, however, has completely displaced it. It remains the most satisfactory explanation of the origin of adaptations, although its all-sufficiency is no longer accepted. The initial step in evolution is the appearance of individual variations which are perpetuated by heredity, rather than the selection of variations after they have appeared. The interest of investigators has shifted to problems of variation and heredity, as exemplified by the rise of the science of genetics.

As a result of this situation there has been much discussion among scientists regarding the adequacy of what is often referred to as the Darwinian theory, meaning natural selection. In condemning selection as an inadequate explanation of the problem, biologists have often seemed to condemn evolution itself. It is not strange that the layman, for whom Darwinism and evolution are synonymous terms, believes that evolution has been rejected when he hears that belief in Darwinism is on the wane. He does not understand that what is thus meant by Darwinism is not the historic fact of evolution, but the proposed cause of evolution—natural selection. This point may not seem vital, but those interested in biological science frequently find the situation used to support claims that the entire concept of organic evolution has fallen into disrepute. There are many, even today, who rejoice at anything that appears to weaken this major generalization of biology.

Such then is the more strictly scientific status of the doctrine of evolution as a whole. The origin, by evolution, of the heavenly bodies and of our earth is evidenced by facts of astronomy and geology, as set forth in any elementary treatise on these sciences. Inorganic evolution, or the modification of nonliving matter, is thus supported by science and does not find serious opposition in the public mind. Organic evolution, or the origin of animal and plant life, receives a similiar support from the facts of biology. If the origin of man were not involved, there would be presumably little serious opposition from nonscientific sources of the present day.

Human Evolution.

But with the evolution of all other living things, both animal and plant, overwhelmingly attested by the facts, it is not only impossible, but puerile to separate man from the general course of events. Moreover, the evidence for man's origin is becoming clearer year by year. Comparative anatomy, embryology, classification, physiology, geographical distribution, fossils and the existing races of mankind tell the same story for man as for the rest of the animal world.

Huxley's essay, entitled "Man's Place in Nature," presents in a masterful manner the anatomical evidence for our kinship with the four species of tailless apes—the gibbon, gorilla, orang and chimpanzee—and his most significant conclusions are even more strongly established at the present day. If creation occurred at 9:00 a.m. on October 23rd of the year 4,004 B. C., as part of the divine plan, it is amazing that such success should have dogged the steps of the students of human skeletal and cultural remains during the last half century. The skeletons, in part or in whole, are known for a number of subhuman races and a vast array of implements and other remains, all showing a progressive advancement. By another fifty years it seems safe to expect that much more of the story will be unveiled. It is further amazing that investigations in Egypt show the existence of a flourishing civilization in the Nile valley as early as 5,000 B. C., and back of this a gradual development from the barbarism of the stone age.

On man's intellectual side, psychology is making increasingly evident the essential animal foundation of human intelligence. Man's claim to importance in the universe, revealed by science, lies not in the pretense that this planet was created for his convenience, but in the claim that he transcends the material universe in so far as he comprehends it. And the method of such comprehension that dominates modern thought is the method of science, not that of theology.

The question of human beginnings is one that is open to investigation, like any other historic or prehistoric event. In this connection a quotation from a famous essay by Herbert Spencer, published in 1852, is appropriate: "Those who cavalierly reject the theory of evolution ," writes Spencer, "as not adequately supported by facts seem quite to forget that their own theory is supported by no facts at all. Like the majority of men who are born to a given belief, they demand the most rigorous proof of any adverse belief, but assume that their own needs none. Here we find, scattered over the globe, vegetable and animal organisms numbering, of the one kind (according to Humboldt), some 320,000 species, and of the other some 2,000,000 species (see Carpenter); and if to these we add the numbers of animal and vegetable species that have become extinct, we may safely estimate the number of species that have existed, and are existing on the earth, at not less than 10,000,000. Well, which is the most rational theory about these 10,000,000 of species? Is it most likely that there have been 10,000,000 of special creations? Or is it most likely that by continual modifications, due to a change of circumstances, 10,000,000 of varieties have been produced, as varieties are being produced still?

And, one might add, if the evidence indicates that all other species have arisen by evolution, is it probable that man, whose bodily structure and functions are so nearly identical with those of the mamalia and particularly the primates—that man arose in a different fashion. We have, moreover, as above indicated, the positive evidence to support this general presumption

Having outlined the evidence for human evolution and stated the presumption in its favor, let us turn to the evidence for special creation, as found in Genesis. Science and common sense alike inquire regarding the nature and sources of this account, if it be regarded as a true statement of the facts. Science faces the matter squarely, desiring only the right to investigate and draw unprejudiced conclusons. The results of such investigations are not in doubt. It appears that the races about the eastern Mediterranean, like other primitive peoples, had their traditions of the origin of the world. The story in Genesis apparently descended to the early Hebrews and to their neighbors in Mesopotamia from a source far antedating the appearance of the Jews as a people and their sacred writings. Archeology and ethnology most reasonably indicate that in its origin this Hebrew-Babylonian tradition may be compared with the stories of many primitive peoples. We take the story in Genesis seriously as an account of prehistorical facts, because it is our story of creation passed down by tradition from our fathers. It is, and will remain, sacred and interesting, because it has been woven into the thought of western culture for almost 2,000 years and because of its intrinsic literary and moral qualities.

But the past history of events, whether of human or animal origins, is subject matter for scientific inquiry, and the answer of science is evolution. The very great antiquity of man, the existence at an earlier period of beings, manlike, but intermediate between man and primates, together with the facts of man's anatomy, his embryology, his physiological reactions, even his mentality, all point to his bodily kinship with the rest of living nature. It is not true men came from monkeys, but that men, monkeys and apes all came from a common mammalian ancestry millions of years in the past.

It is more reasonable to believe that the Bible is a human document representing the history of an advance from the concept of a barbarous and vengeful Jehovah of the earlier Old Testament, through the God of righteousness and justice of the later prophets, and culminating in the concept of a Father as preached by Jesus of Nazareth.

In the foregoing statement we have considered the intellectual aspects of the doctrine of organic evolution. There remains its social aspects. Evolution is one of the basic concepts in modern thought. Suppression of a doctrine established by such overwhelming evidence is a serious matter. From the standpoint of the teacher the situation has more than academic interest.

Evolution has been generally accepted by the intellectually competent who have taken the trouble to inform themselves with an open mind. The following letter was written in response to a request to state his position, it having been alleged that he was not a believer in organic evolution:

"Washington, D. C.,
"August 29, 1922.

"My dear Prof. Curtis:

"May it not suffice for me to say, in reply to your letter of August 25th, that, of course, like every other man of intelligence and education, I do believe in organic evolution. It surprises me that at this late date such questions should be raised.

Sincerely yours,
"WOODROW WILSON."

Prof. W. C. Curtis,
Columbia, Mo.

In view of all the facts, may we not say that the present storm against organic evolution is but an expression of malign influences of prejudice and ignorance, hostile to what we may envision is the high destiny of our western world.

By Prof. Horatio Hackett Newman,

Zoologist, University of Chicago.

(Biography—He was dean of the colleges of science at that university for nearly seven years, having charge especially of premedical and medical students. He has been teaching zoology since 1898. He received his bachelor's degree at McMaster university, and his doctor's degree at Chicago university. He has memberships and fellowships in the American Association for the Advancement of Science, the American Society of Zoologists, etc. He has attracted widespread attention in the scientific world by his studies of experimental embryology and in other zoological subjects. He was among the earliest in this country to organize large classes in various universities for the study of evolution and heredity. His publications include many technical monographs and the following books: "Evolution, Genetics and Eugenics;" "Vertebrate Zoology;" "Outline of General Zoology;" "The Biology of Twins;" "The Physiology of Twinning.")

The evolutionist stands for and believes in a changing world. Evolution is merely the philosophy of change as opposed to the philosophy of fixity and unchangeability. One must choose between these alternate philosophies, for there is no intermediate position; once admit a changing world and you admit the essence of evolution. The particular courses of change or the causes of any particular kinds of change are matters that the expert alone is in a position intelligently to discuss. We know with certainty some few things about the course of evolution, and we believe that we have discovered some important phases of the mechanism of evolution, but these are controversial matters and in no way effect the question as to the validity of the principle. Whether or not evolution may lay claim to rank as a law of nature depends upon the strength, the coherency and the abundance of the so-called evidences of evolution.

The Nature of the Proof of Organic Evolution.

There are two distinct types of evidences of evolution, one of which has to do with changes that have occurred during past ages, the other with changes that are going on at the present time. The evidences of changes that have taken place in the remote past must, in their very nature, be indirect and to some extent circumstantial, for there are no living eye-witnesses of events so far removed from the present and there are no documentary records written in human language. Records of past events are written, however, for him who has learned the language, in the rocks, in the anatomical details of modern species, in the development of animals and plants, in their classification, and in their geographical distribution, past and present. Evidences that species are changing today are quite direct in character, for more or less radical hereditary changes have been seen in the act of taking place, though, as yet, we have little knowledge of the causes responsible for them. The discovery that species are changing at a noticeable rate at the present time is in itself strong evidence that they have changed in the past, and doubtless in the same ways and at the same rates of speed as those observable today; for even the convinced special creationist would hardly claim that species have remained immutable since their creation only to begin change during the present era. Little can be learned about the large changes involved in organic evolution by observing relatively small changes of the present, for it takes immense periods of time for the larger waves of change to run their course and reach their culmination. For the study of past evolutionary events we use the historical method so successfully employed in archaeology and ancient history; for the study of present evolution we make use of the methods of direct observation and experiment. The findings in one field strongly support and supplement the other.

When we admit that the evidences of past evolution are indirect and circumstantial, we should hasten to add that the same is true of all other great scientific generalizations. The evidences upon which the law of gravity are based are no less indirect than are those supporting the principle of evolution. Like all other great scientific generalizations, the law of gravity has acquired its validity through its ability to explain, unify and rationalize many observed facts of physical nature. If certain facts entirely out of accord with the law of gravity were to come to light, physicists would be forced either to modify the statement of the law so as to bring it into harmony with the newly-discovered facts, or else to substitute a new law capable of meeting the situation. Laws of nature are no more or less than condensed statements about the facts of nature and therefore are valid only in so far as they agree with the facts. The nebular hypothesis and its modern rival, the planetesimal hypothesis, are both deductions from facts; they both seem to agree with many of the obscured data, but neither of them is as yet fully adequate for all. In the field of physical chemistry we had first the molecular theory, then the atomic theory, then the ionic theory and now the electron theory; each of those has appeared in direct response to the necessity of explaining new sets of facts, and none of them is so well founded as is the theory of evolution. No one has ever seen a molecule, an atom, an ion or an electron; the existence of and the properties of these entities have been deduced from the behaviors of various chemical substances when subjected to experimental conditions.

The principle of evolution stands in the first rank among natural laws not only in its range of applicability, but in the degree of its validity, the extent to which it may lay claim to rank as an established law. It is the one great law of life. It depends for its validity, not upon conjecture or philosophy, but upon exactly the same sorts of evidence as do other laws of nature.

Evolution has been tried and tested in every conceivable way for considerably over half a century. Vast numbers of biological facts have been examined in the light of this principle and without a single exception they have been entirely compatible with it. Think what a sensation in the scientific world might be created if some one were to discover even one well-authenticated fact that could not be reconciled with the principle of evolution. If the enemies of evolution ever expect to make any real headway in their campaign they should devote their energies toward the discovery of such a fact.

The exact nature of the proof of the principle of evolution is that when great masses of scientific data such as are involved in those branches of biology known as taxonomy, comparative anatomy, embryology, serology, paleontology and geographic distribution, are looked upon as the result of evolutionary processes, they take on orderliness, reasonableness, unity and coherency. Not only this, but each subscience becomes more closely linked with the others and all turn out to be but different aspects of the one great process. No other explanation of biological phenomena that in any sense rivals the evolution principle has ever been offered to the public. This principle cannot be abandoned until one more satisfactory comes forth to take its place. To revert to the thoroughly discredited and unscientific idea of special creation would be as utterly impossible as to revert to the ancient geocentric conception of the universe, according to which a flat earth was thought to occupy the center of the universe and the sun, moon and stars to revolve about it.

Let us reiterate that a theory or a principle is acceptable only so long as it accords with the facts already known and leads to the discovery of new facts and principles. Whether or not the principle of evolution meets these requirements the reader must judge for himself after a perusal of the facts that lie at the basis of the principle.

The evidences of evolution that we shall investigate are contained within the following fields of biology:

First—Comparative anatomy or morphology, the science of structure.

Second—Taxonomy, the science of classification.

Third—Serology, the science of blood tests.

Fourth—Embryology, the science of development.

Fifth—Paleontology, the science of extinct life.

Sixth—Geographic distribution, the study of the horizontal distribution of species upon the earth's surface.

Seventh—Genetics, the analytic and experimental study of evolutionary processes going on today.

A careful study of the situation reveals that the entire fabric of evolutionary evidences is woven about a single broad assumption: That fundamental structure resemblance signifies blood relationship; that, generally speaking, the closeness of structural resemblance runs essentially parallel with closeness of kinship. Most biologists would say that this may once have been only an assumption, but that it is now so amply supported by facts that it has become axiomatic. However obvious the validity of this assumption may be, it is the plain duty of one who attempts to justify the evolutionary principle to avoid taking steps that are in the least open to serious criticism. If we cannot rely upon this principle we can make no sure progress toward the proof of evolution.

The assumption we are now discussing is tantamount to an affirmation of the principle of heredity; that like tends to produce like. We continually employ the principle in every day life. We fully expect the offspring of sparrows to be sparrows, of robins to be robins; and if we should ever find an instance to the contrary, we would be greatly surprised and shocked. Futhermore, we have learned by experience that offspring not only belong to the same species as the parents, but resemble the parents more closely than they do other people. Whenever we see two people whose resemblance is closer than usual we immediately come to the conclusion that such persons are relations, probably offspring of the same parents. Every one has had the experience of meeting two persons so strikingly alike that it is almost impossible to distinguish them apart, and the natural assumption is that such persons are duplicate or identical twins. Twins of this sort are vastly more closely related than are brothers or sisters, or even than are fraternal twins who are usually no more alike than are brothers and sisters of closely similar ages. It is practically established that duplicate twins are products of the early division of a single germ cell. No closer degree of kinship can well be imagined than this, for the two individuals bear the same relationship to each other as do the two bilateral halves of one body.

The writer has had an exceptional opportunity to determine the exact degrees of resemblance existing between separate offspring derived from a single egg. It so happens that a peculiar species of mammal, the nine-banded armadillo, almost always gives birth to four young at a time. These quadruplets are invariably all the same sex in a litter and are nearly identical in their anatomical details. A study of their embryonic history has proven beyond question that in every case the four embryos are produced by the division of a single normally fertilized egg. Large numbers of advanced sets of quadruplets fetuses were studied statistically with the idea of determining the exact degree or their resemblance. An average or a considerable number of determinations revealed the somewhat startling fact that their coefficient of correlation is .93, which is merely another way of saying that they are 93 per cent identical. The remarkable closeness of this resemblance may be fully appreciated when it is realized that the only structural resemblances belonging to this order of closeness are those existing between the right and left halves of single individuals, and that the next order of resemblance is that between siblings (brothers or sisters) who are only 50 per cent identical.

This, then, is a crucial test of the validity of the assumption that closeness of resemblance is a function of closeness of kinship, for here we have the closest approach to identity in connection with what is also the closest possible blood relationship.

Employing the principle of heredity in a somewhat broader way, and in a way that is hardly likely to be questioned even by the most captious, we account for the common possession of certain structural peculiarities by all members of a given kind or species of animal by saying that characters have been derived from a common ancester. It is only a short step in logic to conclude that two closely similar kinds or species of animals have been derived one from the other or from a common species. Once having taken this step we are on the road that leads inevitably to an evolutionary interpretation of natural groups. If the principle of heredity holds for fraternities, for races, for species, where are we to draw the line? It does not seem reasonable to admit that structural resemblances within the fraternity, the race, the species, are accounted for as a product of heredity, and to deny that equally plain resemblances among the species of a genus or among the genera of a family have a hereditary basis. It is logically impossible to draw the line at any level of organic classification, and say that fundamental structural resemblance is the product of heredity up to such and such a level, but that beyond some arbitrarily settled point heredity ceases to operate.

Evidences From Comparative Anatomy.

The foundation stones of comparative anatomy are the principles of homology and of analogy. The former implies heredity and the latter variation.

The Principle of Homology.

Any one who has at all seriously studied comparative anatomy must have been impressed with the fact that the animal kingdom exhibits several distinct types of architecture, each of which characterizes one of the grand divisions of the kingdom. Within each of these great assemblages of animals characterized by a common plan of organization there are almost innumerable structural diversities within the scope of the fundamental plan. These major or minor departures from the ideally generalized condition reminded one of variations upon a theme in music; no matter how elaborate the variations may be, the skilled musician recognizes the common theme running through it all. This fundamental unity amidst minor diversity of form or of function is looked upon as a common inheritance from a more or less remote ancestor. In animals belonging to the same group, and therefore having the same general plan of organization, we find many organs having the same embryonic origin and the same general relations to other structures, but with vastly different superficial appearance and playing quite diverse functional roles. Such structures are said to be homologous.

A common example of homologous structures is presented by the forelimbs of various types of backboned animals (vertebrates); such, for example, as that of man, that of the whale, that of the bird and that of the horse. The arm of man is by far the most generalized of these; it is not far from the ideal prototypic land vertebrate fore limb, and in that it is not specialized for any particular function, but is a versatile tool of the brain. The flipper of the whale is a short, broad, paddlelike structure, apparently without digits, wrist, forearm or upper arm; but on close examination it is seen to possess all of these structures in a condition homologous, almost bone for bone and muscle for muscle, with those of the human arm. The wing of the bird, a highly specialized organ of flight, appears superficially to have nothing in common with the arm of man; but a study of its anatomy shows the same bony architecture and muscle complex, modified rather profoundly for a different function and with the thumb and two of the fingers greatly reduced or entirely unrepresented in the adult stage. The fore-leg of the horse is a specialized cursorial appendage, and in accord with this function has but one functional toe with a heavy toenail or hoof. Two other toes are represented by the so-called split bones, mere vestiges of once useful structures. In other respects the horse's leg is quite homologous with that of other land vertebrates. The evolutionary explanation of the fact that these several types of limbs (each playing an entirely different role in nature and each so unlike the other in form and proportions) have the same fundamental architecture, is that they have all inherited these characters from some distant common ancestor. In each case the inheritance has undergone modification in harmony with the life needs of the organism. This, of course, implies descent with modification, which is no more or less than evolution.

An equally significant situation comes to light in connection with the hind-limbs of vertebrates. The leg of man, a specialized walking appendage, is much less versatile than is the arm; yet it is closely homologous with the latter. The hind-limb of the whale is, in some species, entirely wanting in the adult or else is in vestigial condition. The leg of the bird is decidedly reptilian in structure and is believed to have retained, in large measure, the characteristics of that of the supposed reptilian ancestors. The hind-limb of the horse, though somewhat stronger and heavier than the fore-limb, resembles the latter closely both in form and function. Snakes are typically limbless vertebrates, but the python has small but clearly defined hind-limbs, somewhat reduced in number of bones and almost entirely hidden beneath the scaly integument.

No other attempt to explain homologies such as those briefly outlined above has been made except that of special creation, and this implies a slavish adherence to a preconceived ideal plan together with capricious departures from the plan in various instances. A systematic attempt to apply the special creation concept to all cases of homologies involves one in the utmost confusion of ideas and leads almost inevitably to irreverence, which is abhorent to evolutionists as well as to special creationists.

Vestigal Structures.

These may be defined in functionless rudiments of structures whose homologues are found in a functional state in other members of a group with a common architectural plan. Thus the hind-limbs of the whale and of the python, the thumb of the bird, the split bones of the horse, are vestigal homologues of structures well developed in more generalized groups of vertebrates.

The case of the hind limb vestiges in the various species of whales may be emphasized as a crucial one. Several different degrees of rudimentation are found in different types of whales, ranging from a state in which the pelvic bones and those of most of the leg clearly recognizable as such down to one in which these bones are entirely absent in the adult condition. In the cases where the bones are obvious, the situation is just this—deeply buried beneath the thick cushion of blubber in the pelvic region there lies a little handful of bones, ridiculously minute in comparison with the giant proportions of the other parts of the skeleton. These bones are immovable because their muscular connections are atrophied; they do no service in supporting the frame of the animal; in short, they cannot possibly function as bones at all. The somewhat puerile argument of the antievolutionist that these vestigial limb bones play some useful, though unknown role, else they would never have been created, cannot seriously be entertained in this case, for what can they make of the fact that some whales entirely lack these structures? More difficult even than this for the special creationist to explain is the fact that, even in these whales that lack vestigal limb bones in the adult condition, posterior limb buds appear in the early embryonic period and then slowly atrophy. The case just described in in no way exceptional or peculiar. It is, on the contrary, quite typical of a very general phenomenon.

Vestigal Structures in Man.

There are, according to Wiedersheim, no less than 180 vestigal structures in the human body, sufficient to make of a man a veritable walking museum of antiquities. Among these are the vermiform appendix, the abbreviated tail with its set of caudal muscles, a complicated set of muscles homologous with those employed by other animals for moving their ears, but practically functionless in all but a few men; a complete equipment of scalp muscles used by other animals for erecting the hair, but of very doubtful utility in man even in the rare instances when they function voluntarily; gill slits in the embryo the homologues of which are used in aquatic respiration; miniature third eyelids (nictitating membrane), functional in all reptiles and birds, greatly reduced or vestigial in all mammals; the lanugo, a complete coating of ebryonic down or hair, which disappears long before birth and can hardly serve any useful function while it lasts. These and numerous other structures of the same sort can be reasonably interpreted as evidence that man has descended from ancestors in which these organs were functional. Man has never completely lost these characters; he continues to inherit them though he no longer has any use for them. Heredity is stubborn and tenacious, clinging persistently to vestiges of all that the race has once possessed, though chiefly concerned in bringing to perfection the more recent adaptive features of the race.

Homology Versus Analogy.

It is quite common to find different animals with certain structures that look alike and function alike, but are not homologous. The eye of the octopus, a cephaloped mollusk, has a chorion, a lens, a retina, an optic nerve, and a general aspect decidedly like that of a fish. As an optical instrument it must obviously function in the same manner as does the eye of an acquatic vertebrate; but not one part of the eye of a cephalopod is homologous with that of a vertebrate. Because those two types of eye look alike and function alike, but arise from quite different embroyonic primodin adapted to meet a common function, they are known as analogous structures. They are to be sharply contrasted with homologous structures, which may be widely different in form and function so long as they arise from equivalent embryonic primordic. Both homologies and analogies imply changes in relation to the environment, and, therefore plainly favor the idea of descent with modification.

Connecting Links.

If one group of animals has been derived by descent from another there should be some form more or less intermediate between the two and with some characteristics of both groups. Many such connecting links actually exist at the present time. Almost every order of animals possesses some primitive members that have, doubtless, evolved at a slower rate than their relatives and have on that account retained a larger measure of ancestral traits than have the more typical representatives of the group. Thus there is a group of primitive annelid worms, represented by Dinophilus, Protordrillus and Pollygordius, that serve partially to bridge the gap between the two grand divisions, annolide and flatworms. The case of the several species of Dinophilus is especially noteworthy, for these little animals are so evenly balanced between the characteristics of one phylum and those of the other that some authors place them among the flatworms, others among the annelids and still others are inclined to place them in an anomalous group by themselves. There is an interesting genus of primitive centipes, called Peripatus, which possess as many annelid features as anthropoid features. Among vertebrates we have the familiar example of the lung fishes with both the gills of fishes and lungs homologous with those of land vertebrates. And finally, we may mention those curious egg-laying mammals, momotrems, of Australia and New Zealand, which though obviously mammalian in most respects, possess, in addition of laying eggs after the fashion of reptiles, many other decidedly reptilian traits. The reader interested in following up in more detail this interesting branch of comparative anatomy will find the subject skillfully handled by Geoffrey Smith in a volume entitled "Primitive Animals."

Comparative anatomy is a mature and well organzed science and involves a vast amount of technical data. No one but a trained comparative anatomist can reasonably be expected to appreciate the dependence of this subject upon the principle of evolution. Without evolution as a guiding principle, comparative anatomy would be a hopeless mass of meaningless and disconnected facts; with the aid of the principle of homology, an evolutionary assumption, it has grown to be one of the most scientific branches of biology. This may be taken as an illustration of the nature of the proof of organic evolution; that when it is used as a working hypothesis or guiding principle it really works in that it is not only consistent with all of the facts, but lends significance and interest to facts that would otherwise be drab and disconnected.

Evidences From Classification.

The object of classification is to arrange all species of animals and plants in groups of various degrees of inclusiveness which shall express as closely as possible the actual degrees of relationship existing between them. In pursuance of this object we begin by grouping together as one species all animals that are essentially alike in their anatomical details. As an example of the methods of classification we may take the following familiar instance: The European wolf is a particular kind of animal constituting a species called lupus (the Latin word for wolf), all members of which are more like one another than they are like wolves of other sorts, for the reason that they have a common inheritance. There are not a few other species of wolves, each given a Latin name, and all of these wolf species, including dogs (believed to be domesticated and, therefore, highly modified wolves), are placed in one genus, Canis. Several other genera of more or less wolflike animals, such as jackals and foxes, are grouped with the genus Canis, and constitute the family Canidae, the assumption being that they are all the diversified descendants of some common wolflike ancestor. Other families, such as the cat family (felidae), the bear family (urisdae) and several other families of terrestial beasts of prey constitute the suborder fissipedia. These, in turn, are grouped with the marine beasts of prey, such as seals, sea lions, walruses (suborder pinnipedia) to form the mammalian order, carnivora. Several other orders of animals with many characteristics in common are combined to form the class mammalia, which is one of several classes belonging to the subphylum vertebrata, a branch of the phylum chordata. A phylum is one of the grand subdivisions of the animals combined to form the class mammalia, with the same fundamental plan of organization, the common features of which are believed to be derived from a common ancestral type.

The underlying assumption of classification is the same that underlies comparative anatomy; that degrees of resemblance run parallel with degrees of blood relationship, that the most nearly identical individuals are most closely related and that those that bear the least fundamental resemblance to each other are either not genetically related at all or else had a common ancestor far back in the misty past when animal life was in process of origin. We have already shown that this assumption holds good in all cases where it has been possible to put it to the test. No further justification need be offered in this place for making use of the only adequate instrument of classification: the principle of homology.

What is a Species?

The species is the unit of classification, but there is serious doubt as to whether species have any reality outside of the minds of taxonomists. Certainly it is extremely difficult, if at all possible, exactly to draw sharp boundary lines between closely similar species. When we examine a large number of individuals belonging to a given species we find that there are no two exactly alike in all respects. As a rule there is a wide range of diversity within the limits of the group we call a species and the extreme variants are often so unlike the type form that, were it not for the intergrading stops between them, they would often be adjudged distinct species. Moreover, the species of a prosperous genus are so variable that it becomes an almost impossible task to determine where one species ends and another begins, so closely do they intergrade one into another. A species, then, is not a fixed and definite assemblage such as one would expect it to be if specially created as an immutable thing. On the contrary, intensive study of any widely distributed species gives the impression of an intricate network of interrelated individuals changing in a great variety of ways.

The completed classification of any large group, such as the vertebrates, presents itself as an elaborately branching system whose resemblance to a tree is unmistakable. The phylum branches into subphyla, some of the latter into several classes, classes into orders, orders into families, families into genera, genera into species, species into varieties. We may compare the phylum to one of the main branches coming off from the trunk, while the varieties may be thought of as the terminal twigs. This treelike arrangement is exactly what one would expect to find in a group descended from a common ancestry and modified along many different lines. It is in reality a genealogical tree. If this striking arrangement is a part of the plan of special creation it is indeed strangely unfortunate that it speaks so plainly of descent with modification.

Man's Place in the System of Classification:

There is no greater difficulty in connection with the classification of man than in that of any other living species. Indeed there are scores, even hundreds, of species whose exact affinities with other groups are far less obvious than those of the human species. Anatomically the genus homo bears a striking resemblance to the anthropoid apes. Bone for bone, muscle for muscle, nerve for nerve, and in many special details man and the anthrapoid apes are extremely similar. Homologies are so obvious that even the novice in comparative anatomy notes them at a glance. Man is many degrees closer anatomically to the great apes than the latter are to the true monkeys, yet the special creationist insists upon placing man in biological isolation as a creature without affinities to the animal world. If a man is a creature apart from all animals it is extremely difficult to understand the significance of the fact that he is constructed along lines so closely similar to those of certain animals; that his processes of reproduction are exactly those of other animals; that in his development he shows the closest parallelism step for step to the apes, that his modes of nutrition, respiration, excretion, involve the same chemical processes, and that even his fundamental psychological process are of the same kind, though differing in degree of specialization, as are those of lower animals.

Comparative anatomists recognize man as a vertebrate, for he has all of the characteristic features of that group. He is obviously a mammal, for he complies with qualifications of that class in having hair; in giving birth to living young after a period of uterine development; in suckling the young by means of mammary glands; in having two sets of teeth, one suceeding the other; in having the teeth differentiated into incisors, canines and molars; and in many other particulars of skeleton, muscular system, circulatory system, alimentary system, brain and other parts of the central nervous system. Among mammals, man belongs to the well-defined order of Primates, an order anatomically about halfway between the most generalized and the most specialized of the mammalian orders. Apart from his extraordinary nervous specialization, man is a relatively generalized mammal as compared with such highly specialized types as for example, the whales. The older taxonomists placed man and the other primates at the top of the genealogical tree, assigning to him the central tip of the central branch as though the goal of all organic evolution were man. Accordingly, those mammals such as the whales, which are least like man, were considered the lowest members of the class. There has been within recent years a pronounced reversal of this anthropocentric point of view, which has resulted in a complete revision of the arrangement of mammalian orders, with the insectivora the lowest, the cetacea (whales) the highest, and the primates about intermediate in systematic position.

The order primates consists of two suborders—lemuroidae and anthropoidae. The lemurs or half apes are small arborial animals with somewhat squirrel-like habits, but with flat nails and certain other primate characters. They serve to link up the primates with the most primitive of the mammalian orders, the insectivora, which are now believed, on anatomical and paleontological grounds, to be ancestral not only to the primates but to most of the other modern mammalian orders. The anthropoid or man-like primates are divided into four distinct families: The Hapalidae or marmosets; the Cercopithecidate or new world monkeys; the Simiidae or anthropoid apes, and the Hominidae or men. The family Hominidae includes four genera: The genus Pithecanthropus, represented by the fragmentary remains of an extinct Javan ape-man, the genus Paleanthropus, the genus Eanthropus and the genus Homo, including in addition to the existing species, Homo sapiens, several different extinct human species known as the Dawn man, the Neaderthal man, the Rhodesian man, and others.

The species Homo sapiens consists of at least four subspecies or major varieties, each consisting of numerous minor races and admixtures of these. This high degree of diversity within the species is evidence of rapid evolution. If a little over 4,000 years ago, as the special creationists claim, one man was created and has become the ancestor of all men living today, evolution must have gone on at an extremely rapid rate in order to have produced so many widely different races, for there could scarcely have been more than 120 generations in that time. If species are believed to be immutable it is difficult to understand why man should be such a diversified group as he is.

Evidences from Blood Tests

The methods of classifying animals just outlined depend upon relatively gross criteria (homologies) as compared with the refinements characteristic of the serological technique used in blood testing. This latter method of classifying animals depends upon chemical similarities and differences in the bloods of various animals, and the basic assumption is once more that degrees of resemblance parallel degrees of blood relationship. Recent investigation has shown that certain materials in an animal's blood are even more sharply specific than are its visible structural characteristics. Chemical tests of extreme delicacy are used to reveal resemblances in blood. Thus, if we wish to find out what animals are most like man in blood composition we can find it out in the following manner: Human blood is drawn and allowed to clot, a process that separates the solid materials in the blood from the liquid serum. The latter watery fluid contains the specific human blood ingredients. Small doses of it are injected at two-day intervals into the blood vessels of a rabbit. At first the rabbit is sickened by the injection, thus showing a marked reaction to the foreign material. In the course of a short time, however, there is no further reaction, and we may conclude that the rabbit is immunized. What has happened is that some substance has been developed in the rabbit's blood which neutralizes the toxic effects of human blood. It is a sort of antitoxin and may be spoken of as antihuman serum, a material that may now be used as a delicate indicator of blood kinship. When this antihuman serum is mixed with serum taken from the blood of any human being an immediate and definite white precipitate is formed; when mixed with that of any of the anthropoid apes the precipitate is similar to that formed with human serum, but less abundant and somewhat slower in appearing. The tests showed a less prompt and less abundant reaction with the blood of old world monkeys, a slight but definite reaction with that of new world monkeys, and no noticeable reaction with that of lemurs.

The tests further indicated that, if strong enough solutions are used and time enough allowed for the precipitate to settle, there is an unmistakable blood relationship among all mammals and that degrees of relationship run closely parallel with those based upon homologies. Not only this, but not a few affinities, the existence of which had been only vaguely suggested by comparative anatomy, are strongly emphasized by blood tests. One most remarkable revelation is that whales, the most specialized among mammals, are more closely related to the ungulates (hoofed animals), and especially to the swine family, than to any other group of the class mammalia—a diagnosis that had previously been made by several anatomists on what appeared to be rather slender morphological grounds.

At the present time the technique of blood testing for animal affinities is rather difficult and very few workers have attempted to make use of it. The results so far attained, however, are so definite and clean-cut that there is every reason to expect a great future for this new type of evolutionary evidence. Many groups of animals have already been tested and in general the affinities indicated closely parallel those based on homologies. There is, however, no exactness about this parallel; nor could we expect such to be the case. For that matter there is no exact parallelism between the teeth and the feet, between the head and the tail. No two systems of an organism exactly keep pace in their evolution; one may remain relatively conservative while the other may become greatly specialized. Of all systems, the blood appears to have been the most conservative and to have retained most fully its ancestral characters. It is on this account that blood tests are so valuable in revealing relationships that can scarcely be determined in any other way.

Far more important than any information as to animal affinities revealed by blood tests is the fact that the classification of animals based on blood tests is essentially the same as that based on morphology. Suppose, for the sake of argument, that these two modes of classification had revealed quite contrary arrangements, what a blow to our confidence in the validity of evolution! Conversely, what a strong support of the evolution principle is afforded by the fact that the two systems of classification point to the same lines of descent!

Evidences from Embryology

There should be no sharp division between the evidences from comparative anatomy and those from embryology. Those two branches of biology are inseparable; one must be interpreted in the light of the other. Comparative anatomy deals with the adult structures of organisms. Whenever there is any question about homologies of fully developed structures recourse is had to younger and still younger stages, for when structures are really homologous, they tend to be more closely similar the younger they are. Structures that come from the same or similar embryonic are by definition homologous. Therefore the only certain test of homologies is a study of embryology.

It is necessary to bear in mind that an individual is not merely his adult condition; that a species is not fully defined by a description of its adult characteristics. The species characteristics include those of the egg and the sperm, the cleavage pattern, the particular modes of gastrulation and of further differentiation. In brief, the species is fully defined only by a full description of its entire ontogeny. Very closely related species keep step nearly all the way through their ontogenous and diverge only toward the end of their courses. Distantly related forms diverge comparatively early in their developmental paths; while unrelated forms may have little or nothing in common from the beginning.

The most advanced groups of organisms travel a much longer journey before reaching their destination than do organisms of lower status. In many instances certain early stages in the development of an advanced organism resemble in unmistakable ways the end stages of less advanced organisms. There is, in fact, in the long ontogeny of members of high groups, a sort of rough-and-ready repetition of the characteristic features of many lower groups. This fact has so impressed some biologists that they have embodied it into a law, the so-called biogenetic law; that ontogeny recapitulates phylogeny. In less technical language this means that the various stages in the development of the individual are like the various ancestral forms from which the species is descended, the earliest embryonic stages being like the most remote ancestors and the latter stages like the more recent ancestors. In still other words, the concept may be stated as follows: The developmental history of the individual may be regarded as an abbreviated resume of its ancestral history.

In the first place it is obvious that no embryonic stage can be in any real sense the equivalent of any adult ancestor. The most we can affirm is that while some embryonic characters of the higher group strongly remind us of some adult features of lower groups, the tout ensemble of the former is not at all closely similar to that of the latter. In the second place, it should not be forgotten that the embryonic and larval stages of organisms have much more pressing demands upon them than that of recording their ancestral attainments —they must adapt themselves to their surroundings if they are to survive. As a result of this pressing necessity many larvae and even embryos are so profoundly modified in adaptive ways that their ancestral characters are largely obscured. Various larval or fetal organs commonly furnish the outstanding characteristics of developmental histories, and these purely temporary organs not only tell no story of ancestries, but frequently so mask the ancestral story as to make it almost indecipherable. In the third place, different systems of organs develop at different rates, so that when one system has reached an advanced state of differentiation another system may be still in the primordial state. Thus, in the development of fishes the nervous system is far along its course of development before the circulatory system has even begun to differentiate. At such a stage as this the embryo is obviously not equivalent to any adult ancestor, for an organism with so discordant an organization could not survive.

In spite of its faults and limitations, however, the idea that ontogony tends to repeat phylogeny, if used intelligently and not over-applied, is a very useful one. Organisms inherit not only their adult characters from their ancestors, but also their general development patterns. It is therefore inevitable that many features that have been outgrown or subordinated in modern types should be found in a state more nearly ancestral during the embryonic stages. And especially is this the case when particular systems are studied separately. Thus, we find that the human circulatory system develops through a series of stages that are much like the adult conditions of a series of ascending vertebrate classes.

The heart differentiates from a sheet of mesoderm lying beneath the pharynx. It has at first the form of two nearly straight tubes, which soon fuse for part of their length to form a single tube divided at the two ends into two tubes. Later the single tube differentiates lengthwise into two cavities, the auricle and the ventricle, and is now in the stage equivalent to that of an adult fish. The auricle next divides into two chambers, thus resembling that of an amphibian. Finally the ventricle subdivides also, giving rise to the four-chambered heart characteristics of mammals. The main arteries and veins of the head region are at first laid down with reference to what are known as the bronchial arches, the structural framework of the bronchial or gill apparatus of aquatic vertebrates. Later, the whole architecture of this system becomes profoundly modified in adaptation for lung respiration. While the arteries and veins are in the fish-like condition there appear at the anterior end of the body in the prospective neck region four pairs of crevices, gill slits, which in fishes open directly into the pharynx and furnish a surface for gills. In the human embryo, however, these clefts never break through, but, after persisting for some time without playing any useful role, gradually disappear. The only persistent residue of the gill slits is the Eustacean tube, which connects the pharynx with the middle ear. Never at any time do the gill slits function in a respiratory capacity, for they never possess any bronchial tissue. Only one interpretation of these transitory gill slits of man can be seriously entertained, namely, that, although these structures are inherited from the early aquatic ancestry, adaptive demands have caused their suppression in favor of more useful structures. Inheritance causes their appearance; lack of function prevents their development and causes their disappearance or modification.

Nothing is to be gained by a multiplication of parallelisms such as the above. Suffice it to say that the nervous system, the alimentary system, the uregential system and other systems go through stages similar to those described above and that these resemble adult stages of lower classes of vertebrates. The embryology of man is now pretty thoroughly known in spite of the great difficulty of obtaining the early stages. Step for step it is almost precisely like that of other primates, especially like that of the anthropoids, and it is only in the latest stages that it takes on distinctly human characteristics. This is not equivalent to saying that the expert embryologist is in any doubt as to the diagnosis of a human embryo no matter how early the stage, for there are specific features about all embryos from the egg stage on to the end of development that may be distinguished by any one sufficiently versed in the subject. In spite of these specific differences, however, there can be no question that the embryology of man and that of any of the anthropoid apes show the closest of resemblances at every stage and diverge sharply only in the late stages of prenatal life. So close a resemblance in developmental histories is found only in species that are members of the same ancestral stock, for they have both inherited the characteristic features of their development from their common ancestors.

The evidence of human evolution as derived from a study of embryology is in no wise exceptional; in the contrary, it is quite typical and may be taken as indicating that from the developmental standpoint man is at one with other animals.

Evidences from Paleontology

Paleontology is the science of ancient life. Its materials are the more or less complete preserved remains of animals and plants that once lived. We call those remains fossils. Fossils are real; they cannot be explained away. If evolution has taken place and samples of every species that has lived were preserved for study, it would still be a task of immense difficulty to work out the pedigrees of all types of organisms now living, and we might still be largely in the dark as to the causes of the observed changes. As it is, we have fossil remains of perhaps only about one out of each thousand extinct species, a mere random sampling of the types that prevailed during the various past ages. Considering how many factors have been at work to prevent fossilization of large groups of species and how erosion and metamorphosis have worked together to destroy those fossils already preserved, we marvel that our fossil record is sufficiently complete to tell any sort of sequential story. The fact is that the record is surprisingly full and rich.

Age of the Earth

According to the most recent computations based on the rate of radium emanation, 1,000,000,000 years have elapsed since the earth attained its present diameter. Various estimates as to the time since the first life appeared upon the surface of the globe range from 50,000,000 years to about ten times that figure. Even the lowest figure gives ample time for any sort of evolutionary change, no matter how slow.

The Earth's Strata as Time Markers

The crust of the earth is arranged in a series of horizontal strata of varying thickness. The lowest layers are obviously the oldest, except in a few localities where breaks and tilts have occurred. Even in the most disturbed mountainous regions it is an easy task for the geologist to determine the original order of the strata.

First—None of the animals of the past are identical with those of the present. The nearest relationship is between a few species of the past which have been placed in the same genera as those of today.

Second—The animals and plants of each geologic stratum are at least generically different from those of any other stratum.

Third—The animals and plants of the oldest geologic strata represent all of the existing phyla except the vertebrates, but the representatives of the various phyla are relatively generalized as compared with modern representatives of the same phyla.

Fourth—There is a general progression toward more highly specialized forms as one proceeds from lower to higher strata.

Fifth—Many groups of animals reached the climax of their specialization long ages ago and have become extinct.

Sixth—Only the less specialized relatives of those most highly specialized types survived to become the progenitors of the modern representatives of the group.

Seventh—It is common to find a new group arising near the close of some geologic period when vast climatic changes were taking place. Such an incipient group almost regularly becomes the dominant group of the next period, presumably because it arose in response to the new conditions that accompanied the change from one period to another.

Eighth—The evolution of the vertebrate classes is more satisfactorily shown than that of any other group, probably because it arouse within the period which is characterized by an abundant fossil record. Of the vertebrates, the mammals are best represented and show the most complete fossil pedigrees; this, because they are the most recent in origin and their remains have been least disturbed.

Ninth—Many practically complete fossil pedigrees have been worked out, connecting specialized types with simpler and more generalized ancestors. Such pedigrees have been worked out for the horse, the elephant, the camel, the rhinoceros and other equally specialized modern types. A single example of this type of evidence will be given, that of the horse. Many other pedigrees have been worked out that are equally complete and no less significant.

Pedigree of the Horse

As recorded by Dendy, the course of evolution of the horse family (Equidae) "has evidently been determined by the development of extensive, dry, grass-covered, open plains on the American continent. In adaptation of life on such areas structural modification has proceeded chiefly in two directions. The limbs have become greatly elongated and the foot uplifted from the ground, and thus adapted for rapid flight from pursuing enemies, while the middle digit has become more and more important and the others, together with the ulna and the fibula, have gradually disappeared or been reduced to mere vestiges. At the same time the grażing mechanism has been gradually perfected. The neck and head have been elongated so that the animal is able to reach the ground without bending. its legs, and the cheek teeth have acquired complex grinding surfaces and have greatly increased in length to compensate for increased rate of wear. As in so many other groups, the evolution of these special characters has been accompanied by gradual increase in size. Thus Eohippus, of lower eocene times, appears to have been not more than eleven inches high at the shoulder, while existing horses measure about sixty-four inches, and numerous intermediate genera for the most part show regular progress in this respect.

All of these changes have taken place gradually, and a beautiful series of intermediate forms indicating the different stages from Eohippus to the modern horse have been discovered. The sequence of these stages in geological time exactly fits in with the theory that each one has been derived from the one next below it by more perfect adaptation to the conditions of life. Numerous genera have been described, but it is not necessary to mention more than a few."

The first indisputably horse-like animal appears to have been Hyracotherium of the lower eocene of Europe. Another lower eocene genus is Eohippus, which lived in North America, probably having migrated across from Asia by the Alaskan land connection. In Eohippus the forefeet had four hoofed toes of nearly equal size, the homelogue of the thumb having been reduced to a vestige. In the hind foot the great toe had entirely disappeared and the little toe had been reduced to a splint bone. Then came Orehippus of the upper eocene, Mesehippus of the lower miocene, Prothohippus of the lower pliocene, pliohippus of the upper pliocene, and finally, Equus of the quarternary and recent. This history, in so far as it concerns the characters already described, furnishes all of the intermediate conditions and perfectly connects the horses of the past with those of the present. One could hardly ask for a clearer or more conclusive story of evolution than this, and this is only one of many similar cases.

The Fossil Pedigree of Man.

There is nothing peculiar or exceptional about the fossil record of man. It is considerably less complete than that of the horse, the camel, the elephant and other purely terrestrial mammals, but it is far more complete than that of birds, bats and several types of arboeal mammals. Much has been said by the antievolutionists about the fragmentary nature of the fossil record of man, but many other animals have left traces far less readily deciphered and reconstructed.

The outstanding fact brought out by a study of human paleontology is that of man's antiquity. According to the most expert testimony available, the oldest fossil in the human series is about 500,000 years old; and even this estimate makes man a recent product of evolution as compared with many contemporaneous animals. The earliest fossil remains of the present species of man (homo sapiens) have been very conservatively estimated as 25,000 years old, while other species of extinct man date back to a period at least 100,000 years ago. In addition to several species of the genus homo, anthropologists distinguish three other genera of the man family (Hominidae): Pithecanthropus, Paleanthropus and Eoanthropus, all more primitive than any members of the genus Homo. A brief, but frank, statement about each of these links in the human pedigree is all that is necessary for our purposes.

The Java Man.

Pithecanthropus erectus.

This is the so-called Java man, formerly called the ape man or missing link, but now adjudged to be definitely human. The fossil remains consist of a complete calvarium or skull cap, three teeth and a left thigh bone. These were scattered over twenty yards of space and were discovered at different times. There is no proof that these remains belong to the same individual or even to the same species, but they are all human in their anatomical characters and they occurred in fossil-bearing rock about 500,000 years old. Many pages of scientific romance have been written about this species; all sorts of more or less justifiable pictures and models of this hypothetical species have been published. It is then refreshing to read the coldly scientific statement of Gregory:

"The association of gibbon-like, skull-top, modernized human femur and subhuman upper molars with reduced posterior moiety, if correctly assigned to one animal, may, perhaps, define pithecanthropus as an early side branch of the Hominidae, which had already been driven away from the center of dispersal in central Asia, by pressure of higher races. But whatever its precise systematic and phylogenetic position, Pithecanthropus or even its constituent parts, the skull-top, the femur and the molars, severally and collectively testify to the close relationship of the late Tertiary anthropoids with the pleistocene hominidae."

The Heidelberg Man.

Paleanthropus Heidelbergensis.

The genus and species, commonly known as the Heidelberg man, is based solely upon a single lower jaw in an excellent state of preservation, with all teeth in place. The strong points about this find are, first, that it was found in a stratum whose age had been well established; and second, that its discoverer ranks among the leading experts in the field. The age of this venerable relic has been determined as at least 400,000 years, a little more recent than Pithecanthropus. The jaw is very primitive, heavy and clumsily constructed as compared with that of modern man. It lacks the chin prominence, as does the jaw of the gorilla. The teeth are strictly human, though rather larger than those of modern man. This ape-like jaw with human teeth forms an authentic link in the series connecting man with the anthropoids.

Eoanthropus Dawnonis, the Dawn Man.

The most ancient English human relic has been called the dawn man of Piltdown. Owing to the fact that the skull fragments had been badly damaged and scattered by workmen before they came into scientific hands, there has been a great deal of controversy as to their significance. Until the experts arrive at an agreement about this type it might be well for others to reserve judgment. There can be no doubt as to the fact that these remains show a curious admixture of simian and human characteristics, the jaw and teeth being even more simian than that of the Heidelberg man, while the skull, though primitive, is distinctly human. The age of the dawn man is placed at about 200,000 to 300,000 years.

In striking contrast with the fragmentary character of the remains just described are those of three distinct species of the genus homo, which are now to be briefly characterized.

Homo Neanderthalensis, the Neanderthal Man.

The well-established race known as Neanderthal man is represented by many individual skeletons of varying degrees of completeness and showing a considerable range of diversity. Specimens have been found in France, Spain, Belgium, Germany and Austria. This species of primitive man was of low stature about five feet three inches in the males and less in the females.

The posture was somewhat stooping. The relatively large head was long and flat, with ape-like brow ridges and scarcely any forehead, and was borne on an immensely muscular neck in such a way that the face was thrust forward in simian fashion. The lower jaw was heavy and lacked a chin prominence. The teeth were of a type known as taurodont, adapted to a coarse vegetable diet and quite different in structure from those of modern man. The brain of this ancient homo-Neanderthalensis was large and specialized in some parts, but deficient in those parts associated with the higher mental functions.

There can be no question that Neanderthal man was much more primitive, more simian in organization, than modern man. Expert opinion, as expressed by Keith looks upon him as "a separate and peculiar species of man which died out during or soon after the Mousterian period." This dates him back to about 50,000 years ago.

Homo Rhodesiensis, the Rhodesian Man.

Rhodesian man is represented by a perfect skull and a nearly perfect lower jaw, the tibia, both ends of a femur collar bone and parts of the scapula and pelvis. Part of the upper jaw of a second specimen was found in the same locality, the Broken Hill mine in northern Rhodesia. This species is largely of technical interest, and need not be described in detail. Suffice it to say that in some respects it was as primitive as Neanderthal man, but in other respects showed distinct tendencies toward the modern condition. Anthropologists have as yet not reached a decision as to the exact taxonomic status of Rhodesian man, nor has its age been definitely determined.

Homo Sapiens, the Modern Man.

The earliest fossil evidence of the existence of our own species dates back to about 25,000 years ago. At that time there lived a remarkable race, known to us as Cro-Magnons, a race said to be the most perfect physically of which we have any knowledge. Five essentially complete skeletons form the basis of the type description. This tall, strong, obviously intelligent, and artistic race, was different in several important particulars from any modern race. A detailed description of his characteristics would take us too far afield. Our chief interest in this race is that it serves to emphasize the antiquity of our own species.

In conclusion it may be said that the fossil evidences of man's ancestry are neither rich nor poor; that anthropology is a comparatively youthful science, and that new discoveries in the field are being made at a very satisfactory rate.

Evidences from Geographic Distribution

Just as paleontology deals with the vertical distribution or distribution in time of species, so geographic distribution deals with their horizontal distribution upon the earth's surface at any given period of time. Geographic distribution is a sort of cross section of vertical distribution, giving a picture of the complex evolution of organisms at a given moment in the process. Explorers and collectors have amassed a vast amount of data as to the present and past ranges of animals and have mapped out the distribution of the majority of known species. A composite map of the geographic distribution of all known species would be the most intricate picture puzzle imaginable, and it would be almost impossible to make sense of it. A study of the distribution of limited groups, however, should lead to some reasonable explanation of their interrelations. Obviously animals are not distributed strictly according to climatic conditions or habitat complexes, for a given climate in one part of the world is associated with an entirely different fauna from a practically identical climate in another part of the world. Moreover, animals are not always or even very frequently located in those parts of the world that would offer them the best possible life conditions. This is borne out by the fact that not a few animals, when taken out of the normal range and transferred to a distant region, thrive much better than in their native territory. Thus European rabbits, when carried to Australia, throve and multiplied beyond all expectation till they became a pest. Again, as may be easily observed, the English sparrow seems to find America much more congenial than the British Isles.

If animals are not distributed according to habitats, how, then, can we account for their distribution? It is not at all likely that species retain the same ranges for long periods; they are continually changing their locations. We know, also, that the likeliest places to look for two closely similar species are adjacent territories, separated by geographic barriers. A study of the distribution of the species of a large genus usually reveals the fact that the most generalized or type species occupies the central part of the area and that the most specialized species occupy outlying areas adjacent to or connected with the main range of the genus. Taking these and related facts into considereation, we are able to offer as an explanation of the distribution of groups of allied species that a parent species originates in one place, multiplies and tends to migrate centrifugally in all directions, modifying as it goes to fit new conditions. Some of the extreme migrants become isolated from the main body of the species and, no longer interbreeding with them, become at first well-marked local varieties and in time new species. The above is the usual hypothesis employed in explaining geographic distribution, and it obviously implies evolution. When used as a means of unraveling the intricate tangle involved in the distribution of species it has thrown a flood of light upon situations otherwise quite inexplicable. In brief, the evolution hypothesis rationalizes geographic distribution, makes a science of what was formerly a hopeless jumble, and has thus proven itself a valuable scientific agent.

The Inhabitants of Oceanic Islands

Oceanic Islands are small isolated bodies of land of volcanic origin, far from continents. They are the tops of oceanic mountains. All such islands have their inhabitants, and a study of these should furnish a crucial test of the validity of the rival theories of special creation and of evolution. Both creationists and evolutionists agree that these islands must have obtained their populations from continental bodies. If then the island species are identical with those of the continent from which they have been derived, there is no reason to believe that evolution has taken place; if, however, they are different, the degree of difference should be an exact measure of the amount of evolutionary change that has taken place. What are the facts? Practically all species of animals inhabiting oceanic islands are types that are capable of transportation in the air during storms or on floating debris. All species belong to the faunistic groups characteristic of the most available continent, but the species are for the most part peculiar, that is, different from species anywhere else. They may belong to the same genus or family as do those of the continent, but they are at least specificially, frequently generally, different from the latter. Such being the case, we are forced to conclude that new species have originated under island conditions. The extreme case is that of the island of St. Helena, 1,100 miles from Africa. On this little body of land there are 129 species of beetles, all but one of which are peculiar. The species belong to thirty-nine genera, of which twenty-five are peculiar. There are twenty species of land snails, of which seventeen are peculiar. Of twenty-six species of ferns seventeen belong to peculiar genera. The Azores, Bermudas, Galapagos islands, Sandwich islands, all tell much the same story, but their populations are not quite so peculiar.

Evidences from Genetics

Genetics may be defined as to the experimental and analytical study of variation and heredity, the two primary causal factors of organic evolution. As such, genetics aim not so much at furnishing evidence of the fact of evolution as at discovering its causes. Incidentally, however, man takes a hand in controlling evolutionary processes and actually observes new heredity types taking origin from old, he is observing at first hand the actual processes of evolution. We shall merely say that the geneticist is an eye-witness of present-day evolution and is able to offer the most direct evidence that evolution is a fact.

Summary of Evidences

All of the lines of evidence presented point strongly to organic evolution, and none are contrary to this principle. Most of the facts, moreover, are utterly incompatible with the only rival explanation, special creation. Not only do these evidences tell a straightforward story of evolution, but each one is entirely consistent with all of the others. Furthermore, each line of evidence aids in an understanding of the others. Thus embryology greatly illuminates comparative anatomy and classification; geographic distribution is aided by paleontology, and vice-versa; blood tests and classification throw mutual light the one upon the other. The evolution principle is thus a great unifying and integrating scientific conception. Any conception that is so far-reaching, so consistent, and that has led to so much advance in the understanding of nature, is at least an extremely valuable idea and one not lightly to be cast aside in case it fails to agree with one's predjudices.

The Court—Send for the jury.

Mr. Hays—May I have the consent of the other side to fix my record later and see that they are properly marked and introduced?

Mr. Darrow—Your honor, before you send for the jury, I think it my duty to make this motion. Off to the left of where the jury sits a little bit and about ten feet in front of them is a large sign about ten feet long reading, "Read Your Bible," and a hand pointing to it. The word "Bible" is in large letters, perhaps a foot and a half long, and the printing—

The Court—Hardly that long I think, general.

Mr. Darrow—What is that?

The Court—Hardly that long.

Mr. Darrow—Why, we will call it a foot.

The Court—Compromise on a foot.

Mr. Darrow—Well, we will call it a foot, I guess more, but I might be wrong again, judge.

The Court—Well, I believe there will be no insistence.

A Voice—Fourteen inches.

Mr. Darrow—I move that it be removed.

The Court—Yes.

Gen. McKenzie—If your honor please, why should it be removed? It is their defense and stated before the court, that they do not deny the Bible, that they expected to introduce proof to make it harmonize. Why should we remove the sign cautioning the people to read the Word of God just to satisfy the others in the case?

The Court—Of course, you know I stand for the Bible, but your son has suggested that we agree to take it down.

Gen. McKenzie—I do not agree with my son.

Mr. Malone—The house is divided against itself.

Mr. Darrow—The purpose, I do not know why it was put there, but I suggest that it be removed.

The Court—I do not suppose it was put there to influence the trial.

Gen. Stewart—Do I understand you to ask it to be removed?

Mr. Darrow—Yes.

The Court—What do you say about it being removed?

Gen. Stewart—I do not care for it being removed, I will be frank.

J. G. McKenzie—If your honor please, I believe in the Bible as strong as anybody else here but if that sign is objectionable to the attorneys for the defense, and they do not want to be repeatedly reminded of the fact that they should read their Bible, I think this court ought to remove it.

Mr. Malone—May your honor please—

Mr. Hays—May we make our record—

Mr. Malone (continuing)—I do not think that is the statement of the position of the attorneys for the defense. We are trying a case here which we believe has very definite issues, aspects, we believe even though the court has moved downstairs for safety and comfort, that everything which might possibly prejudice the jury along religious lines, for or against the defense, should be removed from in front of the jury. The opinions of the members of the counsel for the defense, our religious beliefs, or Mr. Darrow's nonbelief, are none of the business of counsel for the prosecution. We do not wish that referred to again. The counsel for the defense are not on trial here. Mr. Scopes here is on trial and we are merely asking this court to remove anything of a prejudicial nature that we may try these issues and the court will be taken out of a prejudicial atmosphere. (Applause.)

J. G. McKenzie—If the court please, in reply to the statement of Mr. Malone, I want to withdraw my suggestion in regard to removing the sign, "Read Your Bible," for this reason: I have never seen the time in the history of this country when any man should be afraid to be reminded of the fact that he should read his Bible, and if they should represent a force that is aligned with the devil and his satellites—

Mr. Malone—Your honor, I object to that kind of language.

J. G. McKenzie (continuing)—Finally I say when that time comes that then is time for us to tear up all of the Bibles, throw them in the fire and let the country go to hell.

Mr. Hays—May I ask that our exception to those remarks be put on the record and I should like to move the court to expunge the last remarks.

The Court—Yes, expunge that part of Mr. McKenzie's statement from the record, where he said, if you were satellites of the devil. Any body else want to be heard?

Mr. Malone—Yes, I think it is all right for the individual members of the prosecution to make up their minds as to what forces we represent. I have a right to assume I have as much chance of heaven as they have, to reach it by my own goal, and my understanding of the Bible and of Christianity, and I will be a pretty poor Christian when I get any Biblical or Christian or religious views from any member of the prosecution that I have yet heard from during this trial. (Applause and laughter, with Court Officer Kelso Rice rapping for order.)

Mr. Bryan—If the court, please—

The Court—Col. Bryan, I will hear you.

Officer Rice—People, this is no circus. There are no monkeys up here. This is a lawsuit, let us have order.

Both Sides Swearing By Bible?

Mr. Bryan—May it please the court. Very often in the course of a trial, questions come up which may be decided on one of several grounds. One is the ground as to what is right. There are certain technicalities that are sometimes observed, and then there are decisions made in the spirit of accomodation. I cannot see that there is any inconsistency, even subtechnically, between taking that "Bible" up there off for the defense, if the defense insists that there is nothing in evolution that is contrary to it. (Applause.) If their arguments are sound and sincere, that the Bible can be construed so as to recognize evolution, I cannot see why "Read Your Bible" would necessarily mean partiality toward our side. It seems to me that both of us would want to read the Bible if both of us find in it the basis of our belief. I am going to quote the Bible in defense of our position, and I am going to hold the Bible as safe, though they try to discard it from our wall. Paul said: "If eating meat maketh my brother to offend, I shall eat no meat while the world lasts." I would not go that far, that is, I would not say while the world lasts, but if leaving that up there during the trial makes our brother to offend, I would take it down during the trial.

Mr. Malone—May I make my exception?

Mr. Darrow—Let me say something. Your honor, I just want to make this suggestion. Mr. Bryan says that the Bible and evolution conflict. Well, I do not know, I am for evolution, anyway. We might agree to get up a sign of equal size on the other side and in the same position reading, "Hunter's Biology," or "Read your evolution." This sign is not here for no purpose, and it can have no effect but to influence this case, and I read the Bible myself—more or less—and it is pretty good reading in places. But this case has been made a case where it is to be the Bible or evolution, and we have been informed by Mr. Bryan, who, himself, a profound Bible student and has an essay every Sunday as to what it means. We have been informed that a Tennessee jury who are not especially educated are better judges of the Bible than all of the scholars in the world, and when they see that sign, it means to them their construction of the Bible. It is pretty obvious, it is not fair, your honor, and we object to it.

Mr. Bryan—I am sure the gentleman does not mean to misrepresent me, but if he will get the record he will find that he has misquoted me.

Mr. Darrow—I am sorry if I did. Perhaps I did.

Mr. Bryan—I said any of the scholars whom the defense could or would call—that is different from the statement as made by the gentleman. Besides, the gentleman's statement is not pertinent. He said he would put up "Hunter's Biology." We are not both swearing by Hunter's Biology. We are swearing by the Bible. If we can accept in good faith what the defendant has said.

Mr. Darrow—Oh, no, there is a variance.

The Court Removes Sign.

The Court—The issues in this case, as they have been finally determined by this court, is whether or not it is unlawful to teach that man descended from a lower order of animals. I do not understand that issue involved the Bible. If the Bible is involved, I believe in it and am always on its side, but it is not for me to decide in this case. If the presence of the sign irritates anyone, or if anyone thinks it might influence the jury in any way, I have no purpose except to give both sides a fair trial in this case. Feeling that way about it, I will let the sign come down. Let the jury be brought around.

(The sign was thereupon removed from the courthouse wall.)

Mr. Hays—Your honor, before you bring in the jury we have other matters to introduce which might bring up a question you may not wish for the jury to hear the argument on. It will not take very long.

The Court—What is it, Mr. Hays?

Mr. Hays—Your honor will remember that in my argument the other day I insisted there was no such thing as the Bible, that there are many bibles; but the court took judicial notice that the King James' version was the Bible. The court has the right to take judicial notice of other bibles, and I will ask the court to admit in evidence a translation of the Holy Bible from the Vulgate, which I understand is the Catholic Bible, as evidence in this case.

The Court—Is it in English?

Mr. Hays—Yes, sir.

The Court—Let it be filed.

Mr. Hays—We wish to treat as read the first two pages your honor.

The Court—I was just reading to see if there was any difference, for my own edification.

Mr. Hays—Your honor, we wish to introduce as evidence, likewise, the Hebrew Bible; and we are going to ask that the first two chapters, likewise, be regarded as in evidence. And we believe we can show, through these translations we wish read into the record, that the Bible was not accurately translated into English, and of particular interest on the question of evolution.

The Court—Is that in English?

Mr. Hays—That is not in English.

Mr. Bryan—Well, of course, you want it in English.

Mr. Hays—No, I want the translation of my witness, whose affidavit I have read. I offered it in evidence, but very little of it was in my statement.

The Court—Let it be put in evidence.

Mr. McKenzie—They cannot put that in as proof.

Has Right, Perhaps, to Show Other Bibles.

Mr. Hays—We have a right to do so to this extent: That if it should appear that the Catholic Bible is different in any part from the King James version, or that the Hebrew Bible is different in any part from the King James version, we have a right to show it. We should be permitted, in our argument, to show that there is a difference, and that it is not merely interpretative.

Mr. Bryan—If the Jewish Bible is to be used in this trial, I think we have a right to object to him bringing in some particular translations. We can get a Hebrew Bible translated into English. We have one and will be glad to give it to them; but I do not think they have a right to bring in some individual's private interpretation.

Mr. Hays—Our witness would swear to it on the stand that his translation is correct.

Mr. Bryan—I know, but your witness has not been sworn, and his testimony is for the record. If you are going before the jury with this, I submit that you cannot come with your private interpretation, but you should take the Jewish Bible that is used by the Jews of this country.

Mr. Hays—I think Mr. Bryan is right about that. We have here the Catholic Bible and the King James version, and I offer these for the purposes of the record, and to show what the translation should have been, and for no other purpose.

Mr. Darrow—Mr. Bryan, have you your translation here?

Mr. Bryan—I have it at the house.

Gen. Stewart—Indictment was based on the King James version of the Bible.

The Court—I don't believe it is worth fussing over. I don't think there is any conflict in it. If there is no conflict—

Gen. Stewart—If there is no conflict, there is no use in discussing it.

Mr. Hays—But I say, if there is conflict anywhere in the words in the Bible as it was interpreted there, and in the Bible as it has been translated from time to time, then it is a matter for each individual to determine.

Gen. Stewart—I think that was settled when your honor took judicial notice of the Bible, and I make the point now because there is no use in making it before the jury.

The Court—The question is whether or not Mr. Scopes taught men descended from the lower order of animals.

Mr. Hays—And whether or not that is contrary to the theory in the St. James version.

The Court—No, not—

Mr. Malone—Your honor ruled that we could not go before the jury with it; that Mr. Scopes taught that man descended from a lower order of animals; and you ruled out important testimony for the defense.

Defense Wants Bryan as a Witness.

Mr. Hays—The defense desires to call Mr. Bryan as a witness, and, of course, the only question here is whether Mr. Scopes taught what these children said he taught, we recognize what Mr. Bryan says as a witness would not be very valuable. We think there are other questions involved, and we should want to take Mr. Bryan's testimony for the purposes of our record, even if your honor thinks it is not admissible in general, so we wish to call him now.

The Court—Do you think you have a right to his testimony or evidence like you did these others?

B. G. McKenzie—I don't think it is necessary to call him, calling a lawyer who represents a client.

The Court—If you ask him about any confidential matter, I will protect him, of course.

Mr. Darrow—I do not intend to do that.

The Court—On scientific matters, Col. Bryan can speak for himself.

Mr. Bryan—If your honor please, I insist that Mr. Darrow can be put on the stand, and Mr. Malone and Mr. Hays.

The Court—Call anybody you desire. Ask them any questions you wish.

Mr. Bryan—Then, we will call all three of them .

Mr. Darrow—Not at once?

Mr. Bryan—Where do you want me to sit?

Mr. Bryan Willing.

The Court—Mr. Bryan, you are not objecting to going on the stand?

Mr. Bryan—Not at all.

The Court—Do you want Mr. Bryan sworn?

Mr. Darrow—No.

Mr. Bryan—I can make affirmation; I can say "So help me God, I will tell the truth."

Mr. Darrow—No, I take it you will tell the truth, Mr. Bryan.

Bryan Goes on Witness Stand.

Examination of W. J. Bryan by Clarence Darrow, of counsel for the defense:

Q—You have given considerable study to the Bible, haven't you, Mr. Bryan?

A—Yes, sir, I have tried to.

Q—Well, we all know you have, we are not going to dispute that at all. But you have written and published articles almost weekly, and sometimes have made interpretations of various things?

A—I would not say interpretations, Mr. Darrow, but comments on the lesson.

Q—If you comment to any extent these comments have been interpretations.

A—I presume that my discussion might be to some extent interpretations, but they have not been primarily intended as interpretations.

Q—But you have studied that question, of course?

A—Of what?

Q—Interpretation of the Bible.

A—On this particular question?

Q—Yes, sir.

A—Yes, sir.

Q—Then you have made a general study of it?

A—Yes, I have; I have studied the

Bible for about fifty years, or sometime more than that, but, of course, I have studied it more as I have become older than when I was but a boy.

Q—Do you claim that everything in the Bible should be literally interpreted?

A—I believe everything in the Bible should be accepted as it is given there; some of the Bible is given illustratively. For instance: "Ye are the salt of the earth." I would not insist that man was actually salt, or that he had flesh of salt, but it is used in the sense of salt as saving God's people.

Did Jonah Swallow the Whale?

Q—But when you read that Jonah swallowed the whale—or that the whale swallowed Jonah—excuse me please—how do you literally interpret that?

A—When I read that a big fish swallowed Jonah—it does not say whale.

Q—Doesn't it? Are you sure?

A—That is my recollection of it. A big fish, and I believe it, and I believe in a God who can make a whale and can make a man and make both do what He pleases.

Q—Mr. Bryan, doesn't the New Testament say whale?

A—I am not sure. My impression is that it says fish; but it does not make so much difference; I merely called your attention to where it says fish—it does not say whale.

Q—But in the New Testament it says whale, doesn't it?

A—That may be true; I cannot remember in my own mind what I read about it.

Q—Now, you say, the big fish swallowed Jonah, and he there remained how long—three days—and then he spewed him upon the land. You believe that the big fish was made to swallow Jonah?

A—I am not prepared to say that; the Bible merely says it was done.

Q—You don't know whether it was the ordinary run of fish, or made for that purpose?

A—You may guess; you evolutionists guess.

Q—But when we do guess, we have a sense to guess right.

A—But do not do it often.

Q—You are not prepared to say whether that fish was made especially to swallow a man or not?

A—The Bible doesn't say, so I am not prepared to say.

Q—You don't know whether that was fixed up specially for the purpose?

A—No, the Bible doesn't say.

Q—But do you believe He made them—that He made such a fish and that it was big enough to swallow Jonah?

A—Yes, sir. Let me add: One miracle is just as easy to believe as another.

Q—It is for me.

A—It is for me.

Q—Just as hard?

A—It is hard to believe for you, but easy for me. A miracle is a thing performed beyond what man can perform. When you get beyond what man can do, you get within the realm of miracles; and it is just as easy to believe the miracle of Jonah as any other miracle in the Bible.

Q—Perfectly easy to believe that Jonah swallowed the whale?

A—If the Bible said so; the Bible doesn't make as extreme statements as evolutionists do.

Mr. Darrow—That may be a question, Mr. Bryan, about some of those you have known?

A—The only thing is, you have a definition of fact that includes imagination.

Q—And you have a definition that excludes everything but imagination, everything but imagination?

Gen. Stewart—I object to that as argumentative.

The Witness—You—

Mr. Darrow—The witness must not argue with me, either.

Q—Do you consider the story of Jonah and the whale a miracle?

A—I think it is.

Did the Sun Stand Still?

Q—Do you believe Joshua made the sun stand still?

A—I believe what the Bible says. I suppose you mean that the earth stood still?

Q—I don't know. I am talking about the Bible now.

A—I accept the Bible absolutely.

Q—The Bible says Joshua commanded the sun to stand still for the purpose of lengthening the day, doesn't it, and you believe it?

A—I do.

Q—Do you believe at that time the entire sun went around the earth?

A—No, I believe that the earth goes around the sun.

Q—Do you believe that the men who wrote it thought that the day could be lengthened or that the sun could be stopped?

A—I don't know what they thought.

Q—You don't know?

A—I think they wrote the fact without expressing their own thoughts.

Q—Have you an opinion as to whether or not the men who wrote that thought—

Gen. Stewart—I want to object, your honor; it has gone beyond the pale of any issue that could possibly be injected into this lawsuit, except by imagination. I do not think the defendant has a right to conduct the examination any further and I ask your honor to exclude it.

The Court—I will hear Mr. Bryan.

The Witness—It seems to me it would be too exacting to confine the defense to the facts; if they are not allowed to get away from the facts, what have they to deal with?

The Court—Mr. Bryan is willing to be examined. Go ahead.

Mr. Darrow—Have you an opinion as to whether—whoever wrote the book, I believe it is, Joshua, the Book of Joshua, thought the sun went around the earth or not?

A—I believe that he was inspired.

Mr. Darrow—Can you answer my question?

A—When you let me finish the statement.

Q—It is a simple question, but finish it.

The Witness—You cannot measure the length of my answer by the length of your question.

(Laughter in the courtyard.)

"I Believe Bible Inspired."

Mr. Darrow—No, except that the answer be longer.

(Laughter in the courtyard.)

A—I believe that the Bible is inspired, an inspired author, whether one who wrote as he was directed to write understood the things he was writing about, I don't know.

Q—Whoever inspired it? Do you think whoever inspired it believed that the sun went around the earth?

A—I believe it was inspired by the Almighty, and He may have used language that could be understood at that time.

Q—Was—

The Witness—Instead of using language that could not be understood until Darrow was born.

(Laughter and applause in the courtyard.)

Q—So, it might not, it might have been subject to construction, might it not?

A—It might have been used in language that could be understood then.

Q—That means it is subject to construction?

A—That is your construction. I am answering your question.

Q—Is that correct?

A—That is my answer to it.

Q—Can you answer?

A—I might say, Isaiah spoke of God sitting upon the circle of the earth.

Q—I am not talking about Isaiah.

The Court—Let him illustrate, if he wants to.

Mr. Darrow—Is it your opinion that passage was subject to construction?

A—Well, I think anybody can put his own construction upon it, but I do not mean that necessarily that is a correct construction. I have answered the question.

Q—Don't you believe that in order to lengthen the day it would have been construed that the earth stood still?

A—I would not attempt to say what would have been necessary, but I know this, that I can take a glass of water that would fall to the ground without the strength of my hand and to the extent of the glass of water I can overcome the law of gravitation and lift it up. Whereas without my hand it would fall to the ground. If my puny hand can overcome the law of gravitation, the most universally understood, to that extent, I would not set power to the hand of Almighty God that made the universe.

Mr. Darrow—I read that years ago. Can you answer my question directly? If the day was lengthened by stopping either the earth or the sun, it must have been the earth?

A—Well, I should say so.

Q—Yes? But it was language that was understood at that time, and we now know that the sun stood still as it was with the earth.

Q—We know also the sun does not stand still?

A—Well, it is relatively so, as Mr. Einstein would say.

Q—I ask you if it does stand still?

A—You know as well as I know.

Q—Better. You have no doubt about it?

A—No. And the earth moves around.

Q—Yes?

A—But I think there is nothing improper if you will protect the Lord against your criticism.

Q—I suppose He needs it?

A—He was using language at that time the people understood.

Q—And that you call "interpretation?"

A—No, sir; I would not call it interpretation.

Q—I say, you would call it interpretation at this time, to say it meant something then?

A—You may use your own language to describe what I have to say, and I will use mine in answering.

What If Earth Had Stood Still?

Q—Now, Mr. Bryan, have you ever pondered what would have happened to the earth if it had stood still?

A—No.

Q—You have not?

A—No; the God I believe in could have taken care of that, Mr. Darrow.

Q—I see. Have you ever pondered what would naturally happen to the earth if it stood still suddenly?

A—No.

Q—Don't you know it would have been converted into a molten mass of matter?

A—You testify to that when you get on the stand, I will give you a chance.

Q—Don't you believe it?

A—I would want to hear expert testimony on that.

Q—You have never investigated that subject?

A—I don't think I have ever had the question asked.

Q—Or ever thought of it?

A—I have been too busy on things that I thought were of more importance than that.

Q—You believe the story of the flood to be a literal interpretation?

A—Yes, sir.

Q—When was that flood?

A—I would not attempt to fix the date. The date is fixed, as suggested this morning.

Q—About 4004 B. C.?

A—That has been the estimate of a man that is accepted today. I would not say it is accurate.

Q—That estimate is printed in the Bible?

A—Everybody knows, at least, I think most of the people know, that was the estimate given.

Q—But what do you think that the Bible, itself, says? Don't you know how it was arrived at?

A—I never made a calculation.

Q—A calculation from what?

A—I could not say.

Q—From the generations of man?

A—I would not want to say that.

Q—What do you think?

A—I do not think about things I don't think about.

Q—Do you think about things you do think about?

A—Well, sometimes.

(Laughter in the courtyard.)

The Policeman—Let us have order.

Mr. Darrow—Mr. Bryan, you have read these dates over and over again?

A—Not very accurately, I turn back sometimes to see what the time was.

Q—You want to say now you have no idea how these dates were computed?

A—No, I don't say, but I have told you what my idea was. I say I don't know how accurate it was.

Q—You say from the generation of man—

Gen. Stewart—I am objecting to his cross-examining his own witness.

Mr. Darrow—He is an hostile witness.

The Court—I am going to let Mr. Bryan control—

The Witness—I want him to have all the latitude he wants. For I am going to have some latitude when he gets through.

Mr. Darrow—You can have latitude and longitude.

(Laughter.)

The Court—Order.

Gen. Stewart—The witness is entitled to be examined as to the legal evidence of it. We were supposed to go into the origin of the case, and we have nearly lost the day, your honor.

Mr. McKenzie—I object to it.

Gen. Stewart—Your honor, he is perfectly able to take care of this, but we are attaining no evidence. This is not competent evidence.

Bryan Charges Defense With Evil Motive.

The Witness—These gentlemen have not had much chance—they did not come here to try this case. They came here to try revealed religion. I am here to defend it, and they can ask me any question they please.

The Court—All right.

(Applause from the court yard.)

Mr. Darrow—Great applause from the bleachers.

The Witness—From those whom you call "yokels."

Mr. Darrow—I have never called them yokels.

The Witness—That is the ignorance of Tennessee, the bigotry.

Mr. Darrow—You mean who are applauding you?

The Witness—Those are the people whom you insult.

Mr. Darrow—You insult every man of science and learning in the world because he does not believe in your fool religion.

The Court—I will not stand for that.

Mr. Darrow—For what he is doing?

The Court—I am talking to both of you.

Gen. Stewart—This has gone beyond the pale of a lawsuit, your honor. I have a public duty to perform, under my oath and I ask the court to stop it.

Mr. Darrow is making an effort to insult the gentleman on the witness stand, and I ask that it be stopped, for it has gone beyond the pale of a lawsuit.

The Court—To stop it now would not be just to Mr. Bryan. He wants to ask the other gentleman questions along the same line.

Gen. Stewart—It will all be incompetent.

The Witness—The jury is not here.

The Court—I do not want to be strictly technical.

Mr. Darrow—Then your honor rules, and I accept.

Gen. Stewart—The jury is not here.

What About the Flood?

Mr. Darrow—How long ago was the flood, Mr. Bryan?

A—Let me see Usher's calculation about it?

Mr. Darrow—Surely.

(Handing a Bible to the witness.)

A—I think this does not give it.

Q—It gives an account of Noah. Where is the one in evidence, I am quite certain it is there?

The Witness—Oh, I would put the estimate where it is, because I have no reason to vary it. But I would have to look at it to give you the exact date.

Q—I would, too. Do you remember what book the account is in?

A—Genesis.

Mr. Hays—Is that the one in evidence?

Mr. Neal—That will have it; that is King James' version.

Mr. Darrow—The one in evidence has it.

The Witness—It is given here, as 2348 years B. C.

Q—Well, 2348 years B. C. You believe that all the living things that were not contained in the ark were destroyed.

A—I think the fish may have lived.

Q—Outside of the fish?

A—I cannot say.

Q—You cannot say?

A—No, except that just as it is, I have no proof to the contrary.

Q—I am asking you whether you believe?

A—I do.

Q—That all living things outside of the fish were destroyed?

A—What I say about the fish is merely a matter of humor.

Q—I understand.

The Witness— Due to the fact a man wrote up here the other day to ask whether all the fish were destroyed, and the gentleman who received the letter told him the fish may have lived.

Q—I am referring to the fish, too?

A—I accept that, as the Bible gives it and I have never found any reason for denying, disputing, or rejecting it.

Q—Let us make it definite, 2,348 years?

A—I didn't say that. That is the time given there (indicating a Bible) but I don't pretend to say that is exact.

Q—You never figured it out, these generations, yourself?

A—No, sir; not myself.

Q—But the Bible you have offered in evidence, says 2,340, something, so that 4,200 years ago there was not a living thing on the earth, excepting the people on the ark and the animals of the ark and the fishes?

A—There have been living things before that.

Q—I mean at that time.

A—After that.

Q—Don't you know there are any number of civilizations that are traced back to more than 5,000 years?

A—I know we have people who trace things back according to the number of ciphers they have. But I am not satisfied they are accurate.

Q—You are not satisfied there is any civilization that can be traced back 5,000 years?

A—I would not want to say there is because I have no evidence of it that is satisfactory.

Q—Would you say there is not?

Scientists Will Have to Get Closer.

A—Well, so far as I know, but when the scientists differ, from 24,000,000 to 306,000,000 in their opinion, as to how long ago life came here, I want them nearer, to come nearer together before they demand of me to give up my belief in the Bible.

Q—Do you say that you do not believe that there were any civilizations on this earth that reach back beyond 5,000 years?

A—I am not satisfied by any evidence that I have seen.

Q—I didn't ask you what you are satisfied with. I asked you if you believe it?

The Witness—Will you let me answer it?

The Court—Go right on.

No Evidence Satisfying.

The Witness—I am satisfied by no evidence, that I have found, that would justify me in accepting the opinions of these men against what I believe to be the inspired Word of God.

Q—And you believe every nation, every organization of men, every animal, in the world outside of the fishes—

The Witness—The fish, I want you to understand, is merely a matter of humor.

Q—I believe the Bible says so. Take the fishes in?

A—Let us get together and look over this.

Mr. Darrow—Probably we would better, we will after we get through.

Darrow Relentless on Question of Years.

Q—You believe that all the various human races on the earth have come into being in the last 4,000 years or 4,200 years, whatever it is?

A—No, it would be more than that.

Q—1927?

A—Some time after creation, before the flood.

Q—1,927 added to it?

A—The flood is 2,300 and something, and creation, according to the estimate there, is further back than that.

Q—Then you don't understand me. If we don't get together on it, look at the book. This is the year of grace 1925, isn't it? Let us put down 1,925. Have you a pencil?

(One of the defense attorneys hands Mr. Darrow a pencil.)

The Witness—Add to that 4,004?

Mr. Darrow—Yes.

A—That is the date (referring to the Bible) given here on the first page, according to Bishop Usher, which I say I only accept because I have no reason to doubt it. In that page he gives it.

Q—1,925 plus 4,004 is 5,929 years. If a fallible person is right in his addition. Now, then, what do you subtract from that?

A—That is the beginning.

Q—I was talking about the flood.

A—2,348 on that, we said.

Q—Less than that?

A—No; subtract that from 4,000; it would be about 1,700 years.

Q—That is the same thing?

A—No; subtracted it is 2,300 and something before the beginning of the Christian era, about 1,700 years after the creation.

The Policeman—Let us have order.

Mr. Darrow—If I add 2,300 years, that is the beginning of the Christian era?

A—Yes, sir.

Q—If I add 1,925 to that I will get it, won't I?

A—Yes, sir.

Q—That makes 4,262 years. If it is not correct, we can correct it.

A—According to the Bible there was a civilization before that, destroyed by the flood.

Q—Let me make this definite. You believe that every civilization on the earth and every living thing, except possibly the fishes, that came out of the ark were wiped out by the flood?

A—At that time.

Q—At that time. And then, whatever human beings, including all the tribes, that inhabited the world, and have inhabited the world, and who run their pedigree straight back, and all the animals, have come onto the earth since the flood?

A—Yes.

Q—Within 4,200 years. Do you know a scientific man on the face of the earth that believes any such thing?

A—I cannot say, but I know some scientific men who dispute entirely the antiquity of man as testified to by other scientific men.

Q—Oh, that does not answer the question. Do you know of a single scientific man on the face of the earth that believes any such thing as you stated, about the antiquity of man?

A—I don't think I have ever asked one the direct question.

Q—Quite important, isn't it?

A—Well, I don't know as it is.

Q—It might not be?

No Interest in Remote Ancestors.

A—If I had nothing else to do except speculate on what our remote ancestors were and what our remote descendants have been, but I have been more interested in Christians going on right now, to make it much more important than speculation on either the past or the future.

Q—You have never had any interest in the age of the various races and people and civilization and animals that exist upon the earth today? Is that right?

A—I have never felt a great deal of interest in the effort that has been made to dispute the Bible by the speculations of men, or the investigations of men.

Q—Are you the only human being on earth who knows what the Bible means?

Gen. Stewart—I object.

The Court—Sustained.

To which ruling of the court counsel for the defendant duly excepted.

Mr. Darrow—You do know that there are thousands of people who profess to be Christians who believe the earth is much more ancient and that the human race is much more ancient?

A—I think there may be.

Q—And you never have investigated to find out how long man has been on the earth?

A—I have never found it necessary.

Q—For any reason, whatever it is?

A—To examine every speculation; but if I had done it I never would have done anything else.

Q—I ask for a direct answer?

A—I do not expect to find out all those things, and I do not expect to find out about races.

Q—I didn't ask you that. Now, I ask you if you know if it was interesting enough, or important enough for you to try to find out about how old these ancient civilizations were?

A—No; I have not made a study of it.

Q—Don't you know that the ancient civilizations of China are 6,000 or 7,000 years old, at the very least?

A—No; but they would not run back beyond the creation, according to the Bible, 6,000 years.

Q—You don't know how old they are, is that right?

A—I don't know how old they are, but probably you do. (Laughter in the courtyard.) I think you would give the preference to anybody who opposed the Bible, and I give the preference to the Bible.

Q—I see. Well, you are welcome to your opinion. Have you any idea how old the Egyptian civilization is?

A—No.

Any Other Record of the Flood?

Q—Do you know of any record in the world, outside of the story of the Bible, which conforms to any statement that it is 4,200 years ago or thereabouts that all life was wiped off the face of the earth?

A—I think they have found records.

Q—Do you know of any?

A—Records reciting the flood, but I am not an authority on the subject.

Q—Now, Mr. Bryan, will you say if you know of any record, or have ever heard of any records, that describe that a flood existed 4,200 years ago, or about that time, which wiped all life off the earth?

A—The recollection of what I have read on that subject is not distinct enough to say whether the records attempted to fix a time, but I have seen in the discoveries of archaeologists where they have found records that described the flood.

Q—Mr. Bryan, don't you know that there are many old religions that describe the flood?

A—No, I don't know.

Q—You know there are others besides the Jewish?

A—I don't know whether these are the record of any other religion or refer to this flood.

Q—Don't you ever examine religion so far to know that?

A—Outside of the Bible?

Q—Yes.

A—No; I have not examined to know that, generally.

Q—You have never examined any other religions?

A—Yes, sir.

Q—Have you ever read anything about the origins of religions?

A—Not a great deal.

Q—You have never examined any other religion?

A—Yes, sir.

Q—And you don't know whether any other religion ever gave a similar account of the destruction of the earth by the flood?

Christian Religion Sufficient.

A—The Christian religion has satisfied me, and I have never felt it necessary to look up some competing religions.

Q—Do you consider that every religion on earth competes with the Christian religion?

A—I think everybody who does not believe in the Christian religion believes so—

Q—I am asking what you think?

A—I do not regard them as competitive because I do not think they have the same source as we have.

Q—You are wrong in saying "competitive"?

A—I would not say competitive, but the religious unbelievers.

Q—Unbelievers of what?

A—In the Christian religion.

Q—What about the religion of Buddha?

A—I can tell you something about that, if you want to know.

Confucious or Buddha?

Q—What about the religion of Confucious or Buddha?

A—Well, I can tell you something about that, if you would like to know.

Q—Did you ever investigate them?

A—Somewhat.

Q—Do you regard them as competitive?

A—No, I think they are very inferior. Would you like for me to tell you what I know about it?

Q—No.

A—Well, I shall insist on giving it to you.

Q—You won't talk about free silver, will you?

A—Not at all.

Gen. Stewart—I object to him—counsel going any further with this examination and cross-examining his own witness. He is your own witness.

Have Right to Cross-Examine Hostile Witness.

Mr. Darrow—Well, now, general, you understand we are making up a record, and I assume that every lawyer knows perfectly well that we have a right to cross-examine a hostile witness. Is there any doubt about that?

Gen. Stewart—Under the law in Tennessee if you put a witness on and he proves to be hostile to you, the law provides the method by which you may cross-examine him. You will have to make an affidavit that you are surprised at his statement, and you may do that.

Mr. Bryan—Is there any way by which a witness can make an affidavit? That the attorney is also hostile?

Mr. Darrow—I am not hostile to you. I am hostile to your views, and I suppose that runs with me, too.

Mr. Bryan—But I think when the gentleman asked me about Confucius I ought to be allowed to answer his question.

Mr. Darrow—Oh, tell it, Mr. Bryan, I won't object to it.

Recriprocity and the Golden Rule.

Mr. Bryan—I had occasion to study Confucianism when I went to China. I got all I could find about what Confucius said, and then I bought a book that told us what Menches said about what Confucius said, and I found that there were several direct and strong contrasts between the teachings of Jesus and the teaching of Confucius. In the first place, one of his followers asked if there was any word that would express all that was necessary to know in the relations of life, and he said, "Isn't reciprocity such a word?" I know of no better illustration of the difference between Christianity and Confucianism than the contrast that is brought out there. Reciprocity is a calculating selfishness. If a person does something for you, you do something for him and keep it even. That is the basis of the philosophy of Confucius. Christ's doctrine was not reciprocity. We were told to help people not in proportion as they had helped us—not in proportion as they might have helped us, but in proportion to their needs, and there is all the difference in the world between a religion that teaches you just to keep even with other people and the religion that teaches you to spend yourself for other people and to help them as they need help.

Q—There is no doubt about that; I haven't asked you that.

A—That is one of the differences between the two.

How Old is Confucianism?

Q—Do you know how old the Confucian religion is?

A—I can't give you the exact date of it.

Q—Did you ever investigate to find out?

A—Not to be able to speak definitely as to date, but I can tell you something I read, and will tell you.

Q—Wouldn't you just as soon answer my questions? And get along?

A—Yes, sir.

Q—Of course, if I take any advantage of misquoting you, I don't object to being stopped. Do you know how old the religion of Zoroaster is?

A—No, sir.

Q—Do you know they are both more ancient than the Christian religion?

A—I am not willing to take the opinion of people who are trying to find excuses for rejecting the Christian religion when they attempt to give dates and hours and minutes, and they will have to get together and be more exact than they have yet been able, to compel me to accept just what they say as if it were absolutely true.

A—Are you familiar with James Clark's book on the ten great religions?

A—No.

Q—He was a Unitarian minister, wasn't he? You don't think he was trying to find fault, do you?

A—I am not speaking of the motives of men.

Q—You don't know how old they are, all these other religions?

A—I wouldn't attempt to speak correctly, but I think it is much more important to know the difference between them than to know the age.

Q—Not for the purpose of this inquiry, Mr. Bryan? Do you know about how many people there were on this earth at the beginning of the Christian era?

A—No, I don't think I ever saw a census on that subject.

Q—Do you know about how many people there were on this earth 3,000 years ago?

A—No.

Q—Did you ever try to find out?

A—When you display my ignorance, could you not give me the facts, so I would not be ignorant any longer? Can you tell me how many people there were when Christ was born?

Q—You know, some of us might get the facts and still be ignorant.

A—Will you please give me that? You ought not to ask me a question when you don't know the answer to it.

Q—I can make an estimate.

A—What is your estimate?

How Many People 5,000 Years Ago?

Q—Wait until you get to me. Do you know anything about how many people there were in Egypt 3,500 years ago, or how many people there were in China 5,000 years ago?

A—No.

Q—Have you ever tried to find out?

A—No, sir. You are the first man I ever heard of who has been interested in it. (Laughter.)

Q—Mr. Bryan, am I the first man you ever heard of who has been interested in the age of human societies and primitive man?

A—You are the first man I ever heard speak of the number of people at those different periods.

Q—Where have you lived all your life?

A—Not near you. (Laughter and applause.)

Q—Nor near anybody of learning?

A—Oh, don't assume you know it all.

Q—Do you know there are thousands of books in our libraries on all those subjects I have been asking you about?

A—I couldn't say, but I will take your word for it.

Q—Did you ever read a book on primitive man? Like Tyler's Primitive Culture, or Boaz, or any of the great authorities?

A—I don't think I ever read the ones you have mentioned.

Q—Have you read any?

A—Well I have read a little from time to time. But I didn't pursue it, because I didn't know I was to be called as a witness.

Never Interested in Primitive Peoples.

Q—You have never in all your life made any attempt to find out about the other peoples of the earth—how old their civilizations are—how long they had existed on the earth, have you?

A—No, sir, I have been so well satisfied with the Christian religion that I have spent no time trying to find arguments against it.

Q—Were you afraid you might find some?

A—No, sir, I am not afraid now that you will show me any.

Q—You remember that man who said—I am not quoting literally—that one could not be content though he rose from the dead—you suppose you could be content?

A—Well, will you give the rest of it, Mr. Darrow?

Q—No.

A—Why not?

Q—I am not interested.

A—Why scrap the Bible—"they have Moses and the prophets"?

Q—Who has?

A—That is the rest of the quotation you didn't finish.

Q—And so you think if they have Moses and the prophets they don't need to find out anything else?

A—That was the answer that was made there.

Q—And you follow the same rule?

"All the Information I Need."

A—I have all the information I want to live by and to die by.

Q—And that's all you are interested in?

A—I am not looking for any more on religion.

Q—You don't care how old the earth is, how old man is and how long the animals have been here?

A—I am not so much interested in that.

Q—You have never made any investigation to find out?

A—No, sir, I have never.

Q—All right?

A—Now, will you let me finish the question?

Q—What question was that. If there is anything more you want to say about Confucious I don't object.

A—Oh, yes, I have got two more things.

Mr. Darrow—If your honor please I don't object, but his speeches are not germane to my question.

Mr. Hicks (Sue K.)—Your honor, he put him on.

The Court—You went into it and I will let him explain.

Mr. Darrow—I asked him certain specific questions about Confucious.

Mr. Hicks (Sue K.)—The questions he is asking are not germane, either.

Mr. Darrow—I think they are.

Other Differences.

The Witness—I mentioned the word reciprocity to show the difference between Christ's teachings in that respect and the teachings of Confucius. I call your attention to another difference. One of the followers of Confucius asked him "what do you think of the doctrine that you should reward evil with good?" and the answer of Confucius was "reward evil with justice and reward good with good. Love your enemies. Overcome evil with good," and there is a difference between the two teachings—a difference incalculable in its effect and in—The third difference—people who scoff at religion and try to make it appear that Jesus brought nothing into the world, talk about the Golden Rule of Confucius. Confucius said "do not unto others what you would not have others do unto you." It was purely negative. Jesus taught "do unto others as you would have others do unto you." There is all the difference in the world between a negative harmlessness and a positive helpfulness and the Christian religion is a religion of helpfulness, of service, embodied in the language of Jesus when he said "let him who would be chiefest among you be the servant of all." Those are the three differences between the teachings of Jesus and the teachings of Confucius, and they are very strong differences on very important questions. Now, Mr. Darrow, you asked me if I knew anything about Buddha.

Q—You want to make a speech on Buddha, too?

A—No, sir; I want to answer your question on Buddha.

Q—I asked you if you knew anything about him?

A—I do.

Q—Well, that's answered, then.

A—Buddha—

Q—Well, wait a minute, you answered the questions—

The Court—I will let him tell what he knows.

Mr. Darrow—All he knows?

The Court—Well, I don't know about that.

The Witness—I won't insist in telling all I know. I will tell more than Mr. Darrow wants told.

Mr. Darrow—Well, all right, tell it, I don't care.

Buddism is Agnostic.

The Witness—Buddism is an agnostic religion.

Q—To what?—what do you mean by agnostic?

A—I don't know.

Q—You don't know what you mean?

A—That is what "agnosticism" is —I don't know. When I was in Ranggoon, Burma, one of the Buddhists told me that they were going to send a delegation to an agnostic congress that was to be held soon at Rome and I read in an official document—

Q—Do you remember his name?

A—No, sir, I don't.

Q—What did he look like, how tall was he?

As Tall As You, But Not So Crooked.

A—I think he was about as tall as you but not so crooked.

Q—Do you know about how old a man he was—do you know whether he was old enough to know what he was talking about?

A—He seemed to be old enough to know what he was talking about. (Laughter.)

Mr. Darrow—If your honor please, instead of answering plain specific questions we are permitting the witness to regale the crowd with what some black man said to him when he was traveling in Rang—who, India?

The Witness—He was dark-colored, but not black.

The Court—I will let him go ahead and answer.

The Witness—I wanted to say that I then read a paper that he gave me, an official paper of the Buddhist church and it advocated the sending of delegates to that agnostic congress at Rome, arguing that it was an agnostic religion and I will give you another evidence of it. I went to call on a Buddhist teacher.

Objects to Bryan Making Speeches.

Mr. Darrow—I object to Mr. Bryan making a speech every time I ask him a question.

The Court—Let him finish this answer and then you can go ahead.

The Witness—I went to call on a Buddhist priest and found him at his noon meal, and there was an Englishman there who was also a Buddhist. He went over as ship's carpenter and became a Buddhist and had been for about six years and while I waited for the Buddhist priest I talked to the Englishman and I asked him what was the most important thing in Buddhism and he said the most important thing was you didn't have to believe to be a Buddhist.

Q—You know the name of the Englishman?

A—No, sir, I don't know his name.

Q—What did he look like? What did he look like?

A—He was what I would call an average looking man.

Q—How could you tell he was an Englishman?

A—He told me so.

Q—Do you know whether he was truthful or not?

A—No, sir, but I took his word for it.

The Court—Well, get along, Mr. Darrow, with your examination.

Mr. Darrow—Mr. Bryan ought to get along.

Tower of Babel.

Q—You have heard of the Tower of Babel haven't you?

A—Yes, sir.

Q—That tower was built under the ambition that they could build a tower up to heaven, wasn't it? And God saw what they were at and to prevent their getting into heaven he confused their tongues?

A—Something like that, I wouldn't say to prevent their getting into heaven. I don't think it is necessary to believe that God was afraid they would get to heaven—

Q—I mean that way?

A—I think it was a rebuke to them.

Q—A rebuke to them trying to go that way?

A—To build that tower for that purpose.

Q—Take that short cut?

A—That is your language, not mine.

Q—Now when was that?

A—Give us the Bible.

Q—Yes, we will have strict authority on it—scientific authority?

A—That was about 100 years before the flood, Mr. Darrow, according to this chronology. It is 2247—the date on one page is 2218 and on the other 2247 and it is described in here—

Q—That is the year 2247?

A—2218 B. C. is at the top of one page and 2247 at the other and there is nothing in here to indicate the change.

Q—Well, make it 2230 then?

A—All right, about.

Q—Then you add 1500 to that—

A—No, 1925.

Q—Add 1925 to that, that would be 4,155 years ago. Up to 4,155 years ago every human being on earth spoke the same language?

A—Yes, sir, I think that is the inference that could be drawn from that.

Q—All the different languages of the earth, dating from the Tower of Babel, is that right? Do you know how many languages are spoken on the face of the earth?

A—No, I know the Bible has been translated into 500 and no other book has been translated into anything like that many.

Q—That is interesting, if true? Do you know all the languages there are?

A—No, sir, I can't tell you. There may be many dialects besides that and some languages, but those are all the principal languages.

Q—There are a great many that are not principal languages?

A—Yes, sir.

Q—You haven't any idea how many there are?

A—No, sir.

Q—How many people have spoken all those various languages?

A—No, sir.

Q—And you say that all those languages of all the sons of men have come on the earth not over 4,150 years ago?

A—I have seen no evidence that would lead me to put it any further back than that.

Q—That is your belief anyway—that that was due to the confusion of tongues at the tower of Babel. Did you ever study philology at all?

A—No, I have never made a study of it—not in the sense in which you speak of it.

Q—You have used language all your life?

A—Well, hardly all my life—ever since I was about a year old.

Q—And good language, too, and you have never taken any pains to find anything about the origin of languages?

A—I have never studied it as a science.

Q—Have you ever by any chance read Max Mueller?

A—No.

Q—The great German philologist?

A—No.

Q—Or any book on that subject?

A—I don't remember to have read a book on that subject, especially, but I have read extracts, of course, and articles on philology.

How Old is Earth?

Q—Mr. Bryan, could you tell me how old the earth is?

A—No, sir, I couldn't.

Q—Could you come anywhere near it?

A—I wouldn't attempt to. I could possibly come as near as the scientists do, but I had rather be more accurate before I give a guess.

Q—You don't think much of scientists, do you?

A—Yes, sir, I do, sir.

Q—Is there any scientists in the world you think much of?

A—I do.

Q—Who?

A—Well, I think the bulk of the scientists—

Q—I don't want that kind of an answer, Mr. Bryan, who are they?

A—I will give you George M. Price, for instance.

Q—Who is he?

A—Professor of geology in a college.

Q—Where?

A—He was out near Lincoln, Neb.

Q—How close to Lincoln Neb.?

A—About three or four miles. He is now in a college out in California.

Q—Where is the college?

A—At Lodi.

Q—That is a small college?

A—I didn't know you had to judge a man by the size of the college—I thought you judged him by the size of the man.

Q—I thought the size of the college made some difference?

A—It might raise a presumption in the minds of some, but I think I would rather find out what be believed.

Q—You would rather find out whether his belief corresponds with your views or prejudices or whatever they are before you said how good he was?

A—Well, you know the word "prejudice" is—

Q—Well, belief, then.

A—I don't think I am any more prejudiced for the Bible than you are against it.

Q—Well, I don't know?

A—Well, I don't know either, it is my guess.

Q—You mentioned Price because he is the only human being in the world so far as you know that signs his name as a geologist that believes like you do?

A—No, there is a man named Wright, who taught at Oberlin.

Q—I will get to Mr. Wright in a moment. Who publishes his book?

A—I can't tell you. I can get you the book.

Q—Don't you know? Don't you know it is Revell & Co., Chicago?

A—I couldn't say.

Q—He publishes yours, doesn't he?

A—Yes, sir.

Gen. Stewart—Will you let me make an exception. I don't think it is pertinent about who publishes a book.

Says Bryan Quotes Montebank.

Mr. Darrow—He has quoted a man that every scientist in this country knows is a montebank and a pretender and not a geologist at all.

The Court—You can ask him about the man, but don't ask him about who publishes the book.

Q—Do you know anything about the college he is in?

A—No, I can't tell you.

Q—Do you know how old his book is?

A—No, sir, it is a recent book.

Q—Do you know anything about his training?

A—No, I can't say on that.

Q—Do you know of any geologist on the face of the earth who ever recognized him?

A—I couldn't say.

Q—Do you think he is all right? How old does he say the earth is?

A—I am not sure that I would insist on some particular geologist that you picked out recognizing him before I would consider him worthy if he agreed with your views?

Q—You would consider him worthy if he agreed with your views?

A—Well, I think his argument is very good.

Q—How old does Mr. Price say the earth is?

A—I haven't examined the book in order to answer questions on it.

Q—Then you don't know anything about how old he says it is?

A—He speaks of the layers that are supposed to measure age and points out that they are not uniform and not always the same and that attempts to measure age by those layers where they are not in the order in which they are usually found makes it difficult to fix the exact age.

Q—Does he say anything whatever about the age of the earth?

A—I wouldn't be able to testify.

Q—You didn't get anything about the age from him?

A—Well, I know he disputes what you say and has very good evidence to dispute it—what some others say about the age.

Q—Where did you get your information about the age of the earth?

A—I am not attempting to give you information about the age of the earth.

Wright of Oberlin?

Q—Then you say there was Mr. Wright, of Oberlin?

A—That was rather I think on the age of man than upon the age of the earth.

Q—There are two Mr. Wrights, of Oberlin?

A—I couldn't say.

Q—Both of them geologists. Do you know how long Mr. Wright says man has been on the earth?

A—Well, he gives the estimates of different people.

Q—Does he give any opinion of his own?

A—I think he does.

Q—What is it?

A—I am not sure.

Q—What is it?

A—It was based upon the last glacial age—that man has appeared since the last glacial age.

Q—Did he say there was no man on earth before the last glacial age?

A—I think he disputes the finding of any proof—where the proof is authentic—but I had rather read him than quote him. I don't like to run the risk of quoting from memory.

Q—You couldn't say then how long Mr. Wright places it?

A—I don't attempt to tell you.

Q—When was the last glacial age?

A—I wouldn't attempt to tell you that.

Q—Have you any idea?

A—I wouldn't want to fix it without looking at some of the figures.

Q—That was since the tower of Babel, wasn't it?

A—Well, I wouldn't want to fix it. I think it was before the time given in here, and that was only given as the possible appearance of man and not the actual.

Q—Have you any idea how far back the last glacial age was?

A—No, sir.

Q—Do you know whether it was more than 6,000 years ago?

A—I think it was more than 6,000 years.

Q—Have you any idea how old the earth is?

A—No.

Bible Gives Age of Earth?

Q—The book you have introduced in evidence tells you, doesn't it?

A—I don't think it does, Mr. Darrow.

Q—Let's see whether it does; is this the one?

A—That is the one, I think.

Q—It says B. C. 4004?

A—That is Bishop Usher's calculation.

Q—That is printed in the Bible you introduced?

A—Yes, sir.

Q—And numerous other Bibles?

A—Yes, sir.

Q—Printed in the Bible in general use in Tennessee?

A—I couldn't say.

Q—And Scofield's Bible?

A—I couldn't say about that.

Q—You have seen it somewhere else?

A—I think that is the chronology usually used.

Q—Does the Bible you have introduced for the jury's consideration say that?

A—Well, you will have to ask those who introduced that.

Q—You haven't practiced law for a long time, so I will ask you if that is the King James version that was introduced? That is your marking, and I assume it is?

A—I think that is the same one.

Mr. Darrow—There is no doubt about it, is there, gentlemen?

Mr. Stewart—That is the same one.

Q—Would you say that the earth was only 4,000 years old?

A—Oh, no; I think it is much older than that.

Q—How much?

A—I couldn't say.

Q—Do you say whether the Bible itself says it is older than that?

A—I don't think the Bible says itself whether it is older or not.

Q—Do you think the earth was made in six days?

Bryan—"Not Six Days of Twenty-four Hours."

A—Not six days of twenty-four hours.

Q—Doesn't it say so?

A—No, sir.

Gen. Stewart—I want to interpose another objection. What is the purpose of this examination?

Mr. Bryan—The purpose is to cast ridicule on everybody who believes in the Bible, and I am perfectly willing that the world shall know that these gentlemen have no other purpose than ridiculing every Christian who believes in the Bible.

Mr. Darrow—We have the purpose of preventing bigots and ignoramuses from controlling the education of the United States and you know it, and that is all.

Mr. Bryan—I am glad to bring out that statement. I want the world to know that this evidence is not for the view Mr. Darrow and his associates have filed affidavits here stating, the purposes of which I understand it, is to show that the Bible story is not true.

Mr. Malone—Mr. Bryan seems anxious to get some evidence in the record that would tend to show that those affidavits are not true.

Bryan Wants World to Know He Is Not Afraid.

Mr. Bryan—I am not trying to get anything into the record. I am simply trying to protect the word of God against the greatest atheist or agnostic in the United States. (Prolonged applause.) I want the papers to know I am not afraid to get on the stand in front of him and let him do his worst. I want the world to know. (Prolonged applause.)

Mr. Darrow—I wish I could get a picture of these clackers.

Gen. Stewart—I am not afraid of Mr. Bryan being perfectly able to take care of himself, but this examination cannot be a legal examination and it cannot be worth a thing in the world, and, your honor, I respectfully except to it, and call on your honor, in the name of all that is legal, to stop this examination and stop it here.

Mr. Hays—I rather sympathize with the general, but Mr. Bryan is produced as a witness because he is a student of the Bible and he presumably understands what the Bible means. He is one of the foremost students in the United States, and we hope to show Mr. Bryan, who is a student of the Bible, what the Bible really means in connection with evolution. Mr. Bryan has already stated that the world is not merely 6,000 years old and that is very helpful to us, and where your evidence is coming from, this Bible, which goes to the jury, is that the world started in 4004 B. C.

Mr. Bryan—You think the Bible says that?

Mr. Hays—The one you have taken in evidence says that.

Mr. Bryan—I don't concede that it does.

Mr. Hays—You know that that chronology is made up by adding together all of the ages of the people in the Bible, counting their ages; and now then, let us show the next stage from a Bible student, that these things are not to be taken literally, but that each man is entitled to his own interpretaton.

Gen. Stewart—The court makes the interpretation.

Mr. Hays—But the court is entitled to information on what is the interpretation of an expert Bible student.

Stewart Bitterly Opposes Procedings.

Gen. Stewart—This is resulting in a harangue and nothing else.

Mr. Darrow—I didn't do any of the haranging; Mr. Bryan has been doing that.

Gen. Stewart—You know absolutely you have done it.

Mr. Darrow—Oh, all right.

Mr. Malone—Mr. Bryan doesn't need any support.

Gen. Stewart—Certainly he doesn't need any support, but I am doing what I conceive my duty to be, and I don't need any advice, if you please, sir. (Applause.)

The Court—That would be irrelevant testimony if it was going to the jury. Of course, it is excluded from the jury on the point it is not competent testimony, on the same ground as the affidaviting.

Mr. Hicks—Your honor, let me say a word right there. It is in the discretion of the court how long you will allow them to question witnesses for the purpose of taking testimony to the supreme court. Now, we as taxpayers of this county, feel that this has gone beyond reason.

The Court—Well, now, that taxpayers doesn't appeal to me so much, when it is only fifteen or twenty minutes time.

Mr. Darrow—I would have been through in a half-hour if Mr. Bryan had answered my questions.

Gen. Stewart—They want to put in affidavits as to what other witnesses would swear, why not let them put in affidavits as to what Mr. Bryan would swear?

Mr. Bryan—God forbid.

Mr. Malone—I will just make this suggestion—

Gen. Stewart—It is not worth anything to them, if your honor please, even for the record in the supreme court.

Mr. Hays—Is not it worth anything to us if Mr. Bryan will accept the story of creation in detail, and if Mr. Bryan, as a Bible student, states you cannot take the Bible necessarily as literally true?

Mr. Stewart—The Bible speaks for itself.

Mr. Hays—You mean to say the Bible itself tells whether these are parables? Does it?

Gen. Stewart—We have left all annals of procedure behind. This is a harangue between Col. Darrow and his witness. He makes so many statements that he is forced to defend himself.

Mr. Darrow—I do not do that.

Gen. Stewart—I except to that as not pertinent to this lawsuit.

The Court—Of course, it is not pertinent, or it would be before the jury.

Gen. Stewart—It is not worth anything before a jury.

The Court—Are you about through, Mr. Darrow?

Mr. Darrow—I want to ask a few more questions about the creation.

The Court—I know. We are going to adjourn when Mr. Bryan comes off the stand for the day. Be very brief, Mr. Darrow. Of course, I believe I will make myself clearer. Of course, it is incompetent testimony before the jury. The only reason I am allowing this to go in at all is that they may have it in the appellate courts, as showing what the affidavit would be.

Bryan Insists He Is Not Afraid of Agnostics or Atheists.

Mr. Bryan—The reason I am answering is not for the benefit of the superior court. It is to keep these gentlemen from saying I was afraid to meet them and let them question me, and I want the Christian world to know that any atheist, agnostic, unbeliever, can question me any time as to my belief in God, and I will answer him.

Mr. Darrow—I want to take an exception to this conduct of this witness. He may be very popular down here in the hills. I do not need to have his explanation for his answer.

The Court—Yes.

Mr. Bryan—If I had not, I would not have answered the question.

Mr. Hays—May I be heard? I do not want your honor to think we are asking questions of Mr. Bryan with the expectation that the higher court will not say that those questions are proper testimony. The reason I state that is this, your law speaks for the Bible. Your law does not say the literal interpretation of the Bible. If Mr. Bryan, who is a student of the Bible, will state that everything in the Bible need not be interpreted literally, that each man must judge for himself; if he will state that, of course, then your honor would charge the jury. We are not bound by a literal interpretation of the Bible. If I have made my argument clear enough for the attorney-general to understand, I will retire.

Gen. Stewart—I will admit you have frequently been difficult of comprehension, and I think you are as much to blame as I am.

Mr. Hays—I know I am.

Gen. Stewart—I think this is not legal evidence for the record in the appellate courts. King James' versions of the Bible, as your honor says—

The Court—I cannot say that.

Gen. Stewart—Your honor has held the court takes judicial knowledge of King James' version of the Bible.

The Court—No, sir; I did not do that.

Gen. Stewart—Your honor charged the grand jury and read from that.

The Court—I happened to have the Bible in my hand, it happened to be a King James' edition, but I will charge the jury, gentlemen, the Bible generally used in Tennessee, as the book ordinarily understood in Tennessee, as the Bible, I do not think it is proper for us to say to the jury what Bible.

Gen. Stewart—Of course, that is all we could ask of your honor. This investigation or interrogation of Mr. Bryan as a witness, Mr. Bryan is called to testify, was of the counsel for the prosecution in this case, and has been asked something, perhaps less than a thousand questions, of course, not personal to this case, and it has resulted in an argument, and argument about every other question cannot be avoided. I submit your honor, it is not worth anything in the record at all, if it is not legal testimony. Mr. Bryan is willing to testify and is able to defend himself. I accept it, if the court please, and ask your honor to stop it.

Mr. Hays—May I ask a question? If your contention is correct that this law does not necessarily mean that the Bible is to be taken literally, word for word, is not this competent evidence?

Gen. Stewart—Why could you not prove it by your scientists?

Mr. Darrow—We are calling one of the most foremost Bible students. You vouch for him.

Mr. Malone—We are offering the best evidence.

Gen. McKenzie—Do you think this evidence is competent before a jury?

Mr. Darrow—I think so.

The Court—It is not competent evidence for the jury.

Gen. McKenzie—Nor is it competent in the appellate courts, and these gentlemen would no more file the testimony of Col. Bryan as a part of the record in this case than they would file a rattlesnake and handle it themselves.

Messrs. Darrow, Hays and Malone (In Unison)—We will file it. We will file it. File every word of it.

Mr. Bryan—Your honor, they have not asked a question legally, and the only reason they have asked any question is for the purpose, as the question about Jonah was asked, for a chance to give this agnostic an opportunity to criticize a believer in the word of God; and I answered the question in order to shut his mouth so that he cannot go out and tell his atheistic friends that I would not answer his question. That is the only reason, no more reason in the world.

Mr. Malone—Your honor on this very subject, I would like to say that I would have asked Mr. Bryan —and I consider myself as good a Christian as he is—every question that Mr. Darrow has asked him for the purpose of bringing out whether or not there is to be taken in this court only a literal interpretation of the Bible, or whether, obviously, as these questions indicate, if a general and literal construction cannot be put upon the parts of the Bible which have been covered by Mr. Darrow's questions. I hope for the last time no further attempt will be made by counsel on the other side of the case, or Mr. Bryan, to say the defense is concerned at all with Mr. Darrow's particular religious views or lack of religious views. We are here as lawyers with the same right to our views. I have the same right to mine as a Christian as Mr. Bryan has to his, and we do not intend to have this case charged by Mr. Darrow's agnosticism or Mr. Bryan's brand of Christianity. (A great applause.)

The Court—I will pass on each question as asked, if it is objected to.

Mr. Darrow:

Q—Mr. Bryan, do you believe that the first woman was Eve?

A—Yes.

Q—Do you believe she was literally made out of Adam's rib?

A—I do.

Where Did Cain Get His Wife?

Q—Did you ever discover where Cain got his wife?

A—No, sir; I leave the agnostics to hunt for her.

Q—You have never found out?

A—I have never tried to find.

Q—You have never tried to find?

A—No.

Q—The Bible says he got one, doesn't it? Were there other people on the earth at that time?

A—I cannot say.

Q—You cannot say. Did that ever enter your consideration?

A—Never bothered me.

Q—There were no others recorded, but Cain got a wife.

A—That is what the Bible says.

Q—Where she came from you do not know. All right. Does the statement, "The morning and the evening were the first day," and "The morning and the evening were the second day," mean anything to you?

A—I do not think it necessarily means a twenty-four-hour day.

Q—You do not?

A—No.

Q—What do you consider it to be?

A—I have not attempted to explain it. If you will take the second chapter—let me have the book. (Examining Bible.) The fourth verse of the second chapter says: "These are the generations of the heavens and of the earth, when they were created in the day that the Lord God made the earth and the heavens," the word "day" there in the very next chapter is used to describe a period. I do not see that there is any necessity for construing the words, "the evening and the morning," as meaning necessarily a twenty-four-hour day, "in the day when the Lord made the heaven and the earth."

Q—Then, when the Bible said, for instance, "and God called the firmament heaven. And the evening and the morning were the second day," that does not necessarily mean twenty-four hours?

A—I do not think it necessarily does.

Q—Do you think it does or does not?

A—I know a great many think so.

Q—What do you think?

A—I do not think it does.

Q—You think those were not literal days?

A—I do not think they were twenty-four-hour days.

Q—What do you think about it?

A—That is my opinion—I do not know that my opinion is better on that subject than those who think it does.

Q—You do not think that?

A—No. But I think it would be just as easy for the kind of God we believe in to make the earth in six days as in six years or in 6,000,000 years or in 600,000,000 years. I do not think it important whether we believe one or the other.

Q—Do you think those were literal days?

A—My impression is they were periods, but I would not attempt to argue as against anybody who wanted to believe in literal days.

Q—Have you any idea of the length of the periods?

A—No; I don't.

Q—Do you think the sun was made on the fourth day?

A—Yes.

Q—And they had evening and morning without the sun?

A—I am simply saying it is a period.

Q—They had evening and morning for four periods without the sun, do you think?

A—I believe in creation as there told, and if I am not able to explain it I will accept it. Then you can explain it to suit yourself.

Q—Mr. Bryan, what I want to know is, do you believe the sun was made on the fourth day?

A—I believe just as it says there.

Q—Do you believe the sun was made on the fourth day?

A—Read it.

Q—I am very sorry; you have read it so many times you would know, but I will read it again: "And God, said, let there be lights in the firmament of the heaven, to divide the day from the night; and let them be for signs, and for seasons, and for days, and years.

"And let them be for lights in the firmament of the heaven, to give light upon the earth; and it was so.

"And God made two great lights; the greater light to rule the day, and the lesser light to rule the night; He made the stars also.

"And God set them in the firmament of the heaven, to give light upon the earth, and to rule over the day and over the night, and to divide the light from the darkness; and God saw that it was good. And the evening and the morning were the fourth day."

Do you believe, whether it was a literal day or a period, the sun and the moon were not made until the fourth day?

A—I believe they were made in the order in which they were given there, and I think in dispute with Gladstone and Huxley on that point—

Q—Cannot you answer my question?

A— ——I prefer to agree with Gladstone.

Q—I do not care about Gladstone.

A—Then prefer to agree with whoever you please.

Q—Can not you answer my question?

A—I have answered it. I believe that it was made on the fourth day, in the fourth day.

Q—And they had the evening and the morning before that time for three days or three periods. All right, that settles it. Now, if you call those periods, they may have been a very long time.

A—They might have been.

Q—The creation might have been going on for a very long time?

A—It might have continued for millions of years.

Eve and the Serpent.

Q—Yes. All right. Do you believe the story of the temptation of Eve by the serpent?

A—I do.

Q—Do you believe that after Eve ate the apple, or gave it to Adam, whichever way it was, that God cursed Eve, and at that time decreed that all womankind thenceforth and forever should suffer the pains of childbirth in the reproduction of the earth?

A—I believe what it says, and I believe the fact as fully——

Q—That is what it says, doesn't it?

A—Yes.

Q—And for that reason, every woman born of woman, who has to carry on the race, the reason they have childbirth pains is because Eve tempted Adam in the Garden of Eden?

A—I will believe just what the Bible says. I ask to put that in the language of the Bible, for I prefer that to your language. Read the Bible and I will answer.

Q—All right, I will do that: "And I will put enmity between thee and the woman"—that is referring to the serpent?

A—The serpent.

Q—(Reading) "and between thy seed and her seed; it shall bruise thy head, and thou shalt bruise his heel. Unto the woman he said, I will greatly multiply thy sorrow and thy conception; in sorrow thou shalt bring forth children; and thy desire shall be to thy husband, and he shall rule over thee." That is right, is it?

A—I accept it as it is.

Q—And you believe that came about because Eve tempted Adam to eat the fruit?

A—Just as it says.

Q—And you believe that is the reason that God made the serpent to go on his belly after he tempted Eve?

Bryan Insists on Bible Being Quoted Verbatim.

A—I believe the Bible as it is, and I do not permit you to put your language in the place of the language of the Almighty. You read that Bible and ask me questions, and I will answer them. I will not answer your questions in your language.

Q—I will read it to you from the Bible: "And the Lord God said unto the serpent, because thou hast done this, thou art cursed above all cattle, and above every beast of the field; upon thy belly shalt thou go and dust shalt thou eat all the days of thy life." Do you think that is why the serpent is compelled to crawl upon its belly?

A—I believe that.

Q—Have you any idea how the snake went before that time?

A—No, sir.

Q—Do you know whether he walked on his tail or not?

A—No, sir. I have no way to know. (Laughter in audience).

Q—Now, you refer to the cloud that was put in the heaven after the flood, the rainbow. Do you believe in that?

A—Read it.

Q—All right, Mr. Bryan, I will read it for you.

Mr. Bryan—Your honor, I think I can shorten this testimony. The only purpose Mr. Darrow has is to slur at the Bible, but I will answer his question. I will answer it all at once, and I have no objection in the world, I want the world to know that this man, who does not believe in a God, is trying to use a court in Tennessee—

Mr. Darrow—I object to that.

Mr. Bryan—(Continuing) to slur at it, and while it will require time, I am willing to take it.

Mr. Darrow—I object to your statement. I am exempting you on your fool ideas that no intelligent Christian on earth believes.

The Court—Court is adjourned until 9 o'clock tomorrow morning.

CHAPTER VIII.

EIGHTH DAY OF EVOLUTION TRIAL—TUESDAY, JULY 21, 1925.

Court met pursuant to adjournment.

The Bailiff—Is the Rev. Dr. R. C. Camper in the house? (the Rev. Dr. R. C. Camper, of Chattanooga.)

The Court—Let everyone stand up. Dr. Camper will open court with prayer.

Dr. Camper—Oh God, our Heavenly Father, we come into Thy presence this morning, feeling our dependence upon Thee. We pray Thy blessings upon each one that has a part in this court here today. Bless the judge, bless each lawyer, bless each one, Lord, that has a part, and may each and every one do the thing that is good and right here today. Guide us in everything we undertake for good here in this life. We ask it in the name of Jesus Christ. Amen.

The Court—Open court, Mr. Sheriff.

The Bailiff—Oyez, oyez, this honorable circuit court is now open pursuant to adjournment. Sit down.

Judge Not Fast—Just Watch.

Officer Kelso Rice—We opened a little earlier on account of the judge's watch and we are now waiting on counsel. The judge isn't fast, I think it is just his watch. (A short recess was taken) whereupon:

Bryan's Testimony Struck From Records.

The Court—Let's have order. Since the beginning of this trial the judge of this court has had some big problems to pass upon. Of course, there is no way for me to know whether I decided these questions correctly or not until the courts of last resort speak. If I have made a mistake it was a mistake of the head and not the heart. There are two things that may lead a judge into error. One is prejudice and passion, another is an over-zeal to be absolutely fair to all parties. I fear that I may have committed error on yesterday in my over-zeal to ascertain if there was anything in the proof that was offered that might aid the higher courts in determining whether or not I had committed error in my former decrees. I have no disposition to protect any decree that I make from being reversed by a higher court, because, if I am in error, I hope to God that somebody will correct my mistake. I feel that the testimony of Mr. Bryan can shed no light upon any issues that will be pending before the higher courts. The lawsuit now is whether—the issue now is whether or not Mr. Scopes taught that man descended from a lower order of animals. It isn't a question of whether God created man as all complete at once, or it isn't a question as to whether God created man by the process of development and growth. These questions have been eliminated from this court and the only question we have now is whether or not this teacher, this accused, this defendant, taught that man descended from a lower order of animals. As I see it, after due deliberation, I feel that Mr. Bryan's testimony cannot aid the higher court in determining that question. If the question before the higher court involved the issue as to what evolution was or as to how God created man, or created the earth, or created the universe, this testimony might be relevant, but those questions are not before the court, and so taking this view of it, I am pleased to expunge this testimony, given by Mr. Bryan on yesterday, from the records of this court and it will not be further considered.

Mr. Darrow—If your honor, please, we want before this is disposed of—we would like to be heard. I want to say a word, if you please. Of course, I am not at all sure that Mr. Bryan's testimony would aid the supreme court, or any other human being, but he testified by the hour there and I haven't got through with him yet.

Mr. Stewart—I understand the court has ruled on this and I think it is entirely out of order unless you are making an exception.

Mr. Darrow—I want to make my exception.

Gen. Stewart—Make your exception and don't begin an argument about it.

The Court—Confine your remarks to the exception, Mr. Darrow, please.

Mr. Darrow—I will, your honor. I want to except to the ruling of the court and so I might understand perfectly, does your honor mean this will not be certified to in a bill of exceptions containing Mr. Bryan's testimony?

The Court—That is what I mean. I mean I will strike it from the record.

Mr. Darrow—We want to take an exception to that. Of course, when we make up the bill of exceptions we will ask to have it included.

The Court—Yes, sir.

Mr. Darrow—I suppose the only remedy we could have if your honor holds that way is to have the writ sent down if we want it in the record.

The Court—Yes, sir, a writ from the higher court to have it certified up.

Defense Has No Witnesses and No Proof—Asks Verdict of Guilty.

Mr. Darrow—Let me suggest this. We have all been here quite a while and I say it in perfectly good faith, we have no witnesses to offer, no proof to offer on the issues that the court has laid down here, that Mr. Scopes did teach what the children said he taught, that man descended from a lower order of animals—we do not mean to contradict that, and I think to save time we will ask the court to bring in the jury and instruct the jury to find the defendant guilty. We make no objection to that and it will save a lot of time and I think that should be done.

Gen. Stewart—Your honor, may I suggest that the court has not been formerly opened yet.

The Court—Yes, sir, it was before you came in.

Gen. Stewart—Well, I thought your honor came in with us.

The Court—No, sir.

Gen. Stewart—I thought that it might just have been an oversight.

The Court—I thank the attorney-general.

Gen. Stewart—We are pleased to accept the suggestion of Mr. Darrow.

Mr. Hays—Before we do that, may I get my record straight on the offer of proof by having the court rule on this offer of proof and permit me to take an exception? We have offered to prove what was said yesterday, where we have filed statements which were scientific testimony, as well as Biblical testimony, I assume from what your honor has said that you deny us the right to put in that evidence.

The Court—Yes, sir.

Mr. Hays—We take an exception. We further offer to prove—we have offered proof by Biblical students whose statements I read, the real meaning of the Bible and translations into the Bible. We offer to prove by Mr. Bryan that the Bible was not to be taken literally, that the earth was 1,000,000 years old and we had hoped to prove by him further that nothing in the Bible said what the processes were of man's creation. We feel that the statement that the earth was 1,000,000 years old and nothing said about the processes of man's creation that it was perfectly clear that what Scopes taught would not violate the first part of the act. I assume from what your honor said that that is not permitted.

The Court—No, sir.

Mr. Hays—Your honor will permit me an exception?

The Court—Yes, sir.

Mr. Hays—In order to be perfectly clear the evidence that I offer is the evidence of Maynard M. Metcalf, Jacob G. Lipham, Wilbur Nelson,

Dr. Fay-Cooper Cole, Dr. H. H. Newman, Dr. Winterton C. Curtis, Dr. Kirtley F. Mather and proof by Biblical scholars, Dr. Rabbi Rosenwasser and Dr. Whitaker. The proof is in the form of statements that were made, or parts of those statements and I understand then your honor rules out of evidence each and every one of those statements and all and every part of any and all of these statements?

The Court—Yes, sir.

Gen. Stewart—And I understand the court excludes from the record the testimony of Mr. Bryan?

The Court—Yes, I excluded that from the jury.

Mr. Hays—May I have on the record also the statement, that doesn't bear on the issue that has been set down by the court? We would offer to prove it if the issue had been different—we would offer to prove by Mr. White that Mr. Scopes had a contract from Sept. 1 until May 1, at $150 a month, to teach biology in the public schools and that under the law he was obliged to teach biology from the book that was provided by the public schools. I understand then your honor likewise excluded that evidence, because that doesn't bear on the issue that you stated?

The Court—Well, it hasn't been offered.

Mr. Hays—It doesn't bear on the issue stated.

Gen. Stewart—I don't think really there will be any objection to it going in the record. Our view of it is that the law goes into effect from and after its passage and our contention is that it violates the law at any time after the act is passed, which was March 21.

Col. Bryan—May it please the court.

The Court—I will hear you, Mr. Bryan.

Mr. Bryan—At the conclusion of your decision to expunge the testimony given by me upon the record I didn't have time to ask you a question. I fully agree with the court that the testimony taken yesterday was not legitimate or proper. I simply wanted the court to understand that I was not in position to raise an objection at that time myself nor was I willing to have it raised for me without asserting my willingness to be cross-examined. I also stated that if I was to take the witness stand I would ask that the others take the witness stand also, that I might put certain questions to them. Now the testimony was ended and I assume that you expunge the questions as well as the answers.

The Court—Yes, sir.

Mr. Bryan—That it isn't a reflection upon the answers any more than it is upon the questions.

The Court—I expunged the whole proceedings.

Mr. Bryan—Now I had not reached the point where I could make a statement to answer the charges made by the counsel for the defense as to my ignorance and my bigotry.

Mr. Darrow—I object, your honor, now what's all this about.

The Court—Why do you want to make this, Col. Bryan?

Mr. Bryan—I just want to finish my sentence.

Mr. Darrow—Why can't he go outside on the lawn?

Mr. Bryan—I am not asking to make a statement here.

The Court—I will hear what you say.

Bryan Appeals to Justice of Press.

Mr. Bryan—I shall have to trust to the justness of the press, which reported what was said yesterday, to report what I will say, not to the court, but to the press in answer to the charge scattered broadcast over the world and I shall also avail myself of the opportunity to give to the press, not to the court, the questions that I would have asked had I been permitted to call the attorneys on the other side.

Mr. Darrow—I think it would be better, Mr. Bryan, for you to take us out also with the press and ask us the questions and then the press will have both the questions and the answers.

Mr. Bryan—The gentleman who represents the defense, not only differs from me, but he differs from the court very often in the manner of procedure. I simply want to make that statement and say that I shall have to avail myself of the press without having the dignity of its being presented in the court, but I think it is hardly fair for them to bring into the limelight my views on religion and stand behind a dark lantern that throws light on other people, but conceals themselves. I think it is only fair that the country should know the religious attitude of the people who come down here to deprive the people of Tennessee of the right to run their own schools.

Mr. Darrow—I object to that.

The Court—I overrule the objection.

Mr. Bryan—That is all.

Malone Says Will Answer Bryan's Questions?

Mr. Malone—If your honor pleases, I wish to make a statement, if statements are in order. The attorneys for the defense are hiding behind no screen of any kind. They will be very happy at any time in any forum to answer any questions which Mr. Bryan can ask along the lines that were asked him yesterday, if they be ger—

Gen. Stewart—Permit me to suggest—

The Court—All right.

Gen. Stewart—I think the next thing in order is to bring the jury in and charge the jury.

Gen. McKenzie (B. G.)—I suggest that the distinguished gentlemen get together with Col. Bryan; they are all anxious to hear him—and that they have a crowd and have a joint discussion and by that means your views will be reflected.

Mr. Malone—We are not worried about our views. We are in a court of law and our discussion is ended. We are ready for the jury.

Thereupon the jury was brought in and took their seats in the jury box.

And thereupon the following discussion occurred before the court, out of the hearing of the jury and the spectators.

Mr. Darrow—My statement that there was no need to try this case further, and for the court to instruct that the defendant is guilty under the law was not made as a plea of guilty or an admission of guilt. We claim that the defendant is not guilty, but as the court has excluded any testimony, except as to the one issue as to whether he taught that man descended from a lower order of animals, and we cannot contradict that testimony, therefore, there is no logical thing to come except that the jury find a verdict that we may carry to the higher court, purely as a matter of proper procedure. We do not think it is fair to the court or counsel on the other side to waste a lot of time when we know this is the inevitable result and probably the best result for the case. I think that is all right?

Gen. Stewart—I think so; yes.

The Court—You want the jury charged, the regular formal charge?

Mr. Darrow—The general suggested something else that might take its place.

Agree on a Verdict of Guilty.

Gen. Stewart—We suggest—It is, of course, agreed by all that what we want and what we want to get is the case into the appellate court to test the act properly. It was suggested merely to make the record show a verdict of guilty, to show that this jury brought in a verdict of guilty. But, I think the best way to proceed would be to let his honor charge the jury and submit it to them, and I do not think there would be anything improper for you or me to state to the jury, after the jury has been charged—you state to them, if you want to, that you do not object to a verdict of guilty, to be frank about it, what you want is for the case to go before the appellate court. I do not think there will be any exception to that?

Mr. Darrow—You agree with me for the record—you agree then, that if the question might arise here, you will help us see that this case comes before the supreme court?

Gen. Stewart—Yes; anything I can do after you get the record to the supreme court.

Mr. Hays—What about thirty days?

Mr. Darrow—Don't bother about that now.

Gen. Stewart—We want to take the case to the supreme court the first Monday in September in Knoxville. I think you can do that—

Mr. Darrow—What?

Gen. Stewart—You have a daily transcript, that was the purpose of calling the special term. Otherwise it would have to wait, if it don't get to the September term, as we only have one term a year.

Thereupon counsel left the bar of the court and returned to their respective seats.

The Court—Have you gentlemen any statements to make?

Gen. Stewart—We want your honor to proceed to charge the jury.

Mr. Darrow—As long as it is agreed we don't need to talk any longer.

The Court—I suggest no exceptions will be made to my charge, is that true?

Mr. Darrow—I mean on account of not covering other matters.

The Court—I will put it in the formal manner. I will dictate my charge, and it will have to be copied so the court will be at ease a little while.

Thereupon the court left the bench and a short recess ensued, after which the following proceedings occurred, to-wit:

Judge Raulston Charges the Jury.

Gentlemen of the Jury:

This is a case of the State of Tennessee vs. John Thomas Scopes, where it is charged that the accused violated what is commonly known as the antievolution statute, the same being chapter 27 of the acts of the legislature of 1925, the statute providing that it shall be unlawful for any person to teach in any of the universities, normals or other public schools of the state any theory that denies the story of the divine creation of man, as taught in the Bible, and teach instead thereof that man is descended from a lower order of animals.

The indictment in this case is dated at the July special term, 1925, and, in part, charges that John Thomas Scopes, heretofore on the 24th day of April, 1925, did unlawfully teach in the public schools of Rhea county, Tennessee, which said public schools are supported in part and in whole by the public school fund of the state, a certain theory and theories that denied the story of the divine creation of man as taught in the Bible, and did teach instead thereof that man is descended from a lower order of animals, he, the said John Thomas Scopes, being at the time and prior thereto, a teacher in the public school of Dayton, in the county aforesaid, against the peace and dignity of the state.

To this charge the defendant has pleaded not guilty and thus are made up the issue for your determination. Before there can be a conviction the state must make out its case beyond a reasonable doubt as to every essential and necessary element of the case. The court calls the attention of the jury to the wording of the indictment, wherein it is charged that this defendant taught a certain theory or theories that denied the story of the divine creation of man as taught in the Bible, and taught instead thereof that man descended from a lower order of animals. This statute has been before the court during the hearing in this case, upon a motion which made it necessary that the court should construe the statute as to what offense was provided against therein. The court, after due consideration, has held that the proper construction of the statute is that it is made an offense thereby to teach in the public schools of the state of Tennessee which are supported in whole or in part by the public school fund or the state, that man descended from a lower order of animals. In other words, the second clause is explanatory of the first, and interprets the meaning of the legislature; and the court charges you that in order to prove its case the state does not have to specifically prove that the defendant taught a theory that denied the story of the divine creation of man as taught in the Bible, other than to prove that he taught that man descended from a lower order of animals. Therefore, the court charges you that if you find that the proof in this case shows that the defendant did teach in the public schools of Rhea county, the same being supported in whole or in part by the public school fund, subsequent to the passage of this statute, and prior to the finding of this indictment, that man descended from a lower order of animals, and if these facts are shown beyond a reasonable doubt, then the defendant would be guilty and should be so found, and you are not concerned as to whether or not this is a theory denying the story of the divine creation of man as taught, for the issues as they have been finally made up in this case do not involve that question.

By the phrase "beyond a reasonable doubt," I do not mean any possible doubt that might arise, or such a doubt as an ingenuous mind might conjure up, but by reasonable doubt in legal parlance is meant such a doubt as would prevent your mind resting easy as to the guilt of the defendant.

In determining whether or not his guilt is shown beyond a reasonable doubt you must weigh and consider the evidence, and in doing that you would look to the demeanor of the witnesses on the stand, their opportunities to know the facts concerning which they testify, their respectability or want of respectability if such appears, their interest in the result of the lawsuit or want of interest; their bias, prejudice or leaning to one side or the other, if such appears. Their relationship to any of the parties, and all other facts that might enable you to determine what weight should be given their testimony.

You, gentlemen, are the sole and exclusive judges of the facts and the credibility of the witnesses, and judges of the law under the direction of the court.

You enter upon this investigation with the presumption that the defendant is not guilty of any offense, and this presumption stands as a witness for him until it is overcome by competent and credible proof.

There are different methods by which witnesses are impeached. One is by showing that they are unworthy of belief, by those who know them best; another method is by showing that a witness has made contradictory statements as to material facts involved in the case, concerning which he gave testimony. Another is to involve the witness in discrepancies upon the witness stand, by rigid and close cross-examination.

When a witness is once impeached, he stands throughout the trial, but this does not mean that he did not swear the truth. This is a matter for you to determine, but the impeaching process is a circumstance which you will take into consideration in determining what weight you will give this testimony.

If there are conflicts in the statements of the different witnesses, it is your duty to reconcile them if you can, for the law presumes that each witness has sworn the truth. But if you cannot reconcile their testimony, the law makes you the sole and exclusive judges of the credibility of the witnesses and the weight to be given their testimony.

In this case the defendant did not go on the stand. Under our construction and laws he has the right to either testify or not to testify as he sees proper, and his failure to testify creates no presumption of his guilt, but should be considered for no purpose in determining whether or not he is guilty.

Under the provision of the statute in this case, a person who violates the same may be punished by a fine of not less than $100 nor more than $500. If after a fair and honest investigation of all the facts you find the defendant guilty and find that his offense deserves a greater punishment than a fine of $100, then you must impose a fine not to exceed $500 in any event. But if you are content with a $100 fine, then you may simply find the defendant guilty and leave the punishment to the court.

But if the proof fails to show his guilt beyond a reasonable doubt, you should acquit the defendant and your verdict should be not guilty.

Under our constitution and laws the jury can have no prejudice or bias either way, but you should search for and find the truth, and the truth alone, and bring into this court such a verdict you think truth dictates and justice demands.

JOHN T. RAULSTON.

The Court—Any requests.

Mr. Darrow—Your honor, do we have to take exceptions at the time of the charge?

The Court—If you want additional instructions given.

Mr. Darrow—No, I do not, your honor. The only thing is matters you have already passed on as to what the law requires.

The Court—Just on the legal points?

Mr. Darrow—Yes, just on the legal points.

The Court—No, the law imposes on the court the duty to charge the law correctly. You do not have to make exceptions at this time.

Mr. Darrow—In our federal court we have to make them at the time.

The Court—Yes, call them to the judge's attention so as to give him a chance. I wish that was the practice here.

Mr. Darrow—Yes.

The Court—Anything, Mr. Attorney-General?

Gen. Stewart—I think Mr. Darrow has something to say.

Mr. Darrow—May I say a few words to the jury? Gentlemen of the jury, we are sorry to have not had a chance to say anything to you. We will do it some other time. Now, we came down here to offer evidence in this case and the court has held under the law that the evidence we had is not admissible, so all we can do is to take an exception and carry it to a higher court to see whether the evidence is admissible or not. As far as this case stands before the jury, the court has told you very plainly that if you think my client taught that man descended from a lower order of animals, you will find him guilty, and you heard the testimony of the boys on that question and heard read the books, and there is no dispute about the facts. Scopes did not go on the stand, because he could not deny the statements made by the boys. I do not know how you may feel, I am not especially interested in it, but this case and this law will never be decided until it gets to a higher court, and it cannot get to a higher court probably, very well, unless you bring in a verdict. So, I do not want any of you to think we are going to find any fault with you as to your verdict. I am frank to say, while we think it is wrong, and we ought to have been permitted to put in our evidence, the court felt otherwise, as he had a right to hold. We cannot argue to you gentlemen under the instructions given by the court —we cannot even explain to you that we think you should return a verdict of not guilty. We do not see how you could. We do not ask it. We think we will save our point and take it to the higher court and settle whether the law is good, and also whether he should have permitted the evidence. I guess that is plain enough.

Gen. Stewart—That is satisfactory.

The Court—Have you any statement, Mr. Attorney-General?

Gen. Stewart—No, sir; except this, your honor, I want to ask this as a matter of information, did I understand your honor to charge the jury in fixing the fine, if they find guilty, if they were satisfied with the minimum fee, it will not be necessary for the jury to bring in a verdict except simply to say "guilty."

The Court—I will read what I charged: "Under the provision of this statute in this case, a person who violates the same may be punished by a fine of not less than $100 nor more than $500. If after a fair and honest investigation of all the facts you find the defendant guilty and find that his offense deserves a greater punishment than a fine of $100, then you must impose a fine not to exceed $500 in any event. But if you are content with a $100 fine, then you may simply find the defendant guilty and leave the punishment to the court."

Gen. Stewart—Of course, that is a minor matter, but I had it in mind that it would be the duty of the jury to fix whatever fine was imposed.

The Court—As I understand the holding, the court can impose a minimum fine always under the statute, that is our practice in whisky cases, the least fine in a transporting case is $100.

Gen. Stewart—Yes. We have more of that kind than any other in the criminal court.

Mr. Darrow—That is encouraging.

The Court—How is that?

Gen. Stewart—I was telling Mr. Darrow we have more whiskey cases than any other in the criminal court.

Mr. Darrow—I have not even seen a cause for a case since I got down here.

The Court—There is no reason why the jury should not fix the minimum if you prefer. The practice, however, is for the court to impose the minimum.

Gen. Stewart—I am not quite clear on that.

Mr. Darrow—We will not take an exception, either way you want it, because we want the case passed on by the higher court, if you want the jury to fix the fine.

The Court—General, the minimum fine in a transporting case is $100.

Gen. Stewart—Yes.

The Court—Our practice in Tennessee is for the court to impose the fine in a transporting case as like this, the maximum fine of $500. The practice is if the jury thinks it should be greater than the minimum.

Gen. Stewart—I had in mind the general statute on that.

The Court—The general statute is $50.

The Court—If you want us to stipulate—

The jury might fix the fine, they will not be irregular. They will not make any question about that.

Mr. Darrow—No.

Gen. Stewart—I do not think that there is anything that can be said to the jury than what Mr. Darrow said. Of course, the case in its present attitude is that it will be thrashed out by the appellate court, that is what the defense wants, and the state wants. What Mr. Darrow wanted to say to you was that he wanted you to find his client guilty, but did not want to be in the position of pleading guilty, because it would destroy his rights in the appellate court.

The Court—We could not undertake to take the verdict and make up the record before noon, could we?

Gen. Stewart—How is that?

The Court—I say we could not undertake to take the verdict and make up the record before noon, could we?

Gen. Stewart—I take it it will only be a matter of a few moments.

Mr. Hays—Yes, and if your honor will only wait, we ought to be able to get through with the whole matter.

Gen. Stewart—The formal motion, as I suggest, is to just let the counsel treat them as in.

The Court—Mr. Officer, go with the jury and get them a place for deliberation.

Jury Out Nine Minutes.

(The jury thereupon retired for deliberation.)

(The jury returned to the courtroom at 11:23 a. m.)

Officer Rice—Everybody be seated please.

Court—Get your book, Mr. Clerk, so as to poll the jury. Get your seats, gentlemen, and let the jury have their seats. You gentlemen will have to move out a little, so I can see the jury. You may call the jury, Mr. Clerk.

(The clerk calls the roll of the jury.)

Court—Mr. Foreman, will you tell us whether you have agreed on a verdict?

Foreman—Yes, sir. We have, your honor.

Court—What do you find?

Foreman—We have found for the state, found the defendant guilty.

Court—Did you fix the fine?

Foreman—No, sir.

Court—You leave it to the court?

Foreman—Leave it to the court.

Court—Mr. Scopes, will you come around here, please, sir.

(The defendant presents himself before the court.)

Scopes Fined $100.

Court—Mr. Scopes, the jury has found you guilty under this indictment, charging you with having taught in the schools of Rhea county, in violation of what is commonly known as the anti-evolution statute, which makes it unlawful for any teacher to teach in any of the public schools of the state, supported in whole or in part by the public school funds of the state, any theory that denies the story of the divine creation of man, and teach instead thereof that man has descended from a lower order of animals. The jury have found you guilty. The statute makes this an offense punishable by fine of not less than $100 nor more than $500. The court now fixes your fine at $100, and imposes that fine upon you—

Mr. Neal—May it please your honor, we want to be heard a moment.

Scopes Says Fine is Unjust.

Court—Oh—Have you anything to say, Mr. Scopes, as to why the court should not impose punishment upon you?

Defendant J. T. Scopes—Your honor, I feel that I have been convicted of violating an unjust statute. I will continue in the future, as I have in the past, to oppose this law in any way I can. Any other action would be in violation of my ideal of academic freedom—that is, to teach the truth as guaranteed in our constitution, of personal and religious freedom. I think the fine is unjust.

Court—So then the court now imposes on you a fine of $100 and costs, which you will arrange with the clerk.

Mr. Malone—Your honor, what about bail?

Court—Sir?

Mr. Malone—What about bail?

Court—Well, how much bail can Mr. Scopes make?

Mr. Malone—We can arrange any amount your honor demands.

Court—Let him give bond for $500. Well, it is misdemeanor case; he does not have to go before the supreme court and only makes bond for his appearance back here at the next term, following the next term of the supreme court.

Gen. Stewart—I believe the next term—

Court—This county, I believe, goes to Knoxville. My counties are close on the border, some of my counties go to Nashville, and some to Knoxville. This county tries its cases there the first Monday in December—

Gen. Stewart—September.

Court—The first Monday in September.

Mr. Malone—I want to state in that connection that the Baltimore Evening Sun has generousy offered to go bond for Mr. Scopes, and the defense has accepted the offer.

Gen. Stewart—The bond, of course, would have to be given for his appearance back here at the December term of court; the defendant does not appear before the supreme court in a misdemeanor case.

Court—In misdemeanors the defendant does not appear in the supreme court. He can make bond to the term of this court following the August term, which will be the first Monday in December, I believe, so the bond will be made returnable to this court on the first Monday in December, 1925. Now, by that time, I presume, the supreme court will have passed on this case.

Malone Thanks People for Courtesy.

Mr. Malone—Your honor, may I at this time say, on behalf of my colleagues, that we wish to thank the people of the state of Tennessee, not only for their hospitality, but for the opportunity of trying out these great issues here.

(Applause and hand-clapping.)

Mr. Hays—For the purpose of the record, may I make a motion here?

Court—Yes, sir.

Mr. Hays—I should like, if possible, that this term of this court be extended for thirty days, in order to enable us to get up a record.

Court—Make the proper motion.

Gen. Stewart—We have an intervening term of the court in August, a regular term. Of course, it can be kept open.

Mr. Hays—I don't know your procedure; we may want you to tell us—

The Court—Under the statutes of Tennessee, the court can allow sixty days to perfect an appeal. Now, I prefer not to allow the whole sixty days in this instance, because that would carry it beyond the meeting of the supreme court.

Mr. Hays—If your honor will allow us sixty days, we will get it up in ten days if we can do that, if possible, but in the event it is not humanly possible, we do not want to be barred. If your honor will give us plenty of time, I assure you we will do everything we can to get it up immediately.

The Court—I suppose you know, Mr. Hays, that in Tennessee the bill of exceptions is just a copy of the evidence and proceedings and the judge's charge. Of course, the technical record is then gotten up separately—a copy of the warrant, and motions and decrees and orders—that is part of the technical record and doesn't become part of the bill of exceptions; and that may be made up at any time.

Mr. Hays—If your honor will just protect us on our record, we will leave the whole thing to you. We don't want to be put in a position to lose our rights.

The Court—I believe you can have it ready in thirty days, Mr. Hays; you have had a daily transcript.

Mr. Hays—We will have that, but there has been a great deal of confusion, there may be a chance of having some of it missing.

Court—I will give you thirty days, and assure you that if you are not ready—Let's see, that would be the 21st of August. Just so you can get it ready and get the case to the supreme court in time. You see, the clerk of this court will have to make up a bill of exceptions, unless he can be furnished with a carbon copy. If he could, that woud expedite matters. It would take, I suppose, a week or ten days to copy the bill of exceptions.

Mr. Hays—It was a long drawn-out matter, and it took a long time to try it, and we wouldn't want to be foreclosed.

Court—I will give you all the time I can, I assure you.

Mr. Hays—I understand that you give us thirty days?

Court—Yes, sir.

Mr. Darrow—Then another matter, if the court please, is a petition for the certification of Mr. Bryan's testimony.

Court—I believe you can do that, Mr. Darrow, after you file your bill of exceptions, by going to the supreme court for a writ of error, whereby the record in this case will be certified up there.

Mr. Hays—And then, your honor, I wish to make a motion in arrest of judgment, and I suggest that you give us opportunity to file that in original form.

Court—You will raise the same question that the court has ruled on?

Mr. Hays—I want to raise all the questions raised during the trial. I presume the motion will be overruled, but I want to get that in the record.

Court—Mr. Attorney-General, are you willing to that?

Gen. Stewart—Yes, sir, I am willing.

Court—Let the record show that the motion has been filed and overruled.

Hays Moves for a New Trial.

Mr. Hays—Yes, sir, and get my exception in. I want to be certain the record is correct; I wish to

Courtroom and Jury just before Verdict

make a motion for a new trial, on the same grounds.

Court—Let the record show that your motions were filed and overruled.

Mr. Hays—And so instead of stating my grounds, I can file them later.

The Court—Yes, sir.

Mr. Hays—Then your honor will note my exception.

The Court—Yes, sir.

Mr. Hays—I also am advised that I should move for permission to appeal.

The Court—Yes, the record will show that you prayed for an appeal and that it was granted by the court.

Mr. Hays—Yes, sir, thank you. And that makes my record complete.

The Court—I think so—if it doesn't, if there is anything overlooked we will overlook it.

Now let me say—the court has not adjourned. Have any of you gentlemen anything to say—but you are standing up and they are taking your pictures, I imagine. Otherwise, you might sit down. (Laughter.)

Mr. Hays—I presume the filing of it in regular form could be done during the term.

The Court—We could extend the term a few days, yes, sir.

Mr. Hays—Isn't that the same time you have extended in order that we might get our records fixed.

The Court—This is a special term now.

Mr. Hays—Are we to understand that the special term for thirty days is extended?

The Court—No, you get your motion filed and the court allows you ten days to make your bill of exceptions. You know the law requires a motion for a new trial to be filed during the term.

Mr. Hays—Yes, sir.

The Court—And in order to meet the requirements of that statute, if the motion isn't ready we will have to extend this term a few days.

Gen. Stewart—Your honor can leave the minutes of this day open.

The Court—Yes, sir.

Gen. Stewart—And it could be entered on the minutes as of this date.

The Court—Just let the records show that they were entered as of today.

Gen. Stewart—Yes—prepare them right away, of course.

The Court—Does anyone have anything to say? Mr. Muto wants to be heard.

Newspaper Men Express Thanks.

Mr. Tony Muto—May it please, your honor, on behalf of the ladies and gentlemen of the press that came down here to cover this trial for the various newspapers, magazines and syndicates, I wish to thank the court and all the officials, the Dayton Progressive club, for all the courtesies and kindnesses that have been shown us. (Applause.)

Special Writer from Toronto—May I have the privilege, as the only Canadian correspondent present, to express my great appreciation of the extreme courtesy which has been accorded me and my brethren of the press by the court and the citizens of Dayton. I shall take back with me a deeper appreciation of the great republic for which we have felt so kindly, and whose institutions we so magnify and admire. (Applause.)

The Court—Has any other press man anything to say? Let me hear you, Mr. Bell—did you say anything?

Mr. Bryan Bell—No, sir.

A Voice—As one of the public, who has come a number of miles to hear this trial, I desire to express my appreciation of the hospitality and reasonable expenses that have been incurred while here. (Applause.)

The Court—Anyone else? Gordon, did you have something you wished to say?

Gordon McKenzie Speaks for Tennessee People.

Gordon McKenzie—On behalf of Rhea county and Gen. Stewart, and on behalf of the prosecution, I desire to say to the gentlemen who have just made their statements, that we are delighted to have had you with us. We have learned to take a broader view of life since you came. You have brought to us your ideas—your views—and we have communicated to you, as best we could, some of our views. As to whether or not we like those views, that is a matter that should not address itself to us at this time, but we do appreciate your views, and while much has been said and much has been written about the narrow-minded people of Tennessee we do not feel hard toward you for having said that, because that is your idea. We people here want to be more broad-minded than some have given us credit for, and we appreciate your coming, and we have been greatly elevated, edified and educated by your presence. And should the time ever come when you are back near the garden spot of the world, we hope that you will stop off and stay awhile with us here in order that we may chat about the days of the past, when the Scopes trial was tried in Dayton. (Applause.)

A Voice—I feel, as a member of the Tennessee bar, that we should not be remiss in our recognition of the counsel from outside of our state who have appeared in this case, and I want to exonerate them from any accusation of any unfair attitude, and to say that the bar of Tennessee appreciates the distinguished services of these great lawyers who have come to discuss among us a fundamental problem which affects our government, and the government of all the states, and we appreciate from the bottom of our hearts their labors amongst us, and we feel that they have as much right to be heard as our local counsel, and we welcome them to our state on this occasion and on any other occasion when matters of great magnitude involving our national welfare come before us. (Applause.)

Mr. Neal—As one of the Tennessee lawyers that has been connected from the beginning with this case, I want to thank your honor and the gentlemen on the other side for the great courtesy they have shown to my distinguished associates from other states. (Applause.)

The Court—Col. Bryan, I will hear you.

Bryan't Last Court Speech.

"This issue will be settled right whether on our side or the other."

Mr. Bryan—I don't know that there is any special reason why I should add to what has been said, and yet the subject has been presented from so many viewpoints that I hope the court will pardon me if I mention a viewpoint that has not been referred to. Dayton is the center and the seat of this trial largely by circumstance. We are told that more words have been sent across the ocean by cable to Europe and Australia about this trial than has ever been sent by cable in regard to anything else happening in the United States. That isn't because the trial is held in Dayton. It isn't because a schoolteacher has been subjected to the danger of a fine from $100.00 to $500.00, but I think illustrates how people can be drawn into prominence by attaching themselves to a great cause. Causes stir the world. It is because it goes deep. It is because it extends wide, and because it reaches into the future beyond the power of man to see. Here has been fought out a little case of little consequence as a case, but the world is interested because it raises an issue, and that issue will some day be settled right, whether it is settled on our side or the other side. It is going to be settled right. There can be no settlement of a great cause without discussion, and people will not discuss a cause until their attention is drawn to it, and the value of this trial is not in any incident of the trial, it is not because of anybody who is attached to it, either in an official way or as counsel on either side. Human beings are mighty small, your honor. We are apt to magnify the personal element and we sometimes become inflated with our importance, but the world little cares for man as an individual.

He is born, he works, he dies, but causes go on forever, and we who participated in this case may congratulate ourselves that we have attached ourselves to a mighty issue. Now, if I were to attempt to define that issue, I might find objection from the other side. Their definition of the issue might not be as mine is, and therefore, I will not take advantage of the privilege the court gives me this morning to make a statement that might be controverted, and nothing that I would say would determine it. I have no power to define this issue finally and authoritatively. None of the counsel on our side has this power and none of the counsel on the other side has this power, even this honorable court has no such power. The people will determine this is-issue. They will take sides upon this issue, they will state the question involved in this issue, they will examine the information—not so much that which has been brought out here, for very little has been brought out here, but this case will stimulate investigation and investigation will bring out information, and the facts will be known, and upon the facts, as ascertained, the decision will be rendered, and I think, my friends, and your honor, that if we are actuated by the spirit that should actuate everyone of us, no matter what our views may be, we ought not only desire, but pray, that that which is right will prevail, whether it be our way or somebody else's. (Applause.)

Mr. Darrow—May I say a word?

The Court—Colonel, be glad to hear from you.

Darrow Compares Trial with Witchcraft Cases.

Mr. Darrow—I want to say a word. I want to say in thorough sincerity that I appreciate the courtesy of the counsel on the other side from the beginning of this case, at least the Tennessee counsel, that I appreciate the hospitality of the citizens here. I shall go away with a feeling of respect and gratitude toward them for their courtesy and their liberality toward us persons; and that I appreciate the kind, and I think I may say, general treatment of this court, who might have sent me to jail, but did not.

(Laughter in the courtroom.)

Mr. Darrow (Continuing)—And on the side of the controversy between the court and myself I have already ruled that the court was right, so I do not need to go further.

The Court—Thank you.

Mr. Darrow—But, I mean it.

The Court—Yes.

Mr. Darrow (Continuing)—Of course, there is much that Mr. Bryan has said that is true. And nature—nature, I refer to does not choose any special setting for mere events. I fancy that the place where the Magna Charta was wrested from the barons in England was a very small place, probably not as big as Dayton. But events come along as they come along. I think this case will be remembered because it is the first case of this sort since we stopped trying people in America for witchcraft because here we have done our best to turn back the tide that has sought to force itself upon this—upon this modern world, of testing every fact in science by a religious dictum. That is all I care to say.

The Court—Any one else?

A Voice—Yes, your honor.

The Court—Mr. Rappleyea.

Mr. Rappleyea—As Dr. Spencer said a few months ago that big movements make big men, but this the case of the reverse, where big men have made big movements. I especially wish to pay my respects and thanks and take this opportunity, perhaps the last I shall have, to Mr. Bryan for relieving me of the embarrassing position I was in as original prosecutor, and carrying through what he thought was right in spite of the criticisms that he has had. Mr. Bryan, I thank you. (Applause.)

The Court's Farewell Oration.

The Court—My fellow citizens, I recently read somewhere what I think was a definition of a great man, and that was this: That he possesses a passion to know the truth and have the courage to declare it in the face of all opposition. It is easy enough, my friends, to have a passion to find a truth, or to find a fact, rather, that coincides with our preconceived notions and ideas, but it sometimes takes courage to search diligently for a truth, that may destroy our preconceived notions and ideas.

The man that only has a passion to find the truth is not a complete and great man; but he must also have the courage to declare it in the face of all opposition. It does not take any great courage for a man to stand for a principle that meets with the approval of public sentiment around him. But it sometimes takes courage to declare a truth or stand for a fact that is in contravention to the public sentiment.

Now, my friends, the man—I am not speaking in regard to the issues in this case, but I am speaking in general terms—that a man who is big enough to search for the truth and find it, and declare it in the face of all opposition is a big man.

Now, we spoke—Dayton has been referred to. That the law—that something big could not come out of Dayton. Why, my friends, the greatest Man that has ever walked on the face of the earth, the Man that left the portals of heaven, the Man that came down from heaven to earth that man might live, was born in a little town, and He lived and spent His life among a simple, unpretentious people.

We do not measure greatness by the size of the village or the town or the neighborhood from which it came. But greatness depends upon the principles that are involved. Some one recently wrote on this subject, and in referring to this case, that the great Dred Scott bill, one of the most famous lawsuits ever tried in America, a case that drew public attention, perhaps, from the whole world simply involved the liberty of one colored man.

Someone has also referred to a case from the District of Columbia, where the president of the United States appoints the magistrates, and President Adams appointed a magistrate but failed to issue his commission and went out of office; that he later mandamussed a successor of President Adams to compel him to issue a commission to him for the simple office of justice of the peace. John Marshall, the man that ruled and reigned, that presided over the supreme court of the United States for thirty-four long years, and one of the most noted lawyers and judges that ever lived in America, made his fame and laid the foundation for his fame by writing the opinion involving the office of justice of the peace.

Now, my friends, the people in America are great people. We are great in the South, and they are great in the North. We are great because we are willing to lay down our differences when we fight the battle out and be friends. And, let me tell you, there are two things in this world that are indestructible, that man cannot destroy, or no force in the world can destroy.

One is truth. You may crush it to the earth but it will rise again. It is indestructible, and the causes of the law of God. Another thing indestructible in America and in Europe and everywhere else, is the Word of God, that He has given to man, that man may use it as a waybill to the other world. Indestructible, my friends, by any force because it is the word of the Man, of the forces that created the universe, and He has said in His word that "My word will not perish" but will live forever.

I am glad to have had these gentlemen with us. This little talk of mine comes from my heart, gentlemen. I have had some difficult problems to decide in this lawsuit, and I only pray to God that I have decided them right. If I have not, the higher courts will find the mistake. But if I failed to decide them right, it was for the want of legal learning, and legal attainments, and not for the want of a disposition to do everybody justice.

We are glad to have you with us. (Applause.)

Mr. Hays—May I, as one of the counsel for the defense, ask your honor to allow me to send you the "Origin of Species and the Descent of Man," by Charles Darwin?

(Laughter.)

The Court—Yes; yes.

(Laughter and applause.)

The Court—Has anyone else anything to say.

(No response.)

If not—

Officer Kelso Rice—Now, people, when court is adjourned—

The Court—Wait, do not adjourn yet.

(A train whistle blows.)

The Court—Go ahead, officer.

Officer Rice—Do not crowd the aisles. When the court has adjourned move slowly, do not be in a hurry. But move slowly, everybody, when court is adjourned and do not block the aisleways at all. Keep moving.

The Court—We will adjourn. And Brother Jones will pronounce the benediction.

Dr. Jones—May the grace of our Lord Jesus Christ, the love of God and the communion and fellowship of the Holy Ghost abide with you all. Amen.

The Court—The court will adjourn sine die.

At Attorney's Table During Scopes Trial.

TEXT OF BRYAN'S PROPOSED ADDRESS IN SCOPES CASE.

As a member of the counsel of prosecution in the Scopes evolution case in Dayton, William Jennings Bryan had prepared an address in defense of Tennessee's law against the teaching of evolution in the public schools. This address was not delivered during the trial because arguments to the jury by counsel on both sides were dispensed with by agreement. Arrangements for publication of it were made by Mr. Bryan only a few hours before his death. The text of the address follows:

May It Please the Court,
and Gentlemen of the Jury:

Demosthenes, the greatest of ancient orators, in his "Oration on the Crown," the most famous of his speeches, began by supplicating the favor of all the gods and goddesses of Greece. If, in a case which involved only his own fame and fate, he felt justified in petitioning the heathen gods of his country, surely we, who deal with the momentous issues involved in this case, may well pray to the Ruler of the universe for wisdom to guide us in the performance of our several parts in this historic trial.

Let me, in the first place, congratulate our cause that circumstances have committed the trial to a community like this and entrusted the decision to a jury made up largely of the yeomanry of the state. The book in issue in this trial contains on its first page two pictures contrasting the disturbing noises of a great city with the calm serenity of the country. It is a tribute that rural life has fully earned.

I appreciate the sturdy honesty and independence of those who come into daily contact with the earth, who, living near to nature, worship nature's God, and who, dealing with the myriad mysteries of earth and air, seek to learn from revelation about the Bible's wonder-working God. I admire the stern virtues, the vigilance and the patriotism of the class from which the jury is drawn, and am reminded of the lines of Scotland's immortal bard, which, when changed but slightly, describe your country's confidence in you:

"O Scotia, my dear, my native soil!
For whom my warmest wish to Heaven is sent,
Long may thy hardy sons of rustic toil
Be blest with health, and peace, and sweet content!
"And, oh, may Heav'n their simple lives prevent
From luxury's contagion, weak and vile!
Then, howe'er crowns and coronets be rent,
A virtuous populace may rise the while,
And stand, a wall of fire, around their much-loved isle."

Let us now separate the issues from the misrepresentations, intentional or unintentional, that have obscured both the letter and the purpose of the law. This is not an interference with freedom of conscience. A teacher can think as he pleases and worship God as he likes, or refuse to worship God at all. He can believe in the Bible or discard it; he can accept Christ or reject Him. This law places no obligations or restaints upon him. And so with freedom of speech; he can, so long as he acts as an individual, say anything he likes on any subject. This law does not violate any right guaranteed by any constitution to any individual. It deals with the defendant, not as an individual, but as an employee, an official or public servant, paid by the state, and therefore under instructions from the state.

Right of the State to Control Public Schools.

The right of the state to control the public schools is affirmed in the recent decision in the Oregon case, which declares that the state can direct what shall be taught and also forbid the teaching of anything "manifestly inimical to the public welfare." The above decision goes even farther and declares that the parent not only has the right to guard the religious welfare of the child, but is in duty bound to guard it. That decision fits this case exactly. The state had a right to pass this law, and the law represents the determination of the parents to guard the religious welfare of their children.

It need hardly be added that this law did not have its origin in bigotry. It is not trying to force any form of religion on anybody. The majority is not trying to establish a religion or to teach it—it is trying to protect itself from the effort of an insolent minority to force irreligion upon the children under the guise of teaching science. What right has a little irresponsible oligarchy of self-styled "intellectuals" to demand control of the schools of the United States, in which 25,000,000 of children are being educated at an annual expense of nearly $2,000,000,000?

Christians must, in every state of the Union, build their own colleges in which to teach Christianity; it is only simple justice that atheists, agnostics and unbelievers should build their own colleges if they want to teach their own religious views or attack the religious views of others.

The statute is brief and free from ambiguity. It prohibits the teaching, in the public schools, of "any theory that denies the story of divine creation as taught in the Bible," and teaches, "instead, that man descended from a lower order of animals." The first sentence sets forth the purpose of those who passed the law. They forbid the teaching of any evolutionary theory that disputes the Bible record of man's creation and, to make sure that there shall be no misunderstanding, they place their own interpretations on their language and specifically forbid the teaching of any theory that makes man a descendant of any lower form of life.

The evidence shows that defendant taught, in his own language as well as from a book outlining the theory, that man descended from lower forms of life. Howard Morgan's testimony gives us a definition of evolution that will become known throughout the world as this case is discussed. Howard, a 14-year-old boy, has translated the words of the teacher and the textbook into language that even a child can understand. As he recollects it, the defendant said, "A little germ or one cell organism was formed in the sea; this kept evolving until it got to be a pretty good-sized animal, then came on to be a land animal, and it kept evolving, and from this was man." There is no room for difference of opinion here, and there is no need of expert testimony. Here are the facts, corroborated by another student, Harry Shelton, and admitted to be true by counsel for defense. Mr. White, superintendent of schools, testified to the use of Hunters' Civic Biology, and to the fact that the defendant not only admitted teaching evolution, but declared that he could not teach it without violating the law. Mr. Robinson, the chairman of the school board, corroborated the testimony of Superintendent White in regard to the defendant's admissions and declaration. These are the facts; they are sufficient and undisputed. A verdict of guilty must follow.

But the importance of this case requires more. The facts and arguments presented to you must not only convince you of the justice of conviction in this case but, while not necessary to a verdict of guilty, they should convince you of the righteousness of the purpose of the people of the state in the enactment of this law. The state must speak through you to the outside world and repel the aspersions cast by the counsel for the defense upon the intelligence and the enlightenment of the citizens of Tennessee. The people of this state have a high appreciation of the value of education.

The state constitution testifies to that in its demand that education shall be fostered and that science and literature shall be cherished. The continuing and increasing appropriations for public instruction furnish abundant proof that Tennessee places a just estimate upon the learning that is secured in its schools.

Declares Religion Not Hostile to Learning.

Religion is not hostile to learning, Christianity has been the greatest patron learning has ever had. But Christians know that "the fear of the Lord is the beginning of wisdom" now just as it has been in the past, and they therefore oppose the teaching of guesses that encourage godlessness among the students.

Neither does Tennessee undervalue the service rendered by science. The Christian men and women of Tennessee know how deeply mankind is indebted to science for benefits conferred by the discovery of the laws of nature and by the designing of machinery for the utilization of these laws. Give science a fact and it is not only invincible, but it is of incalculable service to man. If one is entitled to draw from society in proportion to the service that he renders to society, who is able to estimate the reward earned by those who have given to us the use of steam, the use of electricity, and enabled us to utilize the weight of water that flows down the mountainside? Who will estimate the value of the service rendered by those who invented the phonograph, the telephone and the radio? Or, to come more closely to our home life, how shall we recompense those who gave us the sewing machine, the harvester, the threshing machine, the tractor, the automobile and the method now employed in making artificial ice? The department of medicine also opens an unlimited field for invaluable service. Typhoid and yellow fever are not feared as they once were. Diphtheria and pneumonia have been robbed of some of their terrors, and a high place on the scroll of fame still awaits the discoverer of remedies for arthritis, cancer, tuberculosis and other dread diseases to which mankind is heir.

Christianity welcomes truth from whatever source it comes, and is not afraid that any real truth from any source can interfere with the divine truth that comes by inspiration from God Himself. It is not scientific truth to which Christians object, for true science is classified knowledge, and nothing therefore can be scientific unless it is true.

Evolution Not Truth; Merely an Hypothesis.

Evolution is not truth; it is merely an hypothesis—it is millions of guesses strung together. It had not been proven in the days of Darwin; he expressed astonishment that with two or three million species it had been impossible to trace any species to any other species. It had not been proven in the days of Huxley, and it has not been proven up to today. It is less than four years ago that Prof. Bateson came all the way from London to Canada to tell the American scientists that every effort to trace one species to another had failed—every one. He said he still had faith in evolution, but had doubts about the origin of species. But of what value is evolution if it cannot explain the origin of species? While many scientists accept evolution as if it were a fact, they all admit, when questioned, that no explanation has been found as to how one species developed into another.

Darwin suggested two laws, sexual selection ad natural selection. Sexual selection has been laughed out of the class room, and natural selection is being abandoned, and no new explanation is satisfactory even to scientists. Some of the more rash advocates of evolution are wont to say that evolution is as firmly established as the law of gravitation or the Copernican theory. The absurdity of such a claim is apparent when we remember that anyone can prove the law of gravitation by throwing a weight into the air, and that anyone can prove the roundness of the earth by going around it, while no one can prove evolution to be true in any way whatever.

Chemistry is an insurmountable obstacle in the path of evolution. It is one of the greatest of the sciences; it separates the atoms—isolates them and walks about them, so to speak. If there were in nature a progressive force, an eternal urge, chemistry would find it. But it is not there. All of the ninety-two original elements are separate and distinct; they combine in fixed and permanent proportions. Water is H20, as it has been from the beginning. It was here before life appeared and has never changed; neither can it be shown that any thing else has materially changed.

There is no more reason to believe that man descended from some inferior animal than there is to believe that a stately mansion has descended from a small cottage. Resemblances are not proof—they simply put us on inquiry. As one fact, such as the absence of the accused from the scene of the murder, outweighs all the resemblances that a thousand witnesses could swear to, so the inability of science to trace any one of the millions of species to another species, outweighs all the resemblances upon which evolutionists rely to establish man's blood relationship with the brutes.

Man's Urge Comes Not From Within, But From Above.

But while the wisest scientists cannot prove a pushing power, such as evolution is supposed to be, there is a lifting power that any child can understand. The plant lifts the mineral up into a higher world, and the animal lifts the plant up into a world still higher. So, it has been reasoned by analogy, man rises, not by a power within him, but only when drawn upward by a higher power. There is a spiritual gravitation that draws all souls toward heaven, just as surely as there is a physical force that draws all matter on the surface of the earth towards the earth's center. Christ is our drawing power; He said, "I, if I be lifted up from the earth, will draw all men unto Me," and His promise is being fulfilled daily all over the world.

It must be remembered that the law under consideration in this case does not prohibit the teaching of evolution up to the line that separates man from the lower forms of animal life. The law might well have gone farther than it does and prohibit the teaching of evolution in lower forms of life; the law is a very conservative statement of the people's opposition to an anti-Biblical hypothesis. The defendant was not content to teach what the law permitted; he, for reasons of his own, persisted in teaching that which was forbidden for reasons entirely satisfactory to the law-makers.

Most of the people who believe in evolution do not know what evolution means. One of the science books taught in the Dayton High school has a chapter on "The Evolution of Machinery." This is a very common misuse of the term. People speak of the evolution of the telephone, the automobile and the musical instrument. But these are merely illustrations of man's power to deal intelligently with inanimate matter; there is no growth from within in the development of machinery.

Equally improper is the use of the word "evolution" to describe the growth of a plant from a seed, the growth of a chicken from an egg or the development of any form of animal life from a single cell. All these give us a circle not a change from one species to another.

Evolution Wrong Word Even in Plant Life.

Evolution—the evolution involved in this case, and the only evolution that is a matter of controversy anywhere—is the evolution taught by defendant, set forth in Hunter's Civic Biology. The author the books now prohibited by the new state law, and illustrated in the diagram printed on page 194 of Hunter's Civic Biology. The author estimates the number of species in the animal kingdom at 518,900. These are divided into eighteen classes, and each class is indicated on a diagram by a circle, proportionate in size to the number of species in each class and attached by a stem to the trunk of the tree. It begins with Protozoa and ends with the mammals. Passing over the classes with which the average is unfamiliar, let me call your attention to a few of the larger and better known groups. The insects are numbered at 360,000, over two-thirds of the total number of species in the animal world. The fishes are numbered at 13,000, the amphibians at 1,400, the reptiles at 3,500, and the birds are 13,000, while 3,500 mammals are crowded together in a little circle that is barely higher than the bird circle. No circle is reserved for man alone. He is, according to the diagram, shut up in the little circle entitled "Mammals," with 3,499 other species of mammals. Does it not seem a little unfair not to distinguish between man and lower forms of life? What shall we say of the intelligence, not to say religion, of those who are so particular to distinguish between fishes and reptiles and birds, but put a man with an immortal soul in the same circle with the wolf, the hyena and the skunk? What must be the impression made upon children by such a degradation of man?

In the preface of this book, the author explains that it is for children, and adds that "the boy or girl of average ability upon admission to the secondary school is not a thinking individual." Whatever may be said in favor of teaching evolution to adults, it surely is not proper to teach it to children who are not yet able to think.

The evolutionist does not undertake to tell us how protozoa, moved by interior and resident forces, sent life up through all the various species, and cannot prove that there was actually any such compelling power at all. And yet, the school children are asked to accept their guesses and build a philosophy of life upon them. If it were not so serious a matter, one might be tempted to speculate upon the various degrees of relationship that, according to evolutionists, exist between man and other forms of life. It might require some very nice calculation to determine at what degree of relationship the killing of a relative ceases to be murder and the eating of one's kin ceases to be cannibalism.

Evolution Casts Doubt Upon Creation Itself.

But it is not a laughing matter when one considers that evolution not only offers no suggestions as to a Creator but tends to put the creative act so far away as to cast doubt upon creation itself. And while it is shaking faith in God as a beginning, it is also creating doubt as to a heaven at the end of life. Evolutionists do not feel that it is incumbent upon them to show how life began or at what point in their long-drawn-out scheme of changing species man became endowed with hope and promise of immortal life. God may be a matter of indifference to the evolutionists and a life beyond may have no charm for them, but the mass of mankind will continue to worship their Creator and continue to find comfort in the promise of their Savior that He has gone to prepare a place for them. Christ has made of death a narrow, star-lit strip between the companionship of yesterday and the reunion of tomorrow; evolution strikes out the stars and deepens the gloom that enshrouds the tomb.

If the results of evolution were unimportant, one might require less proof in support of the hypothesis, but before accepting a new philosophy of life, built upon a materialistic foundation, we have reason to demand something more than guesses; "we may well suppose" is not a sufficient substitute for "Thus saith the Lord."

Darwin's Family Tree Pointed Out By Own Words.

If, your honor, and you, gentlemen of the jury, would have an understanding of the sentiment that lies back of the statute against the teaching of evolution, please consider the facts that I shall now present to you. First, as to the animals to which evolutionists would have us trace our ancestry. The following is Darwin's family tree, as you will find it set forth on pages 180-181 of his "Descent of Man":

"The most ancient progenitors in the kingdom of Vertebrata, at which we are able to obtain an obscure glance, apparently consisted of a group of marine animals, resembling the larvae of existing ascidians. These animals probably gave rise to a group of fishes, as lowly organized as the lancelot; and from these the Ganoids, and other fishes like the Lepidosiren, must have been developed. From such fish a very small advance would carry us on to the amphibians. We have seen that birds and reptiles were once intimately connected together; and the Monotremata now connect mammals with reptiles in a slight degree. But no one can at present say by what line of descent the three higher and related classes, namely, mammals, birds and reptiles, were derived from the two lower vertebrate classes, namely, amphibians and fishes. In the class of mammals the steps are not difficult to conceive which led from the ancient Monotremata to the ancient Marsupials; and from these to the early progenitors of the placental mammals. We may thus ascend to the Lemuridae; and the interval is not very wide from these to the Simiadae. The Simiadae then branched off into two great stems, the New World and Old World monkeys; and from the latter, at a remote period, Man, the wonder and glory of the universe, proceeded. Thus we have given to man a pedigree of prodigious length, but, not, it may be said, of noble quality." (Ed. 1874, Hurst.)

Note the words implying uncertainty; "obscure glance," "apparently," "resembling," "must have been," "slight degree," and "conceive."

Darwin, on page 171 of the same book, tries to locate his first man—that is, the first man to come down out of the trees—in Africa. After leaving man in company with gorillas and chimpanzees, he says, "But it is useless to speculate on this subject." If he had only thought of this earlier the world might have been spared much of the speculation that his brute hypothesis has excited.

On page 79 Darwin gives some fanciful reasons for believing that man is more likely to have descended from the chimpanzee than from the gorilla. His speculations are an excellent illustration of the effect that the evolutionary hypothesis has in cultivating the imagination. Prof. J. Arthur Thomson says that the "idea of evolution is the most potent thought economizing formula the world has yet known." It is more than that; it dispenses with thinking entirely and relies on the imagination.

On page 141 Darwin attempts to trace the mind of man back to the mind of lower animals. On pages 113 and 114 he endeavors to trace man's moral nature back to the animals. It is all animal, animal, animal, with never a thought of God or of religion.

Our first indictment against evolution is that it disputes the truth of the Bible account of man's creation and shakes faith in the Bible as the Word of God. This indictment we prove by comparing the processes described as evolutionary with the text of Genesis. It not only contradicts the Mosaic record as to the beginning of human life, but it disputes the Bible doctrine of reproduction according to kind—the greatest scientific principle known.

Our second indictment is that the evolutionary hypothesis, carried to its logical conclusion, disputes every vital truth of the Bible. Its tendency, natural, if not inevitable, is to lead those who really accept it, first to agnosticism and then to atheism. Evolutionists attack the truth of the Bible, not openly at first, but by using weazel-words like "poetical," "symbolical" and "allegorical" to suck the meaning out the inspired record of man's creation.

We call as our first witness Charles Darwin. He began life a Christian. On page 39, Vol. 1 of the Life and Letters of Charles Darwin, by his son, Francis Darwin, he says, speaking of the period from 1828 to 1831, "I did not then in the least doubt the strict and literal truth of every word in the Bible." On page 412 of Vol. II of the same publication, he says, "When I was collecting facts for 'The Origin' my belief in what is called a personal God was as firm as that of Dr. Pusey himself." It may be a surprise to your honor and to you, gentlemen of the jury, as it was to me, to learn that Darwin spent three years at Cambridge studying for the ministry.

This was Darwin as a young man, before he came under the influence of the doctrine that man came from a lower order of animals. The change wrought in his religious views will be found in a letter written to a German youth in 1879, and printed on page 277 of Vol. I of the Life and Letters above referred to. The letter begins: "I am much engaged, an old man, and out of health, and I cannot spare time to answer your questions fully—nor indeed can they be answered. Science has nothing to do with Christ, except insofar as the habit of scientific research makes a man cautious in admitting evidence. For myself, I do not believe that there ever has been any revelation. As for a future life, every man must judge for himself between conflicting vague probabilities."

Note that "science has nothing to do with Christ, except insofar as the habit of scientific research makes a man cautious in admitting evidence." Stated plainly, that simply means that "the habit of scientific research" makes one cautious in accepting the only evidence that we have of Christ's existence, mission, teaching, crucifixion and resurrection, namely the evidence found in the Bible. To make this interpretation of his words the only possible one, he adds, "For myself, I do not believe that there ever has been any revelation." In rejecting the Bible as a revelation from God, he rejects the Bible's conception of God and he rejects also the supernatural Christ of whom the Bible, and the Bible alone, tells. And, it will be observed, he refuses to express any opinion as to a future

Now let us follow with his son's exposition of his father's views as they are given in extracts from a biography written in 1876. Here is Darwin's language as quoted by his son:

"During these two years (October, 1838, to January, 1839) I was led to think much about religion. Whilst on board the Beagle I was quite orthodox and I remember being heartily laughed at by several of the officers (though themselves orthodox) for quoting the Bible as an unanswerable authority on some point of morality. When thus reflecting, I felt compelled to look for a first cause, having an intelligent mind in some degree analagous to man; and I deserved to be called an atheist. This conclusion was strong in my mind about the time, as far as I can remember, when I wrote the 'Origin of Species'; it is since that time that it has very gradually, with many fluctuations, become weaker. But then arises the doubt, can the mind of man, which has, as I fully believe, been developed from a mind as low as that possessed by the lowest animals, be trusted when it draws such grand conclusions?

"I cannot pretend to throw the least light on such abstruse problems. The mystery of the beginning of all things is insoluble by us; and I for one must be content to remain an agnostic."

Darwin Used Bible As Early Arguments.

When Darwin entered upon his scientific career he was "quite orthodox and quoted the Bible as an unanswerable authority on some point of morality." Even when he wrote "The Origin of Species," the thought of "a first cause, having an intelligent mind in some degree analagous to man" was strong in his mind. It was after that time that "very gradually, with many fluctuations," his belief in God became weaker. He traces this decline for us and concludes by telling us that he cannot pretend to throw the least light on such abstruse problems—the religious problems above referred to. Then comes the flat statement that he "must be content to remain an agnostic"; and to make clear what he means by the word, agnostic, he says that "the mystery of the beginning of all things is insoluble by us"—not by him alone, but by everybody. Here we have the effect of evolution upon its most distinguished exponent; it led him from an orthodox Christian, believing every word of the Bible and in a personal God, down and down and down to helpless and hopeless agnosticism.

But there is one sentence upon which I reserved comment — it throws light upon his downward pathway. "Then arises the doubt, can the mind of man which has, as I fully believe, been developed from a mind as low as that possessed by the lowest animals, be trusted when it draws such grand conclusions?"

Here is the explanation: he drags man down to the brute level, and then, judging man by brute standards, he questions whether man's mind can be trusted to deal with God and immortality?

How can any teacher tell his students that evolution does not tend to destroy his religious faith? How can an honest teacher conceal from his students the effect of evolution upon Darwin himself? And is it not stranger still that preachers who advocate evolution never speak of Darwin's loss of faith, due to his belief in evolution? The parents of Tennessee have reason enough to fear the effect of evolution on the minds of their children. Belief in evolution cannot bring to those who hold such a belief any compensation for the loss of faith in God, trust in the Bible, and belief in the supernatural character of Christ. It is belief in evolution that has caused so many scientists and so many Christians to reject the miracles of the Bible, and then give up, one after another, every vital truth of Christianity. They finally cease to pray and sunder the tie that binds them to their Heavenly Father.

Miracle Should Not Become Stumbling Block.

The miracle should not be a stumbling block to any one. It raises but three questions: First, could God perform a miracle? Yes, the God who created the universe can do anything He wants to with it. He can temporarily suspend any law that He has made or He may employ higher laws that we do not understand.

Second—Would God perform a miracle? To answer that question in the negative one would have to know more about God's plans and purposes than a finite mind can know, and yet some are so wedded to evolution that they deny that God would perform a miracle merely because a miracle is inconsistent with evolution.

If we believe that God can perform a miracle and might desire to do so, we are prepared to consider with open mind the third question, namely, Did God perform the miracles recorded in the Bible? The same evidence that establishes the authority of the Bible establishes the truth of the record of miracles performed.

Now, let me read to the honorable court and to you, gentlemen of the jury, one of the most pathetic confessions that has come to my notice. George John Romanes, a distinguished biologist, sometimes called the successor of Darwin, was prominent enough to be given extended space in both the Encyclopedia Britannica and Encyclopedia Americana. Like Darwin, he was reared in the orthodox faith, and like Darwin, was led away from it by evolution (see "Thoughts on Religion," page 180.) For twenty-five years he could not pray. Soon after he became an agnostic, he wrote a book entitled, "A Candid Examination of

Theism," publishing it under the assumed name, "Physicus." In this book (see page 29, "Thoughts on Religion"), he says:

"And forasmuch as I am far from being able to agree with those who affirm that the twilight doctrine of the 'new faith' is a desirable substitute for the waning splendor of 'the old,' I am not ashamed to confess that with this virtual negation of God the universe to me has lost its soul of loveliness; and although from henceforth the precept to 'work while it is day' will doubtless but gain an intensified force from the terribly intensified meaning of the words that 'the night cometh when no man can work,' yet when at times I think, as think at times I must, of the appalling contrast between the hallowed glory of that creed which once was mine, and the lonely mystery of existence as now I find it—at such times I shall ever feel it impossible to avoid the sharpest pang of which my nature is susceptible."

Do these evolutionists stop to think of the crime they commit when they take faith out of the hearts of men and women and lead them out into a starless night? What pleasure can they find in robbing a human being of "the hallowed glory of that creed" that Romanes once cherished, and in substituting "the lonely mystery of existence" as he found it? Can the fathers and mothers of Tennessee be blamed for trying to protect their children from such a tragedy?

If anyone had been led to complain of the severity of the punishment that hangs over the defendant, let him compare this crime and its mild punishment with the crimes for which a greater punishment is prescribed. What is the taking of a few dollars from one in day or night in comparison with the crime of leading one away from God and away from Christ?

Shakespeare regards the robbing one of his good name as much more grave than the stealing of his purse. But we have a higher authority than Shakespeare to invoke in this connection. He who spake as never man spake, thus describes the crimes that are committed against the young. "It is impossible but that offences will come; but woe unto him through whom they come. It were better for him that a millstone were hanged about his neck, and he cast into the sea, than that he should offend one of these litte ones."

Christ did not overdraw the picture. Who is able to set a price upon the life of a child—a child into whom a mother has poured her life and for whom a father has labored? What may a noble life mean to the child itself, to the parents, and to the world?

And it must be remembered that we can measure the effect on only that part of life which is spent on earth; we have no way of calculating the effect on that infinite circle of life of which existence here is but a small arc. The soul is immortal and religion deals with the soul; the logical effect of the evolutionary hypothesis is to undermine religion and thus affect the soul. I recently received a list of questions that were to be discussed in a prominent eastern school for women. The second question in the list read, "Is religion an obsolescent function that should be allowed to atrophy quietly, without arousing the passionate prejudice of outworn superstition?" The real attack of evolution, it will be seen, is not upon orthodox Christianity, or even upon Christianity, but upon religion—the most basic fact in man's existence and the most practical thing in life.

But I have some more evidences of the effect of evolution upon the life of those who accept it and try to harmonize their thought with it.

Over Half of Scientists Deny Existence of God.

James H. Leuba, a professor of psychology at Bryn Mawr college, Pennsylvania, published a few years ago a book entitled "Belief in God and Immortality." In this book he relates how he secured the opinions of scientists as to the existence of a personal God and a personal immortality. He used a volume entitled "American Men of Science," which, he says, included the names of "practically every American who may properly be called a scientist." There were 5,500 names in the book. He selected 1,000 names as representative of the 5,500, and addressed them personally. Most of them, he said, were teachers in schools of higher learning. The names were kept confidential. Upon the answers received, he asserts that over half of them doubt or deny the existence of a personal God and a personal immortality, and he asserts that unbelief increases in proportions to prominence, the percentage of unbelief being greatest among the most prominent. Among biologists, believers in a personal God numbered less than 31 per cent., while believers in a personal immortality numbered only 37 per cent.

He also questioned the students in nine colleges of high rank and from 1,000 answers received, 97 per cent. of which were from students between 18 and 20, he found that unbelief increased from 15 per cent. in the freshman class up to 40 to 45 per cent. among the men who graduated. On page 280 of this book, we read, "The students' statistics show that young people enter college, possessed of the beliefs still accepted, more or less perfunctorily, in the average home of the land, and gradually abandon the cardinal Christian beliefs." This change from belief to unbelief he attibutes to the influence of the persons "of high culture under whom they studied."

The people of Tennessee have been patient enough; they have acted none too soon. How can they expect to protect society, and even the church, from the deadening influence of agnosticism and atheism if they permit the teachers employed by taxation to poison the minds of the youth with this destructive doctrine? And remember that the law has not heretofore required the writing of the word "poison" on poisonous doctrines. The bodies of our people are so valuable that druggists and physicians must be careful to properly label all poisons; why not be as careful to protect the spiritual life of our people from the poisons that kill the soul?

There is a test that is sometimes used to ascertain whether one suspected of mental infirmity is really insane. He is put into a tank of water and told to dip the tank dry while a stream of water flows into the tank. If he has not sense enough to turn off the stream, he is adjudged insane. Can parents justify themselves if, knowing the effect of belief in evolution, they permit irreligious teachers to inject skepticism and infidelity into the minds of their children?

Do bad doctrines corrupt the morals of students? We have a case in point. Mr. Darrow, one of the most distinguished criminal lawyers in our land, was engaged about a year ago in defending two rich men's sons who were on trial for as dastardly a murder as was ever committed. The older one, "Babe" Leopold, was a brilliant student, 19 years old. He was an evolutionist and an atheist. He was also a follower of Nietzsche, whose books he had devoured and whose philosophy he had adopted. Mr. Darrow made a plea for him, based upon the influence that Nietzsche's philosophy had exerted upon the boy's mind. Here are extracts from his speech:

"Babe took to philosophy. * * * He grew up in this way; he became enamoured of the philosophy of Nietzsche. Your honor, I have read almost everything that Nietzsche ever wrote. A man of wonderful intellect; the most original philosopher of the last century. A man who made a deeper imprint on philosophy than any other man within a hundred years, whether right or wrong. More books have been written about him than probably all the rest of the philosophers in a hundred years. More college professors have talked about him. In a way, he has reached more people, and still he has been a philosopher of what we might call the intellectual cult.

"He wrote one book called 'Beyond the Good and Evil,' which was a criticism of all moral precepts, as we understand them, and a treatise that the intelligent man was beyond good and evil, that the laws for good and the laws for evil did not apply to anybody who approached the superman. He wrote on the will to power.

"I have just made a few short extracts from Nietzsche that show the things that he (Leopold) has read, and these are short and almost taken at random. It is not how this would affect you. It is not how it would affect me. The question is how it would affect the impressionable, visionary, dreamy mind of a boy—a boy who should never have seen it—too early for him."

Mr. Bryan Quotes From Nietzsche's Books.

Quotations from Nietzsche: "Why so soft, oh, my brethren? Why so soft, so unresisting and yielding? Why is there so much disavowal and abnegation in your hearts? Why is there so little fate in your looks? For all creators are hard and it must seem blessedness unto you to press your hand upon millenniums and upon wax. This new table, oh, my brethren, I put over you; Become hard. To be obsessed by moral consideration presupposes a very low grade of intellect. We should substitute for morality the will to our own end, and consequently to the means to accomplish that. A great man, a man whom nature has built up and invented in a grand style, is colder, harder, less cautious and more free from the fear of public opinion. He does not possess the virtues which are compatible with respectability, with being respected, nor any of those things which are counted among the virtues of the herd."

Mr. Darrow says, that the superman, a creation of Nietzsche, has permeated every college and university in the civilized world.

"There is not any university in the world where the professor is not familiar with Nietzsche, not one. * * * Some believe it and some do not believe it. Some read it as I do and take it as a theory, a dream, a vision, mixed with good and bad, but not in any way related to human life. Some take it seriously. * * * There is not a university in the world of any high standing where the professors do not tell you about Nietzsche and discuss him, or where the books are not there.

"If this boy is to blame for this, where did he get it? Is there any blame attached because somebody took Nietzsche's philosophy seriously and fashioned his life upon it? And there is no question in this case but what that is true. Then who is to blame? The university would be more to blame than he is; the scholars of the world would be more to blame than he is. The publishers of the world * * * are more to blame than he is. Your honor, it is hardly fair to hang a 19-year-old boy for the philosophy that was taught him at the university. It does not meet my ideas of justice and fairness to visit upon his head the philosophy that has been taught by university men for twenty-five years."

Transformed Into Murderer By Philosophy of Atheist.

In fairness to Mr. Darrow, I think I ought to quote two more paragraphs. After this bold attempt to excuse the student on the ground that he was transformed from a well-meaning youth into a murderer by the philosophy of an atheist, and on the further ground that this philosophy was in the libraries of all the colleges and discussed by the professors—some adopting the philosophy and some rejecting it—on these two grounds he denies that the boy should be held responsible for the taking of human life. He charges that the scholars in the universities were more responsible than the boy, because they furnished such books to the students, and then he proceeds to exonerate the universities and the scholars, leaving nobody responsible. Here is Mr. Darrow's language:

"Now, I do not want to be misunderstood about this. Even for the sake of saving the lives of my clients, I do not want to be dishonest and tell the court something that I do not honestly think in this case. I do not think that the universities are to blame. I do not think they should be held responsible. I do think, however, that they are too large, and that they should keep a closer watch, if possible, upon the individual.

"But you cannot destroy thought because, forsooth, some brain may be deranged by thought. It is the duty of the university, as I conceive it, to be the great storehouse of the wisdom of the ages, and to have its students come there and learn and choose. I have no doubt but what it has meant the death of many; but that we cannot help."

This is a damnable philosophy, and yet it is the flower that blooms on the stalk of evolution. Mr. Darrow thinks the universities are in duty bound to feed out this poisonous stuff to their students, and when the students become stupefied by it and commit murder, neither they nor the universities are to blame. I am sure, your honor and gentlemen of the jury, that you agree with me when I protest against the adoption of any such philosophy in the state of Tennessee. A criminal is not relieved from responsibility merely because he found Nietzsche's philosophy in a library which ought not to contain it. Neither is the university guiltless if it permits such corrupting nourishment to be fed to the souls that are entrusted to its care. But, go a step farther, would the state be blameless if it permitted the universities under its control to be turned into training schools for murderers? When you get back to the root of this question, you will find that the legislature not only had a right to protect the students from the evolutionary hypothesis but was in duty bound to do so.

While on this subject, let me call your attention to another proposition embodied in Mr. Darrow's speech. He said that Dickey Loeb, the younger boy, had read trashy novels, of the blood and thunder sort. He even went so far as to commend an Illinois statute which forbids minors reading stories of crime. Here is what Mr. Darrow said; "We have a statute in this state, passed only last year, if I recall it, which forbids minors reading story of crime. Why? There is only one reason; because the legislature in its wisdom thought it would have a tendency to produce these thoughts and this life in the boys who read them."

If Illinois can protect her boys, why cannot this state protect the of Illinois any more precious than boys of Tennessee? Are the boys yours?

Quotes Darrow's Plea For Richard Loeb's Life.

But to return to philosophy of an evolutionist. Mr. Darrow said: "I say to you seriously that the parents of Dickey Loeb are more responsible than he, and yet few boys had better parents." * * * Again, he says, "I know that one of two things happened to this boy; that this terrible crime was inherent in his organism, and came from some ancestor, or that it came through his education and his training after he was born." He thinks the boy was not responsible for anything; his guilt was due, according to his philosophy, either to heredity or to environment.

But let me complete Mr. Darrow's philosophy based on evolution. He says: "I do not know what remote ancestor may have sent down the seed that corrupted him, and I do not know through how many ancestors it may have passed until it reached Dickey Loeb. All I know is, it is true, and there is not a biologist in the world who will not say I am right."

Psychologists who build upon the evolutionary hypothesis teach that man is nothing but a bundle of characteristics inherited from brute ancestors. That is the philosophy which Mr. Darrow applied in this celebrated criminal case. "Some remote ancestor"—he does not know how remote—"sent down the seed that corrupted him." You cannot punish the ancestor—he is not only dead but, according to the evolutionists, he was a brute and may have lived a million years ago. And he says that all the biologists agree with him. No wonder so small a per cent of the biologists, according to Leuba, believe in a personal God.

This is the quintessence of evolution, distilled for us by one who follows that doctrine to its logical conclusion. Analyze this dogma of darkness and death. Evolutionists say that back in the twilight of life a beast, name and nature unknown, planted a murderous seed and that the impulse that originated in that seed throbs forever in the blood of the brute's descendents, inspiring killings innumerable, for which the murderers are not responsible because coerced by a fate fixed by the laws of heredity! It is an insult to reason and shocks the heart. That doctrine is as deadly as leprosy; it may aid a lawyer in a criminal case, but it would, if generally adopted, destroy all sense of responsibiity and menace the morals of the world. A brute, they say, can predestine a man to crime, and yet they deny that God incarnate in the flesh can release a human being from this bondage or save him from ancestral sins. No more repulsive doctrine was ever proclaimed by any man; if all the biologists of the world teach this doctrine—as Mr. Darrow says they do—then may heaven defend the youth of our land from their impious babblings.

Minds Are Diverted to Trifling Speculation.

Our third indictment against evolution is that it diverts attention from pressing problems of great importance to trifling speculation. While one evolutionist is trying to imagine what happened in the dim past, another is trying to pry open the door of the distant future. One recently grew eloquent over ancient worms, and another predicted that 75,000 years hence every one will be bald and toothless. Both those who endeavor to clothe our remote ancestors with hair and those who endeavor to remove the hair from the heads of our remote descendants ignore the present with its imperative demands. The science of "How to Live" is the most important of all the sciences. It is desirable to know the physical sciences, but it is necessary to know how to live. Christians desire that their children shall be taught all the sciences, but they do not want them to lose sight of the Rock of Ages while they study the age of rocks; neither do they desire them to become so absorbed in measuring the distance between the stars that they will forget Him who holds the stars in His Hand.

While not more than two per cent of our population are college graduates, these, because of enlarged powers, need a "heavenly vision" even more than those less learned, both for their own restraint and to assure society that their enlarged powers will be used for the benefit of society and not against the public welfare.

Evolution is deadening the spiritual life of a multitude of students. Christians do not desire less education, but they desire that religion shall be entwined with learning so that our boys and girls will return from college with their hearts aflame with love of God and love of fellowmen, and prepared to lead in the altruistic work that the world so sorely needs. The cry in the business world, in the industrial world, in the professional world, in the political world—even in the religious world—is for consecrated talents; for ability plus a passion for service.

Our fourth indictment against the evolutionary hypothesis is that, by paralyzing the hope of reform, it discourages those who labor for the improvement of man's condition. Every upward-looking man or woman seeks to lift the level upon which mankind stands, and they trust that they will see beneficent changes during the brief span of their own lives. Evolution chills their enthusiasm by substituting aeons for years. It obscures all beginnings in the mists of endless ages. It is represented as a cold and heartless process, beginning with time and ending in eternity, and acting so slowy that even the rocks can not preserve a record of the imaginary changes through which it is credited with having carried an original germ of life that appeared sometime from somewhere. Its only program for man is scientific breeding, a system under which a few supposedly superior intellects, self-appointed, would direct the mating and the movements of the mass of mankind—an impossible system! Evolution, disputing the miracle, and ignoring the spiritual in life, has no place for the regeneration of the individual. It recognizes no cry of repentance and scoffs at the doctrine that one can be born again.

Prodigal Son Story Contradicts Evolution.

It is thus the intolerant and unrelenting enemy of the only process that can redeem society through the redemption of the individual. An evolutionist would never write such a story as the Prodigal Son; it contradicts the whole theory of evolution. The two sons inherited from the same parents and, through their parents, from the same ancestors, proximate and remote. And these sons were reared at the same fireside and were surrounded by the same environment during all the days of their youth; and yet they were different. If Mr. Darrow is correct in the theory applied to Loeb (namely, that his crime was due either to inheritance or to environment, how will he explain the difference between the elder brother and the wayward son? The evolutionist may understand from observation, if not by experience, even though he cannot explain, why one of these boys was guilty of every immorality, squandered the money that the father had laboriously earned, and brought disgrace upon the family name; but his theory does not explain why a wicked young man underwent a change of heart, confessed his sin, and begged for forgiveness. And because the evolutionists cannot understand this fact, one of the most important in the human life, he cannot understand the infinite love of the heavenly Father, who stands ready to welcome home any repentant sinner, no matter how far he has wandered, how often he has fallen, or how deep he has sunk in sin.

Your honor has quoted from a wonderful poem written by a great Tennessee poet, Walter Malone. I venture to quote another stanza which puts into exquisite language the new opportunity which a merciful God gives to every one who will turn from sin to righteousness.

"Though deep in mire, wring not your hands and weep;
I lend my arm to all who say, 'I can.'
No shame-faced outcast ever sank so deep
But he might rise and be again a man."

There are no lines like these in all that evolutionists have ever written. Darwin says that science has nothing to do with the Christ who taught the spirit embodied in the words of Walter Malone, and yet this spirit is the only hope of human progress. A heart can be changed in the twinkling of an eye and a change in the life follows a change in the heart. If one heart can be changed, it is possible that many hearts can be changed, and if many hearts can be changed, it is possible that all hearts can be changed—that a world can be born in a day. It is this fact that inspires all who labor for man's betterment. It is because Christians believe in individual regeneration and in the regeneration of society through the regeneration of individuals that they pray, "Thy kingdom come, Thy will be done in earth as it is in heaven." Evolution makes a mockery of the Lord's Prayer!

Evolution Only Defers Hope of All Mankind.

To interpret the words to mean that the improvement desired must come slowly through unfolding ages—a process with which each generation could have little to do—is to defer hope, and hope deferred maketh the heart sick.

Our fifth indictment of the evolutionary hypothesis is that, if taken seriously and made the basis of a philosophy of life, it would eliminate love and carry man back to a struggle of tooth and claw. The Christians who have allowed themselves to be deceived into believing that evolution is a beneficent, or even a rational process, have been associating with those who either do not understand its implications or dare not avow their knowledge of these implications. Let me give you some authority on this subject. I will begin with Darwin, the high priest of evolution, to whom all evolutionists bow.

On pages 149 and 150, in "The Descent of Man," already referred to, he says:

"With savages, the weak in body or mind are soon eliminated; and those that survive commonly exhibit a vigorous state of health. We civilized men, on the other hand, do our utmost to check the process of elimination; we build asylums for the imbecile, the maimed and the sick; we institute poor laws; and our medical men exert their utmost skill to save the life of everyone to the last moment. There is reason to believe that vaccination has preserved thousands who, from a weak constitution, would formerly have succumbed to smallpox. Thus the weak members of civilized society propagate their kind. No one who has attended to the breeding of domestic animals will doubt that this must be highly injurious to the race of man. It is surprising how soon a want of care, or care wrongly directed, leads to the degeneration of a domestic race; but, excepting in the case of man himself, hardly anyone is so ignorant as to allow his worst animals to breed.

"The aid which we feel impelled to give to the helpless is mainly an incidental result of the instinct of sympathy, which was originally acquired as part of the social instincts, but subsequently rendered in the manner previously indicated more tender and more widely diffused. Nor could we check our sympathy, even at the urging of hard reason, without deterioration in the noblest part of our nature . . . We must, therefore, bear the undoubtedly bad effects of the weak surviving and propagating their kind."

Barbarous Sentiment Expressed by Darwin.

Darwin reveals the barbarous sentiment that runs through evolution and dwarfs the moral nature of those who become obsessed with it. Let us analyze the quotation just given. Darwin speaks with approval of the savage custom of eliminating the weak so that only the strong will survive and complains that "we civilized men do our utmost to check the process of elimination." How inhuman such a doctrine as this! He thinks it injurious to "build asylums for the imbecile, the maimed, and the sick," or to care for the poor. Even the medical men come in for criticism because they "exert their utmost skill to save the life of everyone to the last moment." And then note his hostility to vaccination, because it has "preserved thousands who, from a weak constitution would, but for vaccination, have succumbed to smallpox!" All of the sympathetic activities of civilized society are condemned because they enable "the weak members to propagate their kind." Then he drags mankind down to the level of the brute and compares the freedom given to man unfavorably with the restraint that we put on barnyard beasts.

The second paragraph of the above quotation shows that his kindly heart rebelled against the cruelty of his own doctrine. He says that we "feel impelled to give to the helpless," although he traces it to a sympathy which he thinks is developed by evolution; he even admits that we could not check this sympathy "even at the urging of hard reason, without deterioration of the noblest part of our nature." "We must therefore bear" what he regards as "the undoubtedly bad effects of the weak surviving and propagating their kind." Could any doctrine be more destructive of civilization? And what a commentary on evolution! He wants us to believe that evolution develops a human sympathy that finally becomes so tender that it repudiates the law that created it and thus invites a return to a level where the extinguishing of pity and sympathy will permit the brutal instincts to again do their progressive (?) work.

Darrow Says Nietzsche Was Gloriously Wrong.

Let no one think that this acceptance of barbarism as the basic principle of evolution died with Darwin. Within three years a book has appeared whose author is even more frankly brutal than Darwin. The book is entitled, "The New Decalogue of Science" and has attracted wide attention. One of our most reputable magazines has recently printed an article by him defining the religion of a scientist. In his preface he acknowledges indebtedness to twenty-one prominent scientists and educators, nearly all of them "doctors" and "professors." One of them, who has recently been elevated to the head of a great state university, read the manuscript over twice "and made many invaluable suggestions." The author describes Nietzsche who, according to Mr. Darrow, made a murderer out of Babe Leopold, as "the bravest soul since Jesus." He admits that Nietzsche was "gloriously wrong," not certainly," but "perhaps," "in many details of technical knowledge," but he affirms that Nietzsche was "gloriously right in his fearless questioning of the universe and of his own soul."

In another place, the author says, "Most of our morals today are jungle products," and then he affirms that "it would be safer, biologically, if they were more so now." After these two samples of his views, you will not be surprised when I read you the following (see page 34):

"Evolution is a bloody business, but civilization tries to make it a pink tea. Barbarism is the only process by which man has ever organically progressed, and civilization is the only process by which he has ever organically declined. Civilization is the most dangerous enterprise upon which man ever set out. For when you take man out of the bloody, brutal, but beneficent hand of natural selection you place him at once in the soft, perfumed, daintily gloved, but far more dangerous hand of artificial selection. And, unless you call science to your aid and make this artificial selection as efficient as the rude methods of nature, you bungle the whole task."

This aspect of evolution may amaze some of the ministers who have not been admitted to the inner circle of the iconoclasts whose theories menace all the ideals of civilized society. Do these ministers know that "evolution is a bloody business"? Do they know that "barbarism is the only process by which man has ever organically progressed"? And that "civilization is the only process by which he has ever organically declined"? Do they know that "the bloody, brutal hand of natural selection" is "beneficent"? And that the "artificial selection" found in civilization is "dangerous"? What shall we think of the distinguished educators and scientists who read the manuscript before publication and did not protest against this pagan doctrine?

To show that this is a world-wide matter, I now quote from a book issued from the press in 1918, seven years ago. The title of the book is "The Science of Power," and its author, Benjamin Kidd, being an Englishman, could not have any national prejudice against Darwin. On pages 46 and 47, we find Kidd's interpretation of evolution:

"Darwin's presentation of the evolution of the world as the product of natural selection in never-ceasing war—as a product, that is to say, of a struggle in which the individual efficient in the fight for his own interests was always the winning type — touched the profoundest depths of the psychology of the West.

The idea seemed to present the whole order of progress in the world as the result of a purely mechanical and materialistic process resting on force. In so doing it was a conception which reached the springs of that heredity born of the unmeasured ages of conquest out of which the western mind has come. Within half a century the origin of species had become the Bible of the doctrine of the omnipotence of force."

Kidd goes so far as to charge that "Nietzsche's teaching represented the interpretation of the popular Darwinism delivered with the fury and intensity of genius." And Nietzsche, be it remembered, denounced Christianity as the "doctrine of the degenerate," and democracy as "the refuge of weaklings."

Kidd says that Nietzsche gave Germany the doctrine of Darwin's efficient animal in the voice of his superman, and that Bernhardi and the military textbooks in due time gave Germany the doctrine of the superman translated into the national policy of the superstate aiming at world power. (Page 67.)

And what else but the spirit of evolution can account for the popularity of the selfish doctrine, "Each one for himself, and the devil take the hindmost," that threatens the very existence of the doctrine of brotherhood.

In 1900—twenty-five years ago—while an international peace congress was in session in Paris, the following editorial appeared in L'Univers:

"The spirit of peace has fled the earth because evolution has taken possession of it. The plea for peace in past years has been inspired by faith in the divine nature and the divine origin of man; men were then looked upon as children of one Father, and war, therefore, was fratricide. But now that men are looked upon as children of apes, what matters it whether they are slaughtered or not?"

When there is poison in the blood, no one knows on what part of the body it will break out, but we can be sure that it will continue to break out until the blood is purified. One of the leading universities of the South (I love the state too well to mention its name) publishes a monthly magazine entitled "Journal of Social Forces." In the January issue of this year, a contributor has a lengthy article on "Sociology and Ethics," in the course of which he says:

"No attempt will be made to take up the matter of the good or evil of sexual intercourse among humans aside from the matter of conscious procreation, but as an historian, it might be worth while to ask the exponents of the impurity complex to explain the fact that, without exception, the great periods of cultural efflorescence have been those characterized by a large amount of freedom in sex-relations, and that those of the greatest cultural degradation and decline have been accompanied with greater sex repression and purity."

No one charges or suspects that all or any large percentage of the advocates of evolution sympathize with this loathsome application of evolution to social life, but it is worthwhile to inquire why those in charge of a great institution of learning allow such filth to be poured out for the stirring of the passions of its students.

Just one more quotation: The Southeastern Christian Advocate of June 25, 1925, quotes five eminent college men of Great Britain as joining in an answer to the question, "Will civilization survive?" Their reply is that:

"The greatest danger menacing our civilization is the abuse of the achievements of science. Mastery over the forces of nature has endowed the twentieth century man with a power which he is not fit to exercise. Unless the development of morality catches up with the development of technique, humanity is bound to destroy itself."

Can any Christian remain indifferent? Science needs religion to direct its energies and to inspire with lofty purpose those who employ the forces that are unloosened by science. Evolution is at war with religion because religion is supernatural; it is, therefore, the relentless foe of Christianity, which is a revealed religion.

Let us, then, hear the conclusion of the whole matter. Science is a magnificent material force, but it is not a teacher of morals. It can perfect machinery, but it adds no moral restraints to protect society from the misuse of the machine. It can also build gigantic intellectual ships, but it constructs no moral rudders for the control of storm-tossed human vessels. It not only fails to supply the spiritual element needed but some of its unproven hypotheses rob the ship of its compass and thus endangers its cargo.

Science Has Made War More Terrible Than Ever.

In war, science has proven itself an evil genius; it has made war more terrible than it ever was before. Man used to be content to slaughter his fellowmen on a single plane—the earth's surface. Science has taught him to go down into the water and shoot up from below and to go up into the clouds and shoot down from above, thus making the battlefield three times a bloody as it was before; but science does not teach brotherly love. Science has made war so hellish that civilization was about to commit suicide; and now we are told that newly discovered instruments of destruction will make the cruelties of the late war seem trivial in comparison with the cruelties of wars that may come in the future. If civilization is to be saved from the wreckage threatened by intelligence not consecrated by love, it must be saved by the moral code of the meek and lowly Nazarene. His teachings, and His teachings, alone, can solve the problems that vex the heart and perplex the world.

"The world needs a Savior more than it ever did before, and there is only one Name under heaven given among men whereby we must be saved." It is this Name that evolution degrades, for, carried to its logical conclusion, it robs Christ of the glory of a virgin birth, of the majesty of His deity and mission and of the triumph of His resurrection. It also disputes the doctrine of the atonement.

It is for the jury to determine whether this attack upon the Christion religion shall be permitted in the public schools of Tennessee by teachers employed by the state and paid out of the public treasury. This case is no longer local, the defendant ceases to play an important part. The case has assumed the proportions of a battle-royal between unbelief that attempts to speak through so-called science and the defenders of the Christian faith, speaking through the legislators of Tennessee. It is again a choice between God and Baal; it is also a renewal of the issue in Pilate's court. In that historic trial—the greatest in history—force, impersonated by Pilate occupied the throne. Behind it was the Roman government, mistress of the world, and behind the Roman government were the legions of Rome. Before Pilate, stood Christ, the Apostle of Love. Force triumphed; they nailed Him to the tree and those who stood around mocked and jeered and said, "He is dead." But from that day the power of Caesar waned and the power of Christ increased. In a few centuries the Roman government was gone and its legions forgotten; while the crucified Lord has become the greatest fact in history and the growing figure of all time.

Again force and love meet face to face, and the question, "What shall I do with Jesus?" must be answered. A bloody, brutal doctrine—Evolution—demands, as the rabble did nineteen hundred years ago, that He be crucified. That cannot be the answer of this jury representing a Christian state and sworn to uphold the laws of Tennessee. Your answer will be heard throughout the world; it is eagerly awaited by a praying multitude. If the law is nullified, there will be rejoicing wherever God is repudiated, the Savior scoffed at and the Bible ridiculed. Every unbeliever of every kind and degree will be

happy. If, on the other hand, the law is upheld and the religion of the school children protected, millions of Christians will call you blessed and, with hearts full of gratitude to God, will sing again that grand old song of triumph:

"Faith of our fathers, living still,
In spite of dungeon, fire and sword;
O how our hearts beat high with joy
Whene'er we hear that glorious word—
Faith of our fathers—holy faith;
We will be true to thee till death!"

www.ingramcontent.com/pod-product-compliance
Lightning Source LLC
LaVergne TN
LVHW091516200926
848483LV00005B/431/J